Earthquake Disaster Reduction: Masonry Building, Design and Construction

Earthquake Disaster Reduction: Masonry Building, Design and Construction

Dr. Anand Swarup Arya
Professor Emeritus of Earthquake Engineering
Indian Institute of Technology, Roorkee

nidm
Towards a disaster free India
National Institute of Disaster Management
New Delhi

in association with

KW
Knowledge World

KW Publishers Pvt. Ltd.
New Delhi

KNOWLEDGE WORLD

A KW Publishers Book
www.kwpublishers.com

Published in India by

Kalpana Shukla
KW Publishers Pvt. Ltd.
5A/4A, Ansari Road, Daryaganj, New Delhi 110002
email: knowledgeworld@vsnl.net / online.kw@gmail.com

ISBN: 978-81-87966-50-0

Typeset by Dharana, New Delhi; studiodharana@gmail.com
Printed at Chaman Offset Printer, New Delhi

Contents

Chapter 5: Concepts and Techniques for Seismic Base-Isolation of Structures

Chapter 6: Isolator Parameters for Earthquake Response Reduction

Chapter 7: Vulnerability and Risk of Damage to Masonry Buildings in India

Chapter 22: Damage Scenario of a Hypothetical 8.0 Magnitude Earthquake in Kangra Region of Himachal Pradesh

Chapter 23: Repair, Restoration and Retrofitting of Masonry Buildings

Chapter 24: Repair and Strengthening of Damaged Stone Houses After Dhamar Earthquake of 1982

Chapter 1

Earthquake Hazard in India[1]

1.1 Earthquake Occurrences in India and the Neighbourhood

India has a large part of its land area liable to a wide range of probable maximum seismic intensities where shallow earthquakes of magnitudes of 5.0 or more on the Richter scale have been known to occur in the historical past or recorded in the last about 100 years. Fig.1 shows the Map of India with the epicentres of such earthquakes. In the Kutch and Andaman Island regions, besides the Himalayas, giant earthquakes of M > 8.0 have also taken place. The largest earthquake magnitude in India has been 8.7 which had its origin in the Shillong Plateau in 1897. This and the 1950 quake of M = 8.6 in Sadiya region have been so intense that the rivers changed their courses, ground elevations got changed permanently and stones were thrown upward. A list of better known damaging earthquakes in India is given in *Table-1.* For ready reference, the terms Magnitude & Intensity are defined here:

1.1.1 Earthquake Magnitude

The magnitude M of an earthquake is denoted by a number which is a measure of energy released during the earthquake occurrence. It is now measured in different ways, the most commonly used being the Richter Scale according to which "the magnitude of an earthquake is the logarithm to the base 10 of the maximum trace amplitude, expressed in microns, with which the standard short period torsion seismometer (with a period of 0.8 second, magnification of 2800 and damping nearly critical) would register the earthquake at an epicentral distance of 100 km". The scale being logarithmic, the energy of earthquake magnitude 'm+1' is about 31 times the energy released in an earthquake of magnitude 'm'. Magnitude scale is open ended, denoted numerically to one place of decimal (5.6, 8.3, etc.).

1.1.2 Intensity of Earthquake Shaking

"The intensity of an earthquake at a place is a measure of the effects of the earthquake".

---

1 *Dr. A.S.Arya, Professor Emeritus, IIT-Roorkee, National Seismic Advisor, GoI-UNDP, DRM Programme, New Delhi*

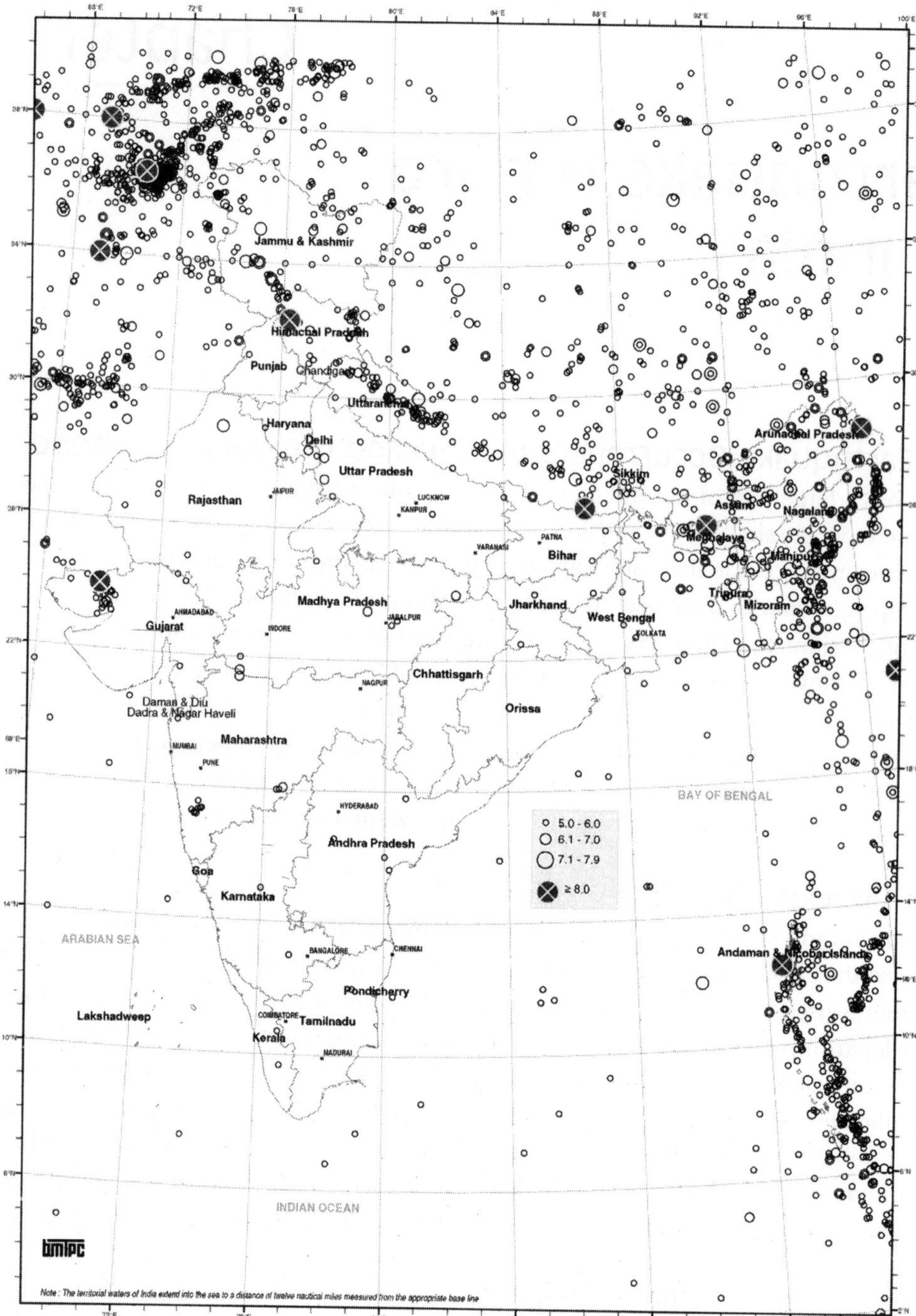

BMTPC : Vulnerability Atlas 2nd Edition; Peer Group, MoH&UPA; Map is Based on digitised data of SOI, GOI; Seismic Zones of India Map IS:1893 - 2002, BIS, GOI, Seismotectonic Atlas of India and its Environs, GSI, GOI

Fig.1: Earthquake Epicentral Map of India

Table 1: Some Better Known Damaging Earthquakes in India

Year	Area	Date	(I.S.T) Time hr:m:s	Latitude degrees North	Longitude degrees East	Magnitude M	Max. MM Int.	Deaths
1819	Gujarat (Kutch)	Jan.16	Mid Night	-	-	8.0	XI	Thousands
1833	Bihar -Nepal	Aug. 26	-	27.5	86.5	7.7	XI	Hundreds
1897	Assam (Shillong)	Jun.12	16:36:-	25.9	91.0	8.7	XII	1600
1900	Kerala (Palghat)	Feb. 8	-	10.7	76.7	6.0	-	
1905	Himachal Pradesh (Kangra)	Apr. 4	06:20:-	32.5	76.5	8.0	XI	20000
1930	Assam (Dhubri)	Jul. 3	02:33:34	25.8	90.2	7.1	IX	Many*
1934	Bihar -Nepal	Jan. 15	14:13:26	26.6	86.8	8.3	XI	14000
1941	Andamans	Jun. 26	-	12.4	92.5	8.0	X	Many
1943	Assam (NE)	Oct. 23	22:53:17	26.8	94.0	7.2	X	
1950	Assam (NE)	Aug. 15	19:39:28	28.7	96.6	8.6	XII	1500
1956	Gujarat (Anjar)	Jul. 21	21:02:36	23.3	70.0	7.0	VIII	Hundreds
1958	UP (Kapkote)	Dec. 28	-	30.0	80.0	6.3	VIII	Many
1960	Delhi	Aug. 27	21:28:59	28.3	77.4	6.0	VII	
1963	Kashmir (Badgam)	Sep 2	07:04:32	33.9	74.7	5.5	VII	Hundreds
1966	Western Nepal	Jun. 27	-	29.5	81.0	6.3	VIII	
1966	UP (Moradabad)	Aug. 15	-	28.0	79.0	5.3	VII	

1967 Nicobar	Jul. 2	-	9.0	93.4	6.2	-	
1967 Maharashtra (Koyna)	Dec. 11	04:21:19	17.4	73.7	6.5	VIII	200
1970 AP (Bhadrachalam)	Apr. 13	-	17.6	80.6	6.5	VII	
1970 Gujarat (Broach)	Mar 23	07:23:03	21.7	72.9	5.7	VII	
1975 Himachal Pradesh	Jan. 19	-	32.5	78.4	6.5	VIII	
1988 Bihar - Nepal	Aug. 21	04:39:10	26.76	86.62	6.6	VIII	1003
1991 UP (Uttarkashi)	Oct. 20	02:53:-	30.75	78.86	6.6	VIII	715
1993 Maharashtra (Killari)	Sep. 30	03:55:47	18.07	76.62	6.3	VIII	7928
1997 MP (Jabalpur)	May 22	04:22:31	23.1	80.1	6.0	VII+	38
1999 UP (Chamoli)	May 29	00:35:13	30.3	79.56	6.5	VIII	63
2001 Gujarat (Kachchh)	Jan. 26	08:46:00	23.6	69.8	Mw = 7.7	X	13805
2005 J&K (Kashmir)	Oct. 08	09:20:00	34.5	73.6	Mw = 7.6	X	Ind-1400 POK-73276

A number of intensity scales have been in vogue in different times, namely the Rossi-Forel (RF), Modified Mercalli (MM), MSK 1964 and Japan Meteorological Agency (JMA) scales. All the scales are close-ended stepped scales, RF having 10 points (I to X), MM and MSK with 12 points (I to XII) and JMA with 7 points (I to VII). Presently MSK 12 point scale is the most used, JMA being used in Japan. In historical earthquakes in India such as the 1905 Kangra earthquake and 1934 Bihar-Nepal earthquake, the RF intensity scale was used for drawing the Isoseismal map, that is equal Intensity distribution map of the affected region.

While for a given earthquake, the Magnitude has one unique value and its epicentral location, the Intensity varies from the maximum in the epicentral area to smaller values at increasing distances from the epicentre. Isoseismals derived from the observed damages in an earthquake as per the Intensity scale show the intensity distribution caused in the earthquake.

The relationship between the earthquake Magnitude and the maximum Intensity caused is not precise. Approximate relationship between them is shown in Table-2 as a general guide.

Table 2: Approximate Relationships Between M, MM Intensity and Felt Area

Earthquake magnitude Richter,	Expected global annual number	Maximum expected Intensity MM	Radius area (km)	felt area $(km)^2$
4.0 - 4.9	6,200	IV - V	50	7,700
5.0 - 5.9	800	VI - VII	110	38,000
6.0 - 6.9	120	VII - VIII	200	125,000
7.0 - 7.9	18	IX - X	400	500,000
8.0 - 8.7	1	XI - XII	800	2,000,000

Source: Earthquakes by Don de Nevi, Celestial Arts, Calif., May 1977, p.102

1.2 Main Seismic Regions

For considering the regional distribution of earthquakes in the Indian subcontinent, the whole area can be divided into the following seismic regions:

(a) Kashmir and Western Himalayas
(b) Central Himalayas (including Nepal Himalayas)
(c) Northeast India
(d) Indo-Gangetic Basin and Rajasthan
(e) Cambay and the Rann of Kutch
(f) Peninsular India
(g) Andaman & Nicobar Islands

The seismic activity of these regions is summarised below.

1.2.1 Kashmir and Western Himalayas

The region covers the States of Jammu and Kashmir, Himachal Pradesh and sub-mountain parts of Punjab. More than 180 earthquakes of M>5 are known to have occurred in the region, the maximum magnitude reaching 8.0 in the great Kangra earthquake of April 4, 1905, which had caused large scale destruction in the region with loss of life exceeding 20,000, mainly due to the collapse of brick, stone and earthen dwellings. More recent damaging earthquakes are the Budgam earthquake of Sept. 2, 1963, Anantnag earthquake of February 20, 1987, Dharmshala earthquake of April 26, 1986, and Kashmir earthquake of October 8, 2005. In the J&K earthquake, more than 75,000 people died due to the collapse of stone buildings, many more on the Pakistan side of the LoC.

1.2.2 Central Himalayas

This region includes the mountain and sub-mountain regions of Uttar Pradesh, the whole of Nepal and the sub-mountain regions of Bihar. The western side of the region has given rise usually to earthquakes of magnitude 6 but maximum reaching 7.5 during the Dharchula quake of 1916. The eastern side of the region has given evidence of very high seismicity and mention may be made of the Bihar-Nepal earthquake of M = 8.4, occurring on January 15, 1934, in which about 13,000 lives were lost and thousands of homes were completely destroyed, rails were bent and sand fountains appeared on the ground. More than 135 earthquakes of magnitude 5 and more are known to have occurred in this region, mostly in the western and eastern parts, the central part being relatively free. The most recent damaging earthquake is of M = 6.7, occurring on August 20, 1988, affecting northern Bihar and eastern Nepal.

1.2.3 North-East India

The region comprises the whole of Indian territory to the east of north Bengal. This (also other neighbouring countries like Bhutan and Myanmar) forms the most severe seismic region of the world, having experienced more than 520 known earthquakes of Magnitude 5 and over, among which 24 had magnitude 7 or more on the Richter scale. The Assam quake of 1897 (M = 8.7) is the largest earthquake ever known and the Sadiya earthquake of 1950 (M = 8.6) was one which has rarely been repeated in the world. These earthquakes were so large that even topographical changes of levels, etc. took place but the loss of life was not so great since in 1897 the population was not so large and the 1950 earthquake occurred in a rather unpopulated region. The Assam type construction using bamboo posts and Ekra-walling was light as well as strong which remained undamaged during these earthquakes. But the present situation is vastly different since the population has increased tremendously and non-engineered masonry and concrete constructions are replacing the Assam type construction. Hence, the danger to life and property in future earthquakes is on the increase with each passing year. Recently, on August 6, 1988, a M = 7.2 earthquake occurred in the Kachhar region.

1.2.4 Indo-Gangetic Basin and Rajasthan

This region is located immediately to the south of the Himalayas, covering Rajasthan, the plains of Punjab, Haryana, Uttar Pradesh, Bihar and Bengal lying north of the Vindhyas. Seismic activity has been noticed on the Moradabad fault, the Sohna fault near Delhi and the Lucknow and Patna faults. More than 110 earthquakes of $M > 5.0$ are known to have occurred in this region, the maximum Magnitude being 6.7. Hence, this region may be seen to have moderate to minor seismic activity. The areas close to the Himalayas are, of course, liable to ground shake during earthquake occurrences in the Himalayan region.

1.2.5 Cambay and the Rann of Kutch

This small region has also suffered from a number of earthquakes, more than 20 of $M \geq 5.0$, including two of $M \geq 7.0$ and another of M = 8.0 which occurred in the Rann of Kutch in 1819, destroying the town of Bhuj and killing 2,000 people. It can be taken as having severe to moderate seismicity. The city of Anjar was destroyed by an earthquake in 1956. The recent Kachchh earthquake of January 26, 2001 of M = 7.7 has been the most destructive in India, killing about 14,000 persons, destroying about 230,000 buildings and damaging more than 800,000 buildings. In this earthquake, for the first time in India, many reinforced concrete frame buildings were totally destroyed in Bhuj, Gandhidham, Anjar and 250 km away in the city of Ahemdabad due to bad design and construction practices.

1.2.6 Peninsular India Including the Islands of Lakshadweep

This region is a pre-Cambrian shield and, hence, generally stable, except some intra-shield seismic activity in its north, that is, along the Narmada Rift, and in the Western Margin. In the whole region, not more than 32 earthquake occurrences of M > 5.0 are known which include the maximum magnitude 6.5 Koyna earthquake of 1967, M 6.4 Marathwada earthquake of 1993 which killed about 8,000 persons and M 6.0 earthquake in Jabalpur area in 1997 and their aftershocks. Thus, the region may be seen to have less seismicity, except for a few small subregions, where maximum Intensity VII on the MSK Intensity scale may be considered probable.

1.2.7 The Andaman - Nicobar Islands

This region is highly seismic having witnessed about 190 earthquakes of M > 5.0 among which may be mentioned the 1941 giant earthquake of Magnitude 8.1. The main town of Port Blair had also experienced damage besides other civil and military installations. The risk of damage here is increasing with the rapid rise of population and larger use of un-reinforced masonry. The other danger is from under-sea occurrence of large (> 7.0 Magnitude, which may cause a tsunami such as the one that was caused by a M = 9.3 earthquake off Sumatra coast on December 24, 2004

1.3 The Seismic Hazard Zoning Map

It is thus seen that the seismic hazard problem in the country covers a substantial part of the land. A seismic zoning map has been standardised by the Bureau of Indian Standards (see IS:1893-Part 1, 2002, Fig.2) which has classified four seismic zones based on experience as well as expected probable intensities on 12 point MSK Intensity scale. Zone V with probable intensities of IX and larger covers an area of 10.8%, Zone IV with probable intensity VIII covers 17.5% and Zone III with probable Intensity VII covers 30.8% of the land area of India. In these three zones, destruction and substantial damage to structures and systems is probable. Thus, an area of about 59% is liable to earthquake damage. Zone II based on MM VI or less could also have minor damage in rural buildings built in field stone, that is, random rubble and adobe and clay mud, but collapse of housing in Zone II is not considered probable.

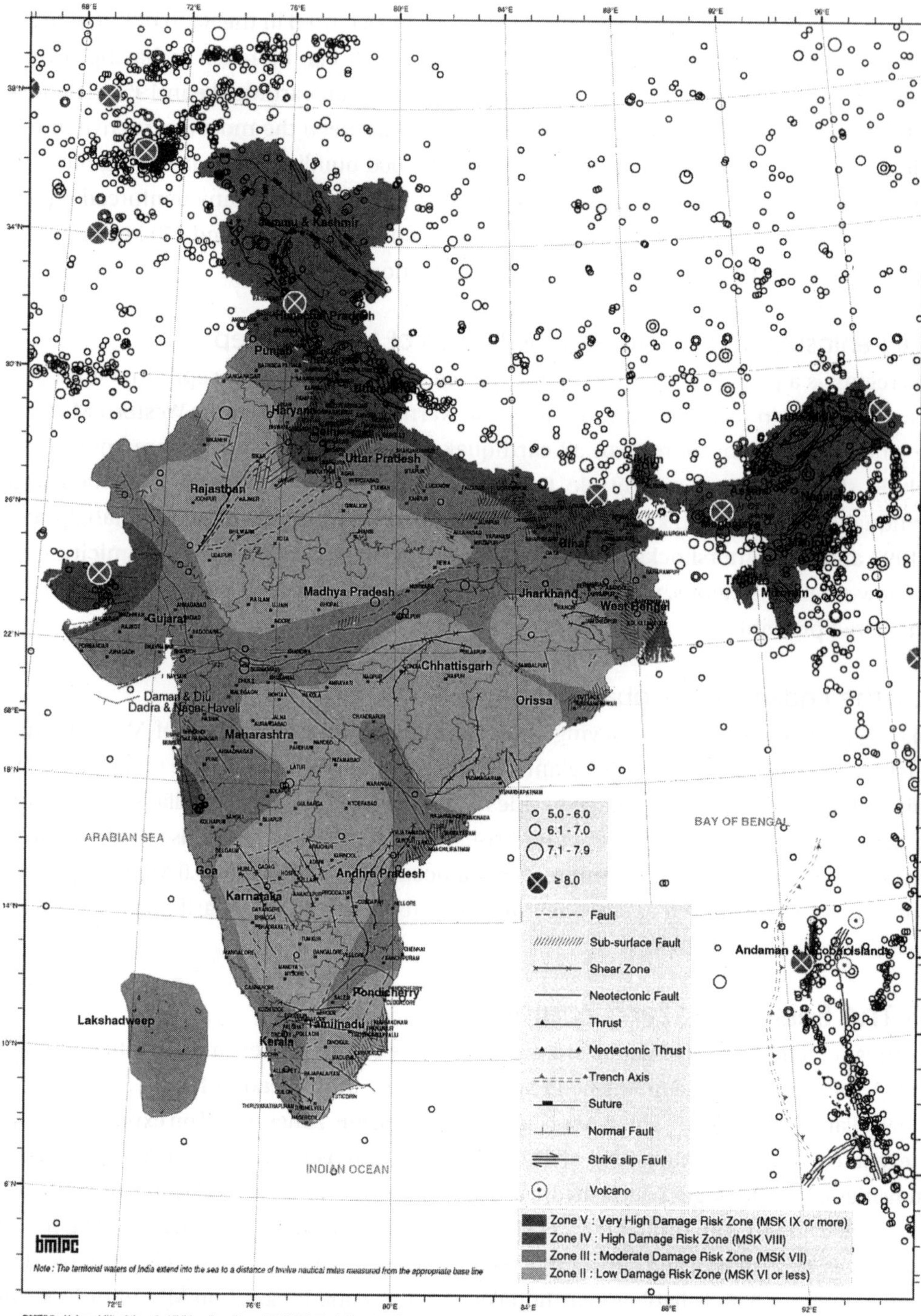

BMTPC : Vulnerability Atlas - 2nd Edition; Peer Group, MoH&UPA; Map is Based on digitised data of SOI, GOI; Seismic Zones of India Map IS:1893 - 2002, BIS, GOI, Seismotectonic Atlas of India and its Environs, GSI, GOI

Fig. 2 Seismic Zoning Map of India

1.4 Damaging Effects of Earthquakes

Earthquakes are the greatest destroyers of man-made buildings, small or big, as well as bridges, power plants, industrial structures and the like. During the 1934 Bihar-Nepal earthquake, railway lines were bent, twisted and shifted, bridge foundations and abutments were badly tilted and superstructures thrown down, numerous rock slides and rock falls occurred in the hills of Nepal, the ground liquefied and subsided in an area 250 km long and 100 km wide and sand fountains occurred in thousands

The overall earthquake effects are indicated in Fig.3 The intense ground vibrations caused by the seismic waves primarily cause damage to soils and structures. But sometimes, secondary effects of these damages may become even more disastrous. Fires caused as a result of the 1923 Kwanto earthquake in Japan caused the death of more than 100,000 people and consumed more buildings than those damaged by the earthquake. Huge floods were caused in the Brahmaputra as a result of the 1950 Sadiya earthquake which destroyed the town of Dibrugarh more than the direct effect of the shaking. Earthquake disaster mitigation planning must take both the primary and secondary effects into consideration.

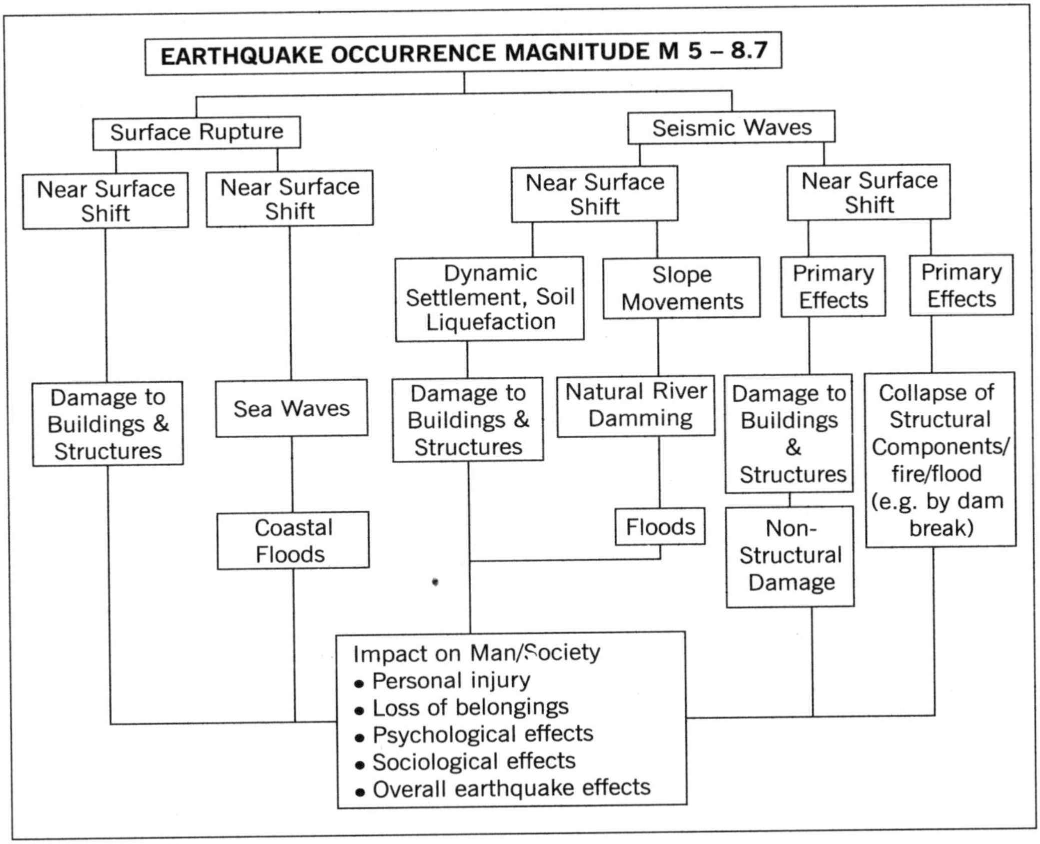

Fig.3: Economical Effects

1.5 Earthquake Prediction

An earthquake prediction, to be meaningful, involves three necessary elements; first, the location of the focal region within an area of about 50-100 km dia; second, the likely Magnitude of the earthquake or the likely damaging Intensity in the epicentral area and third, the time when the event is likely to occur (years, long range, months, medium range or days, short range, etc.). A prediction which does not specify any of these three elements accurately is completely meaningless so far as its usefulness to society is concerned.

Table 3 gives the recorded data of earthquakes in India with M=5. The regions of the country are seen to have an average return period of Magnitude 5.0 or larger earthquakes. Thus, a long range prediction referring to a region taken as a whole, can easily be made, but without useful meaning of the prediction.

Table 3: Earthquake Occurrence in India, M = ≥ 5.0 (Richter Scale) (July 1505 to September 1996)*

S.No.	Seismic Region	No. of Earthquake having M=				Max MSK Intensity	Average Return Period Observed for M≥5.0
		5-5.9	6-6.9	7.79	8.0 or more		
1	Kashmir & western Himalayas (J&K, Himachal Pradesh, Sub-mountain parts of Punjab)	135	35	11	1	X	7 months
2	Central Himalayas (UP & Nepal Himalayas, North Bihar)	104	28	4	1	XI	9 months
3	North-East India	339	157	20	4	XII	2 ½ months
4	Indo-Gangetic basin & Rajasthan (Rajasthan, Punjab, Haryana, Delhi, Plains of UP, Bihar & Bengal)	82	26	5	-	VIII	1 year
5	Cambay and Rann of Kachchh	12	4	1	1	IX -X	6 years
6	Peninsular India	24	8	-	-	VIII	3 years
7	Andaman & Nicobar	136	54	2	1	> IX	6 ½ months
	Whole India, including those before 1897*	**832 (32)**	**312 (40)**	**43 (18)**	**8 (1)**	**XII**	**1 month**

* Total earthquakes as per IMD Catalogue 1195, including 91 earthquakes before 1897 based on historical records.

The point to examine here is whether a successful prediction is feasible at the present juncture of knowledge. Precisely speaking, there has only been one successful very short range prediction in 1975 in Hicheng, China, where the scientists followed the recorded seismic indications a few weeks and few days before (supplemented by other physical observations like animal behaviour a few hours before) the actual occurrence, and issued warnings a couple of days before. The Government took effective action to evacuate the people from their homes. Thus lakhs of lives were saved and only a few were killed even under a M=7.3 earthquake. But none of the houses or other construction could be saved because they were as weak as our constructions and no retrofitting was feasible in the short time available between the prediction and occurrence of the earthquake. The examples of a prediction not coming out true and a prediction not being possible at all also are available from China. An impending earthquake prediction in Beijing kept the people under fear for more than a month before the prediction was called off. In 1976, the city of Tangshan was completely destroyed, killing more than 242,000 people. Here the prediction could not be made at all. Prediction of earthquakes, therefore, at present, cannot be used as a strategy for earthquake safety since either it is not possible at all or is completely unreliable. Moreover, unless prediction be done sufficiently in advance so that some physical safety measures to man-made constructions can be taken, the economic ruination will definitely take place in spite of the prediction.

The seismic zoning of India, based on probable maximum earthquake intensities (see Fig.2)can be looked up as *long range prediction map* based on which an appropriate policy will be to have proper preparedness, using accepted mitigation measures. Again, taking the example of China, after the 1976 Tangshan earthquake, a policy was made to examine existing buildings in most towns and cities which were lying in Intensity VIII and Intensity IX Zones and to retrofit weak buildings such that the collapse and severe damage of these buildings may be avoided in future earthquakes. In a 10-year programme, 120 million square metres of building floor space has already been retrofitted by using measures which are well known to us as flowing from IS:4326-1993 (see IS:13935 First Draft Revision).

1.6 Earthquake Hazard Risk to Urban Areas

In some of the past earthquakes, a number of cities of the world have been adversely impacted. The cities which had to be substantially rebuilt after an earthquake include Tashkent (Uzbekistan, then USSR), Skopje (Macedonia, then Yugoslavia), Anchorage (Alaska, USA) and Tokyo & Kobe (Japan). In India, many towns in North Bihar were destroyed in the 1934 earthquake, Kangra and Dharmsala towns were destroyed in the 1905 earthquake, and more recently, the cities of Jabalpur in the 1997 earthquake and Bhuj, Gandhidham and Ahmedabad were severely damaged in the Kachchh earthquake of 2001 which caused not only deaths but also huge economic losses and trauma to the people. Since urbanisation is occurring now on a large scale, the existing cities are expanding in all directions, even covering those areas which were hitherto considered unsuitable for human habitation. In order to draw the attention toward earthquake hazard risk in urban areas, a brief list of cities situated in Seismic Zones V, IV and III is given in Table 4.

Table 4: Seismic Status of Some Important Towns

(1) Seismic Zone V (Maximum Intensity Expected IX or More)			
Agartala	Aizwal	Bhuj	Darbhanga
Dibrugarh	Dispur	Gwahati	Imphal
Itanagar	Jorhat	Kohima	Madhubani
Mandi (HP)	Sadiya(NE)	Shillong	Srinagar (J&K)
Tezpur			
(2) Seismic Zone IV (Maximum Intensity Expected VIII)			
Almora	Ambala	Amritsar	Bahraich
Barauni	Chandigarh	Darjeeling	Dehradun
Delhi	Faridabad	Gangtok	Gorakhpur
Haridwar	Jammu	Ludhiana	Mathura,
Meerut	Monghyr	Moradabad	Nainital
Patna	Pilibhit	Roorkee	Simla
(3) Seismic Zone III (Maximum Intensity Expected VII)			
Agra	Ahmedabad	Asansol	Bhubaneswar
Bokaro	Chennai	Cochin,	Coimbatore,
Cuttack	Dhanbad	Durgapur	Gaya
Indore	Jabalpur	Kanpur	Kolkata
Lucknow	Mumbai	Nasik	Panaji
Pune	Puri	Rajkot	Surat
Trivandrum	Vadodara	Varanasi	Vijayawada

1.7. Conclusion

From the above presentation, it is seen that the risk of earthquake hazard covers about 59% of the total land area of the country, classified into three seismic zones namely III, IV and V. Damaging earthquakes have occurred in all three zones recently in a short time span of 15 years between 1991and 2005 in which not only the rural areas but also urban centres have been adversely damaged. In view of the rapid urbanisation in the country, the risk to larger populations is increasing which will call for concerted efforts towards prevention and mitigation of the risk to a larger population as compared with the earlier times. ■

1.8 References

1. Arya A.S. et al, *Techno-Legal Aspects of Earthquakes, Windstorm and Flood Hazards and Land Use Zoning* BMTPC, 1998
2. Arya A.S. el al, *(editors) Earthquake Engineering in India,* Jai Krishna 60th Birthday Volume, Ch.1, Sarita Prakashan, Meerut.
3. *Catalogue of Indian Earthquakes,* India Meteorological Department, 1996.
4. IS: 1893 (Part1)-2002, *Criteria for Earthquake Resistant Design of Structures,* B.I.S., 2002.

Chapter 2

Design of Structures Against Earthquake Motions[1, 2]

2.1 General Nature of Earthquake Motions

In an earthquake, the ground moves in all directions in nearly random motion. The total motion may be represented by two horizontal components at right angles to each other and one vertical component. Acceleration records[1] of the Koyna Earthquake of December 11, 1967 in three mutually perpendicular directions are shown in Fig. 1. The time-wise velocities and displacements of the ground can be computed by integrating the acceleration with time. Such curves[2] for the transverse component of the Koyna earthquake are shown in Fig. 2. The general nature of the earthquake motions is indicated by these two figures. It can be seen that the highest intensity peaks of acceleration have a relatively short period, or inversely, high frequency. It is of the order of 12 cycles per second in this earthquake

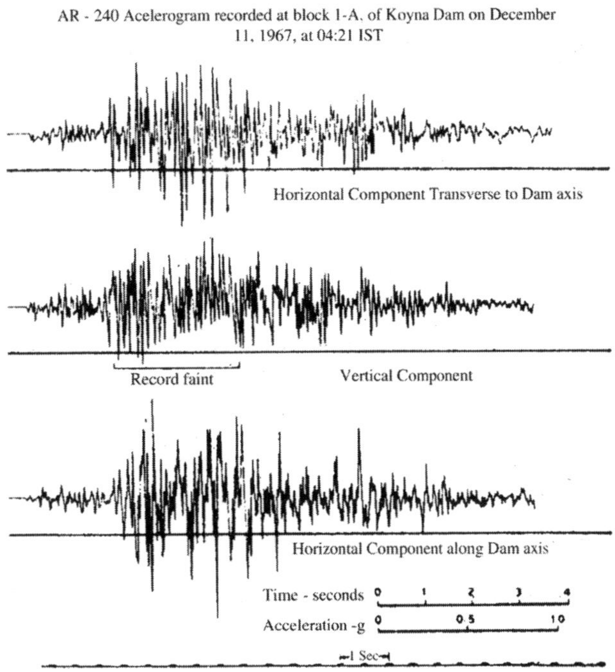

Fig. 1: Accelerograms of Koyna Earthquake (December 11, 1967

1. *Dr. A.S.Arya, Professor & Head, School of Research & Training in Earthquake Engineering, Univ. of Roorkee, Roorkee.*
2. *Civil Engineering Journal, the Institution of Engineers (India), Vol.51, July 1971.*

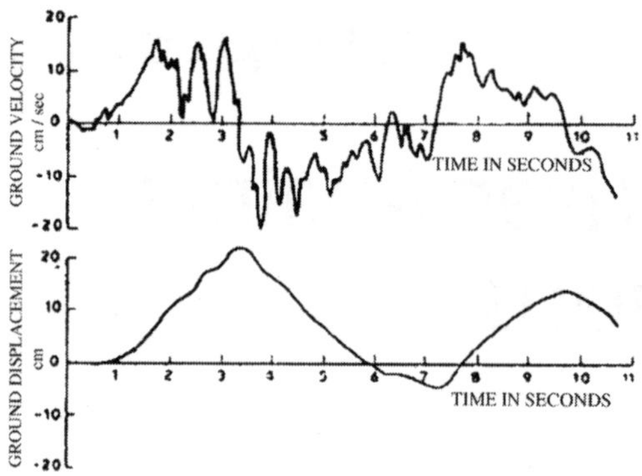

Fig. 2: Ground velocity and displacement for transverse component of Koyna Earthquake

which was recorded on stiff rock. In other earthquake motions recorded on alluvial soil, the frequency has been much lower, of the order of 2-5 cycles per second. The important peaks of velocity have a longer period than the acceleration peaks and the important peaks of ground displacement have a still longer period.

In a given seismically active area, earthquakes of different magnitude occur from time to time, the smaller earthquakes being more numerous than the larger ones. The largest recorded earthquake has naturally the smallest chance of occurrence. At the same time, there is every chance of the largest recorded earthquake being exceeded by a bigger shock in future, the probability of occurrence of which will be still smaller. It is also likely, though with the least probability of occurrence, that an abnormally large, unexpected earthquake may occur which may exceed even the one estimated to be the maximum. Thus, for a structure located in an area of active seismicity and planned to have a long life such as a dam, a major bridge or a monumental building, three earthquake magnitudes will be of interest as described below.

2.2 Frequent Earthquakes

Those earthquakes which are occurring frequently during the life span of the structure are included in this category. For such motions, obviously, the structure must remain undamaged. Therefore, all parts of the structure must remain elastic and the deflections should be small so that even the non-structural parts such as glass panes, window frames, partition walls, etc. are not damaged.

2.3 Probable Earthquake (Design Basis Earthquake)

It is one which may be expected once during the life of the structure. This will generally be taken larger than the maximum recorded earthquake and may be projected from the recorded one to get some idea regarding the peak ground acceleration values, frequencies and total duration. To design the structure for such an earthquake without any damage will be highly uneconomical. Therefore, some minor damage shall have to be accepted in such a case. The structure may be allowed to undergo yielding at *some* sections developing energy absorption in the inelastic deformations but the deformations must be kept within such limits that the structure will not collapse due to secondary effects like instability of columns, crushing of concrete, brittle fracture, elastic or inelastic instability, etc. There should, in fact, be ample factor of safety against collapse, and the permanent deformations after the earthquake should not reduce the functional efficiency of the structure.

2.4 Abnormal Earthquake (Maximum Credible Earthquake)

It is one which is the maximum potential of the seismo-genic features in the area and will exceed all expectations. Obviously, the structure cannot be expected to be designed to have the necessary strength for such an eventuality and will have to undergo considerable damage and deformations. However, it must be designed so as to *preclude collapse* and save human lives and the material contents of the structure. The extent of deformations will now be determined by considerations of safety from collapse rather than by functional efficiency.

The determination of the magnitudes of the earthquakes and the intensities of shaking which will constitute the 'frequent' or 'maximum probable' or the 'abnormal' earthquake at a given site is an extremely complex, rather unsolved, problem in engineering seismology requiring long-time statistical study supported by instrumental measurements of earthquake motions and ground deformations across faults. Such estimates have been made for the California region in the USA and the values[3] are given in Table 1. The recorded motion given in the table has the maximum damaging intensity of all the earthquakes for which strong motion records are available. The maximum probable and abnormal earthquakes indicated in the table are considered as not likely to be exceeded anywhere in the world. For most situations, however, these may be too conservative to use.

Table 1 - Probable Parameters of Max^{m} Ground Motion of Major Earthquakes[3]

Case	Maximum acceleration g	Maximum velocity cm/s	Maximum displacement, cm	Duration of major motion, s
Maximum recorded at El-Centro in California, May 1940, N-S component	0.32	35	30	30
Maximum probable	0.50	60-75	60	90-120
Abnormal^{+}	0.5-0.6	75-90	90-120	120-240

+Lower values for rock and higher values for alluvium.

2.5 Design Acceleration versus Actual Acceleration

As stated earlier, the design accelerations specified in codes are much less than the elastic response accelerations expected in actual earthquakes. Table 2 shows a comparison of code values with those under-actual recorded earthquakes in the Koyna area in India and California in the USA for a few fundamental periods of structures (5,8). A damping value of 5 % which is commonly used for reinforced concrete structure is adopted. The period range considered will cover practically all structures except the extremely tall buildings or long span cable-bridges.

From Table 2, it is clearly seen that the accelerations caused during actual recorded earthquakes are several times larger than the design accelerations specified in the same regions. The maximum probable and abnormal earthquakes will naturally have much higher *intensity* than the recorded earthquake. From the design point of view, this appears to be a hopeless situation but luckily the well designed structures in reinforced concrete and steel have the property of shaking down the load as they deform plastically.

Table 2 - Comparison of Design Acceleration With Elastic Acceleration Under Recorded Earthquake (Ordinary Structures Without Importance Factors, Damping = 5% of critical).

Fundamental period, sec	Design value	Koyna Earthquake 1967 Long.	Ratio (3)/(2)	Design value SEAOC code	El Centre 1940 NS Comp.	Ratio (6)/(5)
(1)	(2)	(3)	4)	(5)	(6)	(7)
0.2	0.076	0.70	9.2	0.086	0.644	7.5
0.5	0.064	0.40	6.2	0.063	0.836	13.3
1.0	0.040	0.17	4.2	0.050	0.518	10.4
1.5	0.030	0.11	3.7	0.044	0.190	4.3
2.0	0.024	0.13	5.4	0.050	0.178	4.5
2.5	0.020	0.12	5.0	0.037	0.176	4.8

2.6 Structural Response

The response of a structure depends upon its own period and damping characteristics besides the nature of the ground motion. For a single degree elastic system, the maximum response of acceleration, relative velocity or displacement may be presented in the form of response spectra such as shown in Fig. 3 for the Koyna earthquake for many values of damping coefficient from 2-20% of critical. For the N-degree elastic system, the response spectra for single-degree elastic system may be used by using the normal mode theory and treating each mode as a single-degree system. The results obtained in various modes may be superimposed to obtain the total response.

2.6.1 Elastic Acceleration Response Spectra

The shape of the elastic response spectra is of interest as it is very much related with the periods associated with the ground motion. It was seen earlier that the periods associated with the ground acceleration are the shortest and those associated with the ground displacements are the longest, the periods of ground velocity being intermediate between them. The spectra show their peaks accordingly. The peaks in the acceleration spectra lie in the low period range and the displacement spectra have the maximum values in the long-period range. The decrease of acceleration spectra with increasing time periods gives the property of shaking down of load to those structures which have the stiffness characteristics of softening spring type, as shown in

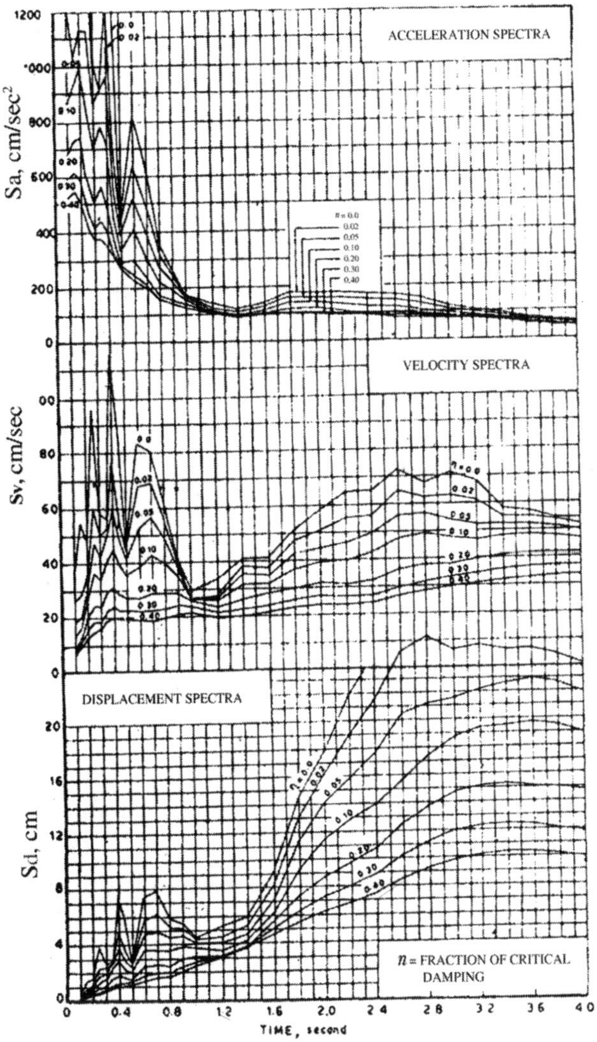

Fig. 3: Response spectra for elastic systems (Koyna Earthquake, December 1967, component along axis of Koyna Dam)

Fig. 4. For such a structure, when the lateral load increases, the deflection will increase more than proportionately; the stiffness will decrease and the time period will increase. Thus, the acceleration response spectrum will decrease. That is, the lateral forces induced in the structure will drop down. Here, the effect of inelasticity has also been considered indirectly.

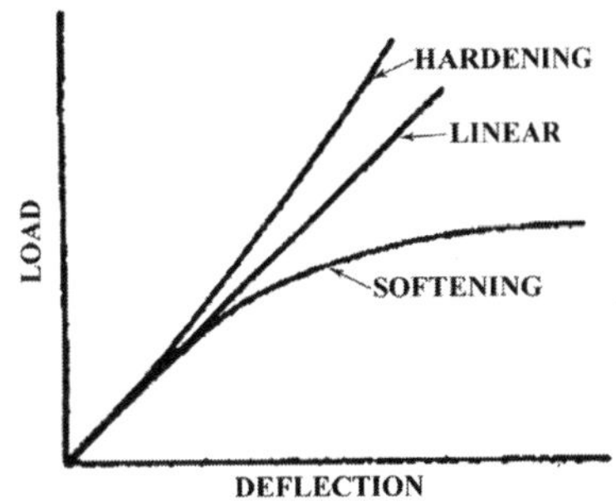

Fig. 4: Spring stiffness characteristics

2.6.2 Inelastic Response

Now the consideration of a non-linear curve in the dynamic analysis is inconvenient, it is customary to represent it by means of a bilinear or elastoplastic curve (Fig. 5). For a single-degree inelastic system, the maximum response can be computed by dividing the values of elastic acceleration spectra by *certain reduction factors corresponding to the extent of plastic deformation.* The plastic deformation is expressed by means of *ductility ratio* which is defined as *the ratio of maximum deflection to the yield point deflection* such as Δ_m/Δ_y shown in Fig. 5.

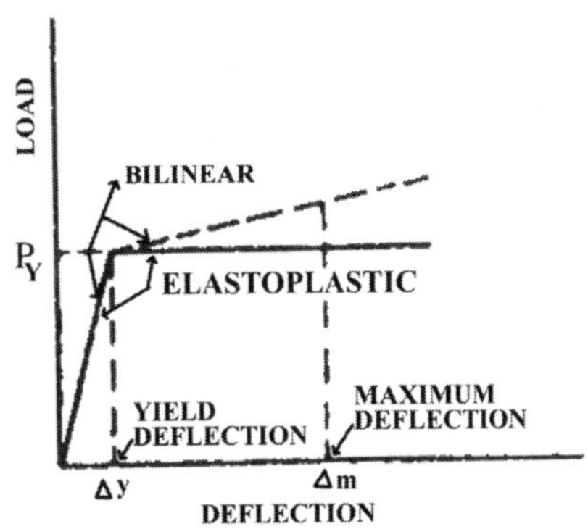

Fig. 5: Idealised load-deflection characteristics

The reduction factors[4] depend upon the earthquake motion, the damping coefficient, the time period and the shape of the load-deflection curve. The most important factors, however, turn out to be the ductility ratio and the time period. For a single degree, elastoplastic systems having damping ratios from 0.02 to 0.05 in the elastic range, the reduction factors given in Table 3 may be used for finding the acceleration that would cause yielding of the structure. In this table, T is the undamped natural period of the structure and μ the ductility ratio. For finding the *design value of acceleration,* the yield acceleration may further be reduced by dividing it by the load factor, that is, the ratio of the yield load to the design load. For steel or concrete structures, the load factor[5] may be taken as 1.4.

Table 3 - Reduction Factors For Single-Degree Elastoplastic Systems

Ductility Ratio μ	Reduction Factors for				
	T=0.5 or less	T = 1.0	T = 1.0	T = 2.0	T = 2.5 or more
2.0	2.9	2.2	2.2	1.7	3.0
3.0	4.7	3.1	2.9	2.4	5.0
4.0	6.4	3.8	3.6	3.6	7.2
5.0	8.4	4.5	4.5	5.0	9.7
6.0	9.3	5.1	5.0	6.2	12.3

We may now examine again, in the light of the above discussion, the structure having T = 0.15 considered in the above section. If a ductility ratio of 3 is permitted, the reduction factor will be 4.7 for yield load and 4.7 x 1.4 = 6.5 for design load. Therefore, a design seismic acceleration of 1.4 g/6.5 = 0.215 g would be adequate. It is easily seen that by permitting a larger amount of ductility ratio, the design coefficient may be further reduced. For □= 5.0 in this case, the design value will become 0.118 g only. The converse of this is also true. That is, a *structure designed for a given seismic coefficient can resist larger shocks by further yielding provided it does not collapse during the plastic deformations* due to other effects like buckling, elastic or inelastic instability, diagonal tension, bond failure, etc.

For multi-storeyed structures, however, such a simple scheme cannot be adopted for obtaining design values in the inelastic case because here the ductility ratios may vary from one storey to the other. In fact, all the ductile deformation may be concentrated in one storey and others may remain elastic or all of them may deform inelastically. An optimum design will perhaps be one in which all storeys have the same ductility ratio. This can, of course, be achieved for only one particular waveform of the earthquake. The important point, however, is that the storeys should be capable of deforming inelastically so that the effective seismic coefficients should be reduced to what the storey could resist. The same arguments may be applied to other multi-degree systems.

2.7 Energy Approach

An alternative approach to the determination of effective seismic coefficients is to consider the *energy absorption by the structure during deformations* for balancing the dynamic energy demand by the earthquake on to the structure. Such an energy balance may be assumed as a necessary requirement for stability of the structure at the instant of time when the structure has the maximum velocity response. Thus, if S_v is the undamped velocity spectrum value and W the weight of the structure, the maximum energy demand will be $WS_v^2/2g$. The structure must be capable of balancing this energy by partly storing as elastic strain energy, partly dissipating it by damping and partly absorbing by plastic deformation at that instant of time. If S_v is the damped velocity spectrum, dissipation of energy by damping is automatically taken care of. Then, the energy absorption by plastic deformation constitutes a much larger share of the remaining energy, as shown by the areas in Fig. 6(i). This also indicates requirement of large plastic deformations for resisting large shocks because when such deformations occur, not only the energy absorption increases, but the increase in energy input also decreases and the reconciliation of the energy takes place[6] at some displacement Δ_m as shown in Fig. 6(ii).

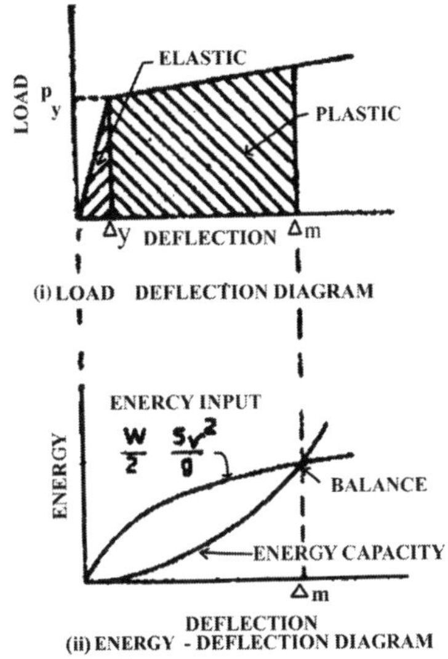

Fig. 6: Energy balance vs plastic deformations

2.8 Design Considerations

From the foregoing discussion, the following design requirements emerge for safety, efficiency and economy:

(i) Choice of suitable earthquake spectra
   - (a) for frequent earthquakes determining the elastic strength of structure,
   - (b) for probable maximum earthquake requiring small amount of ductility,
   - (c) and for abnormal earthquake requiring large amount of plastic deformations.

(ii) Design of structural members and their connections such as to have the necessary design strength and ductility so that the deformations could be sustained without crushing of material, brittle fracture, elastic or inelastic instability of any type in members, their elements or connections.

Whereas the second requirement can be achieved by proper design of members and their connections, the first requirement is subject to lot of uncertainty. However, until more precise information becomes known, a procedure is outlined below which may be adopted for practical designs.

2.9 Design For Frequent Earthquakes

2.9.1 Use of Average Spectra

Where the actual recorded accelerogram is not available, the smoothed average acceleration spectra may be used for finding elastic structural response for the given period and damping. Such spectral values are presented in IS:1893* (Reference 5) for various time periods from 0.2 to 2.8 s and four values of damping ratio, namely, 0.02, 0.05, 0.10 and 0.20. The average spectral value is to be multiplied with a multiplying factor F so as to account for the seismic activity of the region. IS: 1893 draft revision5 specifies the values of factor F for the five seismic zones of India as given in Table 4. For multi-degree systems, the modal analysis using root-mean-square superposition may be adopted for finding the shears and moments at all points of the structure. In the case of multi-storeyed buildings, the base shear may be determined in this manner but the lateral forces at all higher levels may be calculated by the parabolic law as described below.

Table 4 - Multiplying Factor F And Seismic Coefficient For Various Zones

Seismic Zone5	Multiplying Factor, F	Maximum value of seismic coefficient
V	1.000	0.12
IV	0.667	0.08
III	0.500	0.06
II	0.333	0.04
I	0.167	0.02

2.9.2 Use of Seismic Coefficients*

As an alternative to the use of average spectra, the seismic coefficients specified in the Indian standards5 may be used. Of course, the seismic coefficients are to be read along with other provisions in the code such as modifying factors to account for the flexibility and rigidity of structures, and the specified shear and moment distributions are to be adopted along the height of the structure. For instance, in the case of multi-storeyed buildings, the seismic coefficients are to be multiplied by a factor C, which defines the effect of flexibility of building with the increase in the number of storeys and the seismic lateral load is specified to

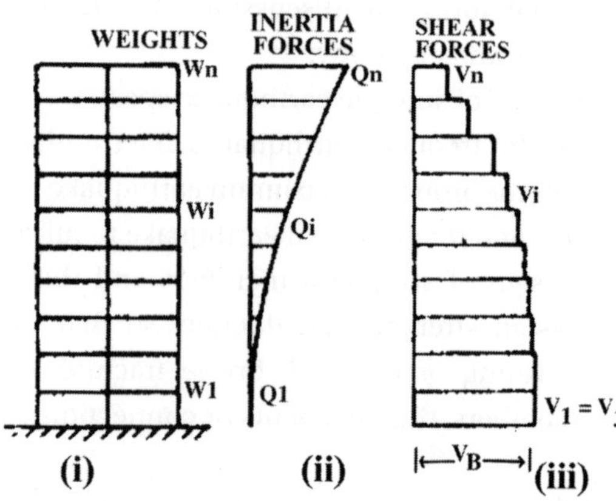

Fig. 7: Earthquake forces and shears in buildings

* *For the current design acceleration spectra and procedure, refer to IS:1893 (Part-1)-2002*

have parabolic distribution along the height with the maximum value at the top of the building, as shown in Fig. 7. It may be mentioned that the parabolic distribution of seismic forces gives considerable uniformity in the values of shear forces along the height of the building which is extremely useful in the behaviour of the structure in the inelastic stage[7].

Alternatively, the ultimate strength method of design may be adopted using a load factor of 1.4 for reinforced concrete and steel structures as follows:

$$U = 1.4 (L + D + E)$$

where U is the ultimate design load or moment, L the live load effect, D the dead load effect, and E the earthquake effect. Thus, the total loads, forces and moments as found for dead, live and earthquake loads together may be increased by the factor 1.4 and the members and their connections designed by the ultimate strength method. If an increase in section or reinforcement is found necessary over the required for the normal loads, it may be incorporated at this stage.

2.10 Design for Larger Shocks - Provision of Ductility

The strength as provided above for 'frequent' earthquakes should be considered as the minimum requirement. Provision for larger earthquakes by increasing the strength will require additional cost due to increased dimensions of the members and connections. Therefore, for economy, yet to have the necessary safety, the secondary effects which may lead to premature collapse, such as crushing of concrete or instability of members and their elements, should be checked and such details of design adopted which will ensure *maintaining the ultimate strength at constant value during further deformations* of the structure. For example, let us consider the load-deflection behaviour of four alternative designs of a member shown in Fig. 8. All of them are intended to have the same ultimate strength but different details of design are adopted leading to different load-deflection curves. Design A presents a brittle behaviour with little ductility. Plain concrete, unreinforced masonry, reinforced concrete columns with largely-spaced lateral ties, slender columns, plate elements with large slenderness ratios, laterally unsupported beams have this type of behaviour and are unsuitable for earthquake resistance. Design **B** where the strength goes on decreasing with deformation is also unsatisfactory because of the progressive failure. Such a characteristic is obtained with some masonry shear walls bounded by a frame of inadequate design. Design C has enough ductility for perhaps the maximum probable earthquake but not for the eventuality of an 'abnormal' earthquake. Design **D** may be adequate in all respects. Therefore, the requirement of ductility is to develop designs of type **D**.

What ductility ratio will be adequate is now the problem. This can be discussed on the basis of the El-Centro earthquake of May, 1940-NS component which was earlier stated to be the most intense earthquake recorded upto date. The multiplying factor F for this earthquake is 2.7 with respect to the average acceleration spectra. In the absence of more precise information, let us assume that for seismic Zone V of IS: 1893 (draft[5] revision), the maximum probable earthquake will be 1.5 times as intense and that the abnormal earthquake may be twice as large as the El-Centro quake. Thus, we get the values of F as 4.05 and 5.40, respectively. Since a value of F = 1.0 has been adopted for Zone V for elastic design and the load factor adopted is 1.4, it may be assumed that the reduction factors required to be achieved by ductility will by 4.05/1.4 = 2.9 for maximum probable and 5.4/1.4 = 3.8 for the abnormal earthquake. Thus, referring to Table 3, the ductility ratios required for various time periods will be as shown in Table 5. It, therefore, follows that a ductility ratio of about 3.0 will be adequate for 'maximum probable' earthquake and about 5.0 for 'abnormal' earthquake. The same factors may be used for other zones where the design earthquake, the 'maximum probable' and the 'abnormal' earthquakes may be assumed to be reduced proportionately to the coefficients for elastic design.

Table 5 - Required Ductility Ratio for Large Earthquakes

Time period, S	Maximum probable earthquake	Abnormal earthquake
0.5 or less	2.0	2.5
1.0	2.9	4.0
1.5	3.0	4.2
2.0	3.4	4.2
2.5 or more	2.0	2.4

2.11 Conclusion

The study shows that the seismic coefficients or F-factors specified with average acceleration spectra are by themselves adequate for small shocks only which may occur rather frequently during the life time of the structure. For large shocks, the strength so provided will be adequate only if the structure has ductile deformation capacity without decrease in strength. It is shown that if a ductility ratio of about 5.0 is available in the structure, it will be able to withstand without collapse even an abnormally large earthquake that may occur rarely in the life of the structure. Of course, premature collapse due to buckling, instability, diagonal tension or bond failure must be prevented by appropriate design details.

2.12 Acknowledgements

The author is thankful to Dr. Jai Krishna, Professor and Director, School of Research and Training in Earthquake Engineering, University of Roorkee, Roorkee for checking the manuscript and making useful suggestions. He is also grateful to Shri Krishen Kumar, Lecturer in Earthquake Engineering for the detailed computations of response spectra. ■

2.13 References

1. *Koyna Earthquake (December 11, 1967)*, Volume 2, Report of the Committee of Experts, Published by the Government of India, New Delhi, 1968.

2. Jai Krishna, et al., "Analysis of Koyna Accelogram of December 11, 1967", Bulletin of the Seismological Society of America, Vol. 59, No. 4, August, 1969, pp. 1719-1731.
3. N.M. Newmark, 'Effects of Earthquakes on Dams and Embankments'. Geotechnique, Vol. 15, No. 2, June 1965, p 141.
4. A.R. Chandrasekaran and S.S. Saini. 'Dynamic Response of Non-linear Systems', Journal of the Institution of Engineers (India), Vol. 48, No. 3, pt CI (SP), November 1967, pp. 645-664.
5. *IS: 1893-Indian Standard Criteria for the Design of Earthquake Resistant Structures, Second Revision (Draft)*, Published by the Indian Standards Institutions, New Delhi, January 1970.
6. J.A. Blume. "A Reserve Energy Technique for the Earthquake Design and Rating of Structures in the Inelastic Range". Proceedings of the Second World Conference on 'Earthquake Engineering', Tokyo, Vol.2, 1960, p.p 1061-1084.
7. K.Muto, "Recent Trends in High-rise Building Design in Japan". Proceedings of the Third World Conference on "Earthquake Engineering", New Zealand, Vol 1, 1964, p 130.
8. *Recommended Lateral Force Requirements and Commentary* pub. By Structural Engineers Association of California, 1968.

Chapter 3

Rational Seismic Coefficients for Earthquake Resistant Design[1,2]

3.1 Introduction

Since the introduction of the response spectrum technique in earthquake resistant design of structures (12-15), considerable progress has been made in the earthquake regulations of the world to include, in one form or the other, the dynamic properties of earthquakes as well as those of the structures. Important differences, however, still exist in the approach for determining the seismic coefficients for design not only from one code to the other in different countries (8), but from region to region or organisation to organisation within the same country (7) and from one structure to the other in the same code (17) also. Usually, base seismic coefficients are specified for different seismic zones, and certain modifying factors are included for taking other parameters into account. The response spectrum technique or dynamic analysis of structures is permissible as an alternative.

The aim of this paper is to state the various parameters affecting the design seismic acceleration at any point of structure; examine, if such parameters are given any weightage in the present codes; and suggest a unified approach to obtain rational seismic coefficients for designing various structures for earthquake motion. Reference will be mainly made to "Recommended Lateral Force Requirements and Commentary" by the Seismology Committee of Structural Engineers Association of California 1968 (22). "Earthquake Resistant Design for Civil Engineering Structures, Earth Structures and Foundations in Japan" compiled by the Japan Society of Civil Engineers, 1968 (7); and 'IS: 1893-1970, Indian Standard Criteria for Earthquake Resistant Design of Structures' by Indian Standards Institution, 1970 (17). These will be briefly referred to as SEAOC, JSCE and IS: 1893 Code respectively in the discussions that follow.

3.2 Earthquake Parameters

For determining motion at a given site, the following eq. parameters are important:

(i) Magnitude of earthquake as a measure of energy release, M;

---

1 A.S.Arya, Professor & Head, Deptt. Of Earthquake Engineering, University of Roorkee, Roorkee

2 Paper no.116, Bulletin, ISET, Vol.8, No. 4, December 1971, pp 178-187.

(ii) Depth of focus of earthquake, h;
(iii) Distance of site from the causative fault, D; and
(iv) The soil on which the structure will stand. This is the intervening medium between the base rock and the structure.

Several expressions are proposed in the literature for computing the peak ground acceleration at the site for the given M, h and D values (10, 20). One such recent expression (20) is as follows:

$$\alpha = \alpha_o \, e^{-x(D/h)^{1.5}} \quad (1)$$

where

α = peak ground acceleration expected at site

α_o = peak ground acceleration expected in epicentral area

x = attenuation factor depending upon the geological conditions in the region including the intervening medium such as soil.

From studies in the Koyna area after the Koyna earthquake of December 11, 1967 (11*), the attenuation factor was found to be 0.1 which is very low as compared with a probable value of α = 0.26 for several earthquakes (20). The reason is a very different geological condition in the Koyna area where more energy is transmitted through stiff rocks so that the attenuation is small.

In the resulting ground motion at site, besides the peak acceleration, the predominant frequency or number of zero crossings per second as well as the total duration of the strong motion are also important for the *effective* seismic intensity or *damage potential* of the earthquake. These two factors are, however, still indeterminate and can at best be estimated by comparing the conditions at the site with some recorded accelerogram in similar geological conditions. The basic data of earthquake engineering are the accelerograms which represent the net product of all the earthquake parameters (16).

3.3 Structural Response

The response of a structure to a given accelerogram depends upon the following factors:
(i) Mass distribution in the structure;
(ii) Stiffness distribution in the structure in the elastic range:
(iii) Dissipation of energy by internal friction, that is, damping; and
(iv) The complete resistance vs deflection diagram of the structure under repetitive loading upto ultimate, changes of stiffnesses with load level, hysteretic energy absorption, and deterioration of energy reserve under cyclic loading.

The interaction of the foundation soil with the structure including feed-back of energy from the structure to the ground could also be an important factor in determining structural response. Thus, the foundation soil has two effects: *first,* modifying the ground motion, and, *second,* interacting with the structure in its response to the ground motion.

In the ultimate behaviour of a structure, absorption or dissipation of energy by inelastic deformations plays an extremely important role. A brittle structure like an unreinforced masonry building, masonry retaining wall or a concrete or masonry dam, will depend only upon its strength in flexural and diagonal tension for resisting earthquakes and other forces. But a ductile structure of reinforced concrete, prestressed concrete or steel, in which all factors leading to premature failure have been eliminated by proper design details, can depend to a great extent on its *deformability* during severe ground shocks.

3.4 Approach to Design

From the facts stated in the preceding paragraph, it is seen that there are many uncertainties as well as difficulties in determining the true ground motion and the true structural response upto *ultimate* stage. For design purposes, therefore, simplifying assumptions have to be introduced.

Other factors which should be included in the design criteria are: the probability of occurrence of strong ground shaking, the nature of building damage which may be sustained without collapse or functional deterioration and the cost of repairing damage as compared with the additional cost for making the structure earthquake resistant. The design criteria must, however, lead to approximately a uniform factor of safety for different structures and for different parts of the same structure unless otherwise desired due to the vulnerability of the structure or its part.

The problem of structural design may be divided into two stages: *first,* design based on elastic stage; and *second,* design based on inelastic stage permitting ductile deformations. The design approach in the two stages is given below:

3.4.1 Design Based On Elastic Stage Only

This stage will apply only to brittle structures where no plastic deformations are intended to be permitted for the expected earthquake motion.

Fortunately the structural response in the elastic stage can be appropriately represented by displacement, velocity and acceleration spectrum curves (14) drawn for the given earthquake accelerogram. Since choosing an appropriate accelerogram is also a tedius task, the response spectra may be idealised for general design purposes such as the average spectra (12, 21) presented in IS: 1893-1970 on arithmetic scale and shown in Fig. 1*. While testing such average design spectra, the expected seismic intensity at a site may be accounted for by using the multiplying factor F. If this approach is used, all the earthquake parameters stated earlier reduce to the choice of this multiplying factor only.

For designing brittle structures in the elastic range, the maximum probable earthquake at site should serve as the basis of the factor F. If the record of such an earthquake is not

** The changes brought out in the current IS:1893-2002 Code should be seen as compared with the 1970 version of the code that was then used in the paper.*

available, the factor F may be chosen by comparison. For instance, for Zone IV of the IS Code, which is based upon Modified Mercalli Intensity VIII, the factor F may be taken to be that of the Koyna Earthquake of December 11, 1967 (Longitudinal Component). Its value is 1.35. For Zone V of the IS Code, based upon M.M. IX or more, the factor F may be chosen as 2.7 which is equal to that for EI-Centro Earthquake of May 18, 1940. N-S component.

For making use of the elastic response spectra, either the normal mode shapes and natural periods of vibration should be worked out and distribution of forces, shears and moments at various points may be obtained by the super-position of adequate number of modes. Alternatively an appropriate approximate method (3) may be used to obtain an equivalent single degree of freedom system. Also an appropriate value of damping has to be assumed (see Ref. 17 App. C).

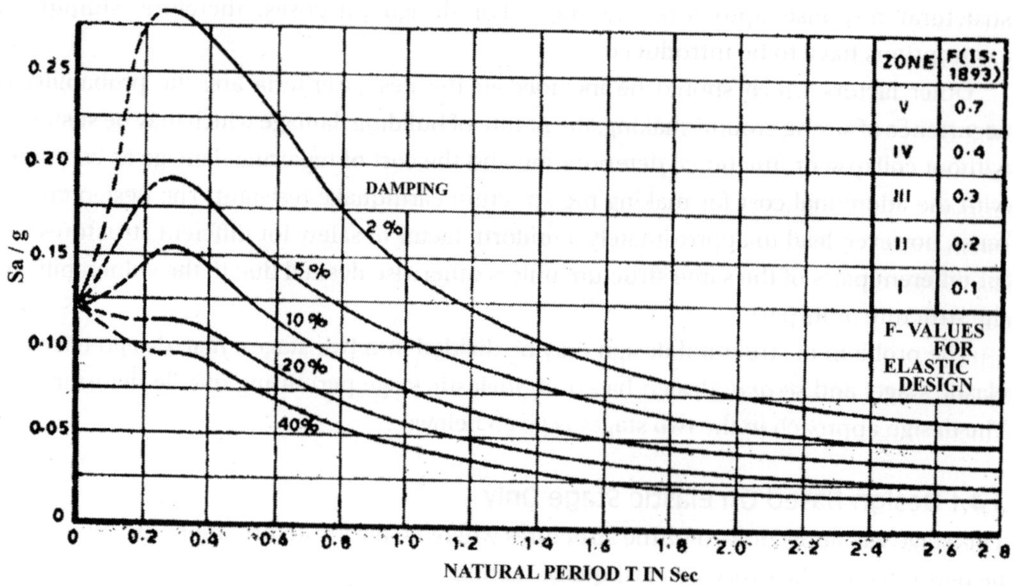

Fig. 1: Average acceleration spectra and Zone-Factors

The allowable stress for such structures must be based upon the cracking strength. If no factor of safety is desired over and above the probable maximum earthquake, stresses may be permitted equal to the cracking strength itself. However, since there is also the probability of occurrence of an abnormal earthquake exceeding the expectations, it will be desirable to adopt a factor of safety with regard to cracking strength.

Now it may be argued that a given seismic zone will neither have the same potential nor the same probability of occurrence of a given seismic intensity at all points in the zone. Moreover, all structures may not require the same immunity from earthquake damage in view of their purpose, expected life, etc., and economic consideration also will not permit the adoption of equal strengthening measures against earthquakes. Like insurance, the premium to be paid for earthquake resistance will depend upon such factors as importance of the structure to the community, its performance during shaking, life expectancy, consequences of its failure, etc. These points merit some discussion.

Since earthquake is an occasional not an every day phenomenon, structures expected to have a long life like dams, large bridges, monumental buildings, etc. should be designed to have a greater factor of safety. Structures which are important to the community before the earthquake as well as in the post-earthquake period like schools, hospitals, water towers, water reservoirs, etc., should be stronger. Finally, structures whose failure may result in a catastrophy such as a dam or nuclear power house, or too much dislocation in the public life such as a major bridge or too long closure of a major power house, will need greater provision against risks. On the other hand, temporary structures used for storage, etc. may not involve much risk, requiring much less or no safety provision against earthquakes.

Rational design for earthquake motion must take the above factors into account.

3.4.2 Design Based On Ductile Deformation

For single degree of freedom systems, it is shown (5,18) that when ductile deformations are permitted at constant strength (elasto-plastic) or somewhat increasing strength (bilinear), the system can resist with the given yield strength much larger ground shocks by undergoing larger plastic deformations under increasing intensity of ground motion. Inversely, it could be stated that for a given large earthquake, an inelastic system will require less yield strength to resist the earthquake as compared with an elastic system having the same natural period and damping as the inelastic system in its elastic range. Hence, when ductile deformations are permissible, a reduction factor may be used to calculate the required strength of the plastic system from that of an equivalent elastic system. In other words, the multiplying factor F of the elastic system could be reduced to take the ductility into account, or, *inelastic response spectra* could be constructed.

For multi-degree systems, however, the situation is more complex. Since a structure behaves non-linearly during inelastic deformations, the principle of superposition cannot be used. The structure must be investigated for ductility at various points by direct integration of the equation of motion (6). Where enough analyses have been carried out such as for multi-storeyed buildings, a reasonable distribution of shear force based on good inelastic behaviour may be adopted. Alternatively, the problem may be analysed by the reserve energy technique (4).

Other concepts in the approach to design are the same as for elastic design. The point mainly to be decided in this approach is: 'for what ductility ratio should the structure be designed?' This question could be posed in a different way as well. What should be the minimum earthquake as a fraction of the maximum probable earthquake for which the structure should be designed elastically? When this is ensured, the ductility requirement can be worked out for the maximum probable as well as the abnormal earthquake and details of the structure designed accordingly.

3.5 Design Seismic Coefficient

From the discussion of various factors presented in the foregoing paragraphs, it becomes evident that besides the zone factor F, other factors are also to be chosen to arrive at the design seismic forces. For most structures, the quantities of interest in design are the shear forces and bending moments at various sections. In a very large variety of structures for which accurate elastic and inelastic analyses have been made in the literature, it is convenient to express the values of forces and moments at other sections as ratios of their values at the base of the structure whereas the base shear and base moment are determined by using appropriate design seismic coefficients. Taking into account the various factors described, the design seismic coefficient may be expressed as

$$\alpha_h = C_D\, C_p\, C_F\, (F\, S_a/g) \qquad (2)$$

where

C_D = Reduction factor due to ductility

C_P = Performance factor to account for structural behaviour, importance, life expectancy and economic consideration

C_F = Soil-foundation factor

F = Seismic intensity factor depending upon seismic zone

S_a = Spectral acceleration from average Response Spectra for given fundamental time period and appropriate damping.

g = Acceleration due to gravity

3.6 Lateral Force for Design

Knowing the overall design seismic coefficient ah, the complete forces and moments in a single degree of freedom system are determined. However, in the case of multi-degree or continuous systems, the internal forces depend upon mode participation factors, mode shapes and spectral acceleration corresponding to higher modes. The design could, however, be simplified by adopting average ratios of forces as determined by dynamic analysis for each structural type, taking a wide range of parameters. In that case, only the fundamental time period is needed for finding S_a in the expression of α_h. Then using the design seismic coefficient α_h, the base shear and base moment may be found as follows:

$$V_B = C_{VB}\, \alpha_h\, W \qquad (3)$$

$$M_B = C_{MB}\, \alpha_h\, W\, \overline{H} \qquad (4)$$

where

C_{VB} = coefficient for shear at base depending on the structure

C_{MB} = coefficient for moment at base depending on the structure

W = total weight of structure including appropriate proportion of live load

$\overline{H}$ = height of centre of gravity of the weight of structure above its base.

Finally, the values of shear and moment along the height can be expressed in terms of the corresponding values at the base as follows:

$$V_x = C_{vx} V_B \quad (5)$$

$$M_x = C_{Mx} M_B \quad (6)$$

where C_{vx} and C_{Mx} are coefficients for shear and moment respectively at any point X of the structure.

The above approach has already been used in IS: 1893 for designing multi-storeyed buildings and stack-like structures and it could be extended to cover other structures like earth and masonry dams by analysing a wide range of these structures.

3.7 Values of Various Factors

The values of various factors in the expression for design seismic coefficient ah are discussed in the following and suggested values given. The other factors C_{VB}, C_{MB}, C_{vx} and C_{Mx} which depend on the type of structure are beyond the scope of the present paper and will not be considered.

3.7.1 Seismic Force Factor (F)

As stated earlier, one approach of fixing the seismic force factor F is with reference to maximum recorded accelerograms in areas of different Intensity Zones. Another approach could be to adopt the factor F for the most severe Intensity Zone and reduce it in other Zones approximately in relation to accelerations associated with various Intensities. Two such approximate scales are shown in Fig. 2 (9, 11). The ratios of accelerations for lower intensities to that of Intensity IX according to the Gutenberg and Richter scale are higher than those according to the Harshberger scale. Adopting the former conservatively, the ratios of average values of acceleration are given in Col. (4) in Table 1. The ratios of F values adopted in IS: 1893-1970 are given in Col. (5) for comparison. It can be seen that the ratios in IS: 1893 are on the conservative side, particularly for zones I to III. Adopting F=2.7 for Zone V and using the ratios as per IS: 1893 given in Col. (5), the values of factor F for various Zones are obtained as shown in Col. (6) which may be used for design.

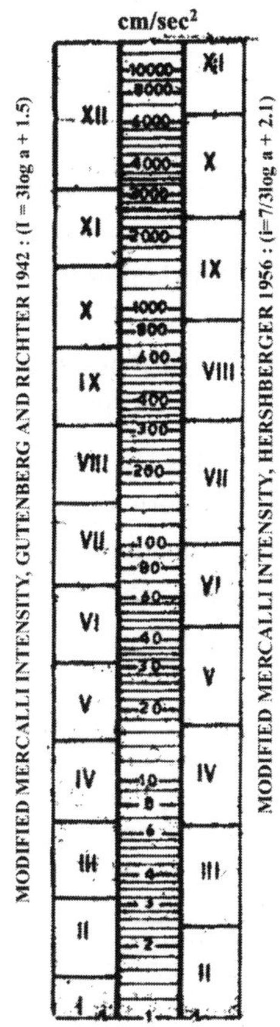

Fig. 2: Intensity vs Acceleration

The value of F for zone IV on the basis of the Koyna earthquakes of December 11, 1967, works out as 1.35 which compares well with the values given in Table 1.

Table 1:- Values of Seismic Zone factor (f)

Zone	M.M. Intensity	Range of acceleration cm/sec2 (Fig. 2)	Ratio of Average Acen. of Zone to that in Zone V	Ratio of values in IS: 1893-1970	Factor F using IS ratios
(1)	(2)	(3)	(4)	(5)	(6)
V	IXor more)	300-700	1.00	1.00	2.7
IV	VIII	150-300	0.50	0.56	1.6
III	VII	70-150	0.25	0.42	1.2
II	VI	30-70	0.10	0.28	0.8
I	V (or less)	15-30	0.05	0.14	0.4

The ratios of coefficients in various Zones in IS, SEAOC and JSCE codes are compared in Table 2. It appears that whereas IS and SEAOC are comparable, ratios in JSCE are on the high side, particularly in view of the acceleration ratios shown in Col.4 of Table 1.

3.7.3 Soil-Foundation Factor (C_F)

To take the influence of the soil-foundation system on the structural response into account, the factor C_F should increase the seismic coefficients for poor soils and poor foundations. The numerical values of C_F are not perhaps possible to find at this stage. The effect of soil has been considered in IS: 1893 Code since its first formulation in 1962. The values are now given more rationally in Cl. 3.4.2, Table 2 of the 1970 Revision taking into account the effect of type of foundation also.

Table 2:- Comparison of Ratios of Seismic Coefficients

Seismic Intensity	IS: 1893		SEAOC		JSCE		
	Zone	Ratio	Zone	Ratio	Zone	Ratio*	
						(1)	(2)
IX or more	V	1.0	3	1.0	High Intensity	1.0	1.0
VII-VIII	III, IV	0.42, 0.56	2	0.5	Medium Intensity	0.9	0.67
V-VI	I,II	0.14, 0.28	1	0.25	Low Intensity	0.8	0.5

*The ratio depends upon the code - (1) Buildings and Foundations; 1.0, 0.9, 0.8 (2) Railways and Harbours 1.0, 0.67, 0.5.

It may be noted that the SEAOC code does not consider the effect of the soil-foundation system on the seismic coefficients. But the design values of seismic coefficients in JSCE show a dependence on soil type. The multiplying factors adopted are 0.8 for hard soil, 1.0 for medium soil and 1.2 for soft soil. Here also, type of foundation is not taken into account. Since the observed behaviour of structures having different types of foundation during past earthquakes has clearly shown that pile and raft foundations are better, the coefficients in IS: 1893 are more rational and may be adopted.

3.7.3 Performance Factor (C_P)

This factor involves structural performance along with importance, life expectancy as well as economic considerations. The value of this factor is intimately linked with the reduction factor due to ductility and will vary from one structure to the other. For those structures which are designed elastically for the probable maximum earthquake, using a factor of safety of about 1.4 over the flexural tensile strength, no increase on account of importance or long life of the structure is necessary. For ordinary structures, on the other hand, like masonry retaining walls, brick buildings, etc. an economy factor less than 1.0 may be used, representing a reduction in the value of design seismic coefficient. That is, such structures may be designed for an earthquake much less than the probable maximum, thus, accepting some damage during a severe shock.

For those structures which can undergo plastic deformations without failure, the design factor will be reduced due to ductility ratio, C_D, discussed in (3.7.4).

Table 3: Values Of Coefficient Cp

Structure	Brittle*	Ductile**
Involving Risk (e.g. Dams, Reservoirs, Refinery Columns, Containers for inflammable gases or liquids, Atomic Power Plants)	1.0	2.0
Unimportant (e.g. storage structures, temporary structures, minor retaining walls along roads, small culverts where diversion easily possible, electric or telephone poles, etc.	0.25	0.5
Ordinary	0.5	1.0
Unimportant (e.g. storage structures, temporary structures, minor retaining walls along roads, small culverts where diversion easily possible, electric or telephone poles, etc.	0.25	0.5

* A factor of safety of 1.4 to be used with respect to tensile strength.
** A load factor of 1.4 and maximum ductility ratio of 4.0 is assumed.
In both the cases, probable maximum earthquake represented by (FSa/g) is to be used with these values.

The IS Code specified an importance factor of 1.5 for buildings, and 2.0 for dams to take the importance of structure into account. According to JSCE factors 0.5, 1.0 and 1.5 are used for unimportant, important and very important structures of ports and harbours. For rail-bridges the corresponding factors are 0.9, 1.0 and 1.1 respectively. For other structures, no specification is laid down to cover importance. In the SEAOC code, the importance factor is not used as such, but it is covered in an overall behaviour factor K. The values in Table 3 could perhaps be adopted for C_P.

3.7.4 Ductility-Reduction Factor (C_D)

It is found (15) that this factor is mainly a function of the time period of the structure and the ductility ratio. Based on the studies with regard to the El-Centro and Taft earthquakes and elasto-plastic and bilinear resistance-deflection characteristics, the values given in Table 4 are suggested for determining the yield strength of the structure at which ductile deformation begins (1,2). For equivalent elastic design permitting 33 $^1/_3$ % increase in allowable stresses or an ultimate load design using 1.4 load factor, the design seismic coefficient may further be reduced in the ratio 1/1.4.

Table 4:- Reduction Factors For Ductility, C_d

Ductility Ratio	Reduction Factor for Time period T equal to				
	0.5 sec. or less	1.0 sec.	1.5 sec.	2.0 sec.	2.5 sec. or more
1.0	1.0	1.0	1.0	1.0	1.0
2.0	0.35	0.45	0.45	0.59	0.33
3.0	0.21	0.32	0.34	0.42	0.20
4.0	0.16	0.26	0.28	0.28	0.14
5.0	0.12	0.22	0.22	0.20	0.10
6.0	0.11	0.20	0.20	0.16	0.08

It is possible to achieve a ductility ratio of about five in reinforced concrete and steel structures by proper design against diagonal tension or buckling failure (2). This value could be adopted for determining the reduction factor for ductile structures.

3.7.5 Shear and Moment Coefficients (C_{VB}, C_{MB}, C_{VX}, C_{MX})

As stated earlier, these coefficients depend mainly upon the type of structure, and its stiffness and mass distribution. These will not be discussed here.

3.8 Illustrative Examples

Example 1. Masonry dam in Zone IV having T = 0.4 sec., damping 10% of critical, Rock foundation.

Since the structure is brittle, no ductility will be considered. The following values are obtained for the structure from the relevant sources:

S_a	= 145 cm/sec2	(Fig. 1 of paper)
F	= 1.6	(Table 1 of paper)
C_F	= 1.0	(Table 2 of IS: 1893-1970)
C_P	= 1.0	(Table 3 of paper)
C_D	= 1.0	(Table 4 of paper)

Hence from Eq. (2), the design seismic coefficient is given by

α_h = 1.9 x 1.0 x 1.0 (1.6 x 145/981)

= 0.236

Using this seismic coefficient, tension may be allowed in the masonry upto (1/1.4) x flexural tensile strength.

Example 2. Reinforced concrete multi-storeyed building for hospital in Zone V having T = 0.6 sec., damping 5% of critical, resting on individual footings with the beams on Type II soil.

It is assumed that the frames are suitably designed to given a ductility ratio of 4.0 only. The following values are obtained for various factors.

S_a	= 148 cm/sec^2	(Fig. 1 of paper)
F	= 2.7	(Table 1 of paper)
C_F	= 1.2	(Table 2 of IS: 1893)
C_P	= 1.5	(Table 3 of paper)
C_D	= 0.18	(Table 4 of paper)

Therefore, design seismic coefficient

α_h = 0.18 x 1.5 x 1.2 (2.7 x 148/981)

= 0.132

When this seismic coefficient is used, critical sections in the structure should be designed to have their ultimate load capacity for the forces caused by the earthquake. If it is desired to use the elastic design method with 33 $^1/_3$ % increase in permissible stresses or use the ultimate load method of design adopting a load factor of 1.4 as recommended in IS: 1893, the seismic coefficient will be further reduced by a factor 1.4. That is, the seismic coefficient for design should be

$$\alpha_h = \frac{0.132}{1.4} = 0.094$$

3.9 Conclusion

It has been shown that the design seismic coefficient for structures can be expressed as a product of five factors, namely, spectral acceleration Sa for appropriate natural period and damping, seismic zone factor F, soil foundation factor CF, performance factor CP and ductility-ratio factor CD. Their values are closely related with permissible ductility in the structure and the factor of safety or load factor specified for design. Suitable values for the various factors are suggested as based upon past experience and current knowledge. ■

3.10 References

1. Arya A.S., "Design of Structures Against Earthquake Motions", Journal of the Institution of Engineers (India), Vol. 51, No. 11, Part Cl. 5, July 1971.
2. Arya A.S., "Efficient Design of Reinforced Concrete Multistoreyed Buildings for Earthquake Resistance", Journal of the Institution of Engineers (India), Vol. 51, No. 7, Part Cl. 4, March 1971, pp.177-184.
3. Biggs J.M., "Introduction to Structural Dynamics", McGraw Hill Book Co., New York, 1964, pp. 169-233.

4. Blume J.A., "A Reserve Energy Technique for the Earthquake Design and Rating of Structures in the Inelastic Range Proceedings of the Second World Conference on Earthquake Engineering, Tokyo, 1969, Vol. II, pp. 1061-1084.
5. Chandrasekaran A.R. and Saini S.S., "Dynamic Response of Non-Linear Systems", Journal of the Institution of Engineers (India) Vol. 48, No. 3, Part Cl. 2(SP), November 1967, pp. 645-664.
6. Clough R.W., Benuska, K.L. and Wilson E.L., "Inelastic Response of Tall Buildings". Procedures of the Third World Conference on Earthquake Engineering, Vol. II, New Zealand 1965.
7. "Earthquake Resistant Design for Civil Engineering Structures and Foundations in Japan", compiled by The Japan Society of Civil Engineers, 1968.
8. "Earthquake Resistant Regulations - A World List", compiled by International Association for Earthquake Engineering, December 1966.
9. Gutenberg, B. and Richter C.F., "Earthquake Magnitude, Intensity, Energy and Acceleration", Bulletin of Seismological Society of America, Vol. 32, July 1942.
10. Gutenherg B. and Richter C.F., "Earthquake Magnitude, Intensity, Energy and Acceleration", Bulletin of Seismological Society of America, Vol. 46 (2), 1956.
11. Hershberger J., "A Comparison of Earthquake Acceleration with Intensity Ratings", Bulletin Seismological Society of America, Vol. 46, October 1956.
12. Housner G.W., "Behaviour of Structures During Earthquakes", Journal, Engineering Mechanics Division, ASCE, Vol. 85, No. EM4, October 1959, pp. 109-129.
13. Housner G.W., "Earthquake Resistant Design Based on Dynamic Properties of Earthquakes", Proceedings, American Concrete Institute, Vol. 53, 1956-57, pp.85-98.
14. Housner, G.W., Martel R.R. and Alford J.L., "Spectrum Analysis of Strong Motion Earthquakes", Bulletin of Seismological Society of America, Vol. 43, April 1953.
15. Hudson D.E., "Response Spectrum Techniques in Engineering Seismology", Proceedings, World Conference on Earthquake Engineering, Berkeley, Calif, 1956, pp. 4.1, to 4.12.
16. Hudson D.E., "Some Problems in the Application of Spectrum Techniques to Strong Motion Earthquake Analysis", Bulletin Seismological Society of America, Vol. 53, No. 2, 1963.
17. "IS: 1893-1970, Indian Standard Criteria for Earthquake Resistant Design of Structures" by BDC: 39, Indian Standard Institution, 1970.
18. Jennings P.C., "Response of Yielding Structures to Statically Generated Ground Motion", Proceedings of Third World Conference on Earthquake Engineering, Vol. II, New Zealand 1965.
19. Krishna J., Arya A.S., Kumar K., "Importance of Isoforce Lines of an Earthquake with Special Reference to Koyna Earthquake of December 11, 1967", 4th Symposium on Earthquake Engineering, University of Roorkee, November 1970.
20. Krishna J. and Chandra B., "Ground Acceleration during Earthquakes", Bulletin of Indian Society of Earthquake Technology, Vol. VI, No. 1, March 1969, pp. 1-6.
21. "Nuclear Reactors and Earthquakes". United States Atomic Energy Commission, Division of Technical Information, August 1963, p. 39.
22. "Recommended Lateral Force Requirements and Commentary" by Seismology Committee of Structural Engineers Association of California, 1968.

Chapter 4

Determination of Isoacceleration Lines by Sliding and Overturning of Objects[1,2]

4.1 Introduction

How to obtain data on peak ground acceleration near the epicentre and its attenuation with distance, in the absence of strong motion instruments arranged specifically for the purpose, is a common problem. Even in well instrumented areas, instruments spaced close enough which would give the attenuation of the ground motion from the epicenter in all directions are not available. Yet for design purposes, such information is of basic importance. As seen from results published by Hershberger (1), the Modified Mercalli Intensity of any value corresponds to too wide a range of accelerations and does not lead to acceptable design criteria. Since very close network of accelerographs is not possible except in limited areas due to prohibitive cost, it appears that the problem will stay for all time in most countries and an alternative solution is necessary. With this objective in mind, attention has been paid to the earthquake behaviour of thousands of small household objects and others which either slide or overturn or do not move at all. Such information becomes available in abundance after a moderate or severe earthquake and, if scientifically interpreted, can give extremely useful results at small cost. The authors have made such studies after the Koyna earthquake of December 11, 1967 (2) and Broach earthquake of March 23, 1970 (3) in India. This paper presents the results of such observations, their interpretation and scientific evaluation. Particular reference is made here to the Broach earthquake. Data obtained from Koyna earthquake was presented earlier (4).

4.2 Basic Concepts

For determining the peak ground acceleration at a place, several observations of movements and non-movements are required. In the case of movement of an object, the force applied by

---

1 *Jai Krishna, Anand S. Arya & Krtishen Kumar, School of Research & Training in Earthquake Engineering, University of Roorkee, Roorkee, U.P. INDIA*

2 *Fifth World Conference on Earthquake Engineering, Rome, Italy, June 25-29, 1973, Session 4A, paper no.158*

the *earthquake must be more than the minimum force required to cause motion.* On the contrary, the force required to move an object which did not move during the earthquake will give *an upper limit on the earthquake acceleration.* Thus, by determining the forces in the two limits, the lower and upper bounds on the peak ground acceleration are obtained. If a number of observations are possible at a place (village or town), the range between the upper and lower bounds gets narrowed and the peak ground acceleration is estimated with sufficient accuracy.

There are several factors which affect the mechanism of overturning or sliding of objects. The earthquake motion may be considered to have two mutually perpendicular components and a vertical component. The overturning of an object will depends upon that horizontal component against which the stability is the least, including the effect of the vertical component. It is likely that the component in the vertical direction decreases with distance from the epicentre at a different rate compared to the decrease in horizontal component and that the frequency pattern of the accelerations may also change. If the effect of this variation is considered to be small, a constant ratio of the vertical to the horizontal component may be assumed. Where an accelerogram of the earthquake is available in the epicentral area, the critical combination of vertical and horizontal components may be worked out for worst effect on overturning and sliding motions and used for evaluating the observed data.

Other factors which affect the results depend upon the location of the observed object. Objects on the ground floor resting on rigid supports will experience the ground motion but those on higher elevations on first or higher floors, or on relatively less rigid supports, will experience an amplified motion due to the motion of the supporting member itself relative to the ground. Moreover, some objects may be more favourably placed than the others for their movement to occur. Then a small amount of sliding may escape observation whereas for overturning of an object, the chances of wrong information are not high. Therefore, care has to be exercised in selecting an object for determination of peak ground acceleration from its reported behaviour during the earthquake.

4.3 Conditions for Sliding and Overturning

The parameters entering the calculations for sliding and overturning of rigid objects are the coefficient of friction between the contact surfaces and the shape of the object, and the earthquake motion.

a) **Sliding of Objects:** A rigid object resting on even a horizontal surface, is shown in Fig. 1 with the

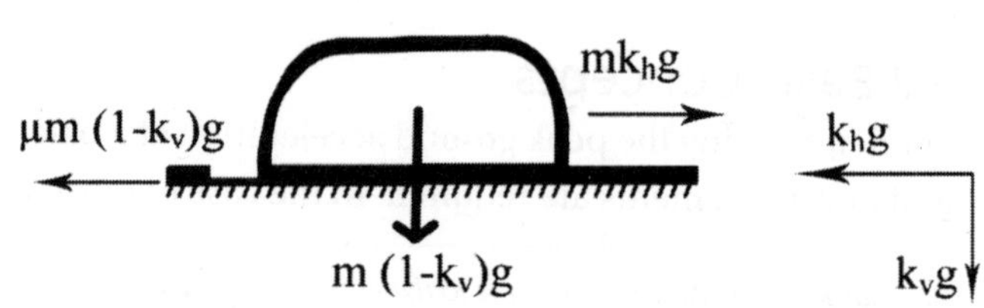

Fig. 1: Forces at threshold of sliding

forces marked on it. For sliding motion, the horizontal inertia force must exceed the frictional force, that is,

$$mk_h g > \mu m (1 - k_v) g$$

or
$$\frac{k_h}{1-k_v} > \mu \qquad (1)$$

where μ is the coefficient of friction, k_h and k_v are acceleration coefficients of the horizontal and vertical components respectively, m is the mass of the object and g is acceleration due to gravity.

b) **Overturning of Objects:** For overturning motion, the moment of the horizontal inertia force about base must exceed the stabilising moment due to modified weight under the vertical acceleration of the ground. From the forces marked in Fig. 2, this condition can be expressed as:

$$mk_h gh > m (1-k_v) ga$$

or
$$\frac{k_h}{1-k_v} > \left(\frac{a}{h}\right) \qquad (2)$$

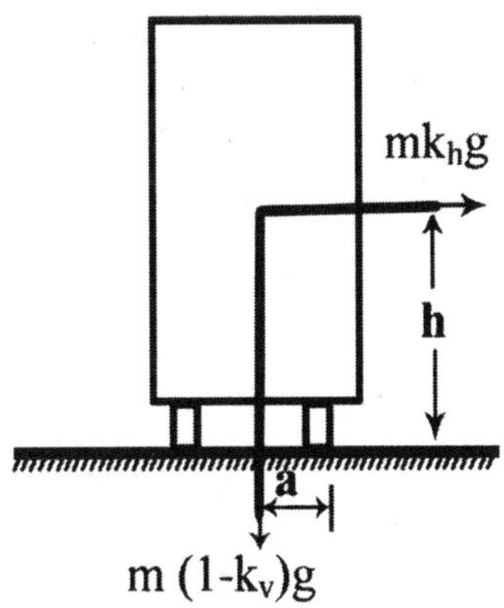

Fig.2: Forces at threshold of overturning

From Eq. (1) and (2), it may be noted that the smaller of the two quantities μ and (a/h) will be the governing condition. That is if μ < (a/h) the object would slide and if μ > (a/h) the object would overturn, provided, of course, that $k_h/(1-k_v)$ exceeds either μ or (a/h) as the case may be.

c) **Effect of Vertical Acceleration:** As indicated by Eq. (1) and (2), the horizontal acceleration required to cause overturning or sliding will be reduced due to the simultaneous action of downward vertical motion of the ground. At a place where an accelerogram is available, the critical combination of the two accelerations can be ascertained for defining the sliding and overturning of rigid objects. For example, for the Koyna and El Centro earthquakes, the following results were obtained by scanning their vertical and horizontal accelerograms:

	Koyna (Long).	El Centro (NS)
Factor $[k_h/(1-k_v)]_{max}$	0.684	0.329
Corresponding horizontal acceleration coefficient, k_h	0.542	0.319
Corresponding vertical acceleration coefficient, k_v	0.207	0.029
Acceleration ratio k_v/k_h	0.382	0.091
Factor $(k_h)_{max}$ as recorded	0.630	0.319

Although $[k_h/(1-k_v)]_{max}$ will always be larger than $(k_h)_{max}$, the difference between them is expected to be small (as seen above for both earthquakes) unless a large positive peak of the vertical component happens to coincide with the largest peak of the horizontal acceleration component. Therefore, for practical purposes, the effect of kv may be neglected in determining $(k_h)_{max}$.

4.4 Dynamic Analysis of Friction Supported Masses

The overturning or extent of sliding displacement of a rigid object under earthquake motion would not only depend on the maximum acceleration but also the duration of the acceleration peaks. For establishing the scope and limitations of Eq. (1) and (2) in predicting the peak ground acceleration, a dynamic analysis is necessary using base motion as input data. The equations of motion, considering one horizontal and one vertical component of ground acceleration acting simultaneously, may be written as follows with the notations shown in Figs. 3 and 4.

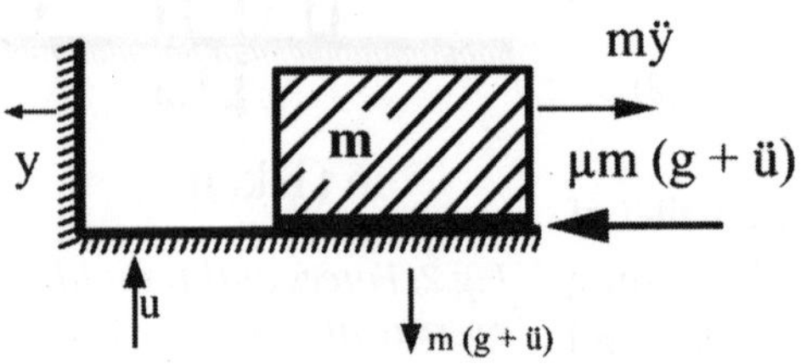

Fig.3: Forces on object in sliding

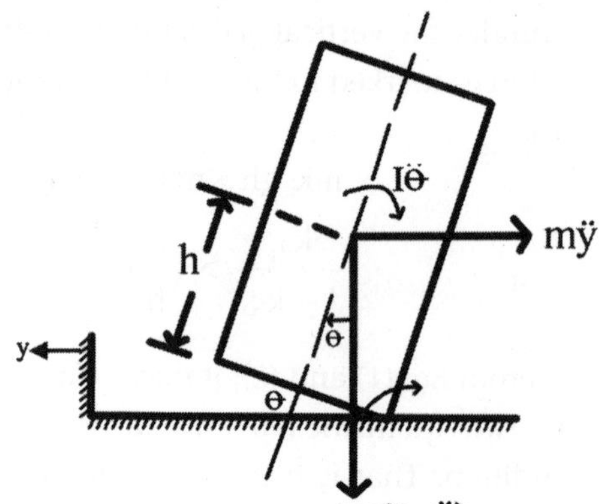

Fig.4: Forces on object in overturning motion

For sliding

$$\dot{v}+\phi(v)=-\ddot{y} \quad (3)$$

where

$$\phi(v) = \mu\,(g+\ddot{u})\,\{\text{sign}(v)\} \quad (4)$$

and $v = x-y; \quad \dot{v} = \dot{x} - \dot{y}$ (5)

For overturning

$$I\,\theta + m\,(g+\ddot{u})(a\cos\theta - h\sin\theta) = m\ddot{y}\,(a\sin\theta + h\cos\theta) \quad (6)$$

In Eq. (6), whenever velocity v changes sign, there is an impact with reduction in velocity so that

θ (after impact) = – v θ (before impact) (7)

where the coefficient of elastic restilution, v, has to be determined experimentally and is a function of the velocity itself besides the nature of the surface of the object and the surface on which it rests.

Results from Eq. (3) and (6) for several pulses as well as Koyna accelerogram are described below:

a) **Sliding of Objects:** For single rectangular and triangular pulses, the sliding motion gives closed-form solutions. Taking a single triangular pulse from an accelerogram for the sake of understanding, the sliding movement is computed for two values of time base, t_o = 0.05 and 0.10 sec, two coefficients of friction, 0.2 and 0.3 and several values of acceleration peak as shown in Table 1. It is noted that the extent of sliding is a function of the square of the duration t_o, provided, of course, that peak acceleration, Ag, exceeds μg. It is also seen that the sliding movement is very small when A exceeds μ slightly. For instance, when A = 0.3, μ = 0.2, and t_o = 0.05 sec, the movement will only be 0.11 mm. This movement is too small to be noticeable and the object may be taken as if it did not slide. Thus, the fact of sliding gives a lower bound but, upper bound may actually be higher than observed.

Table 1:- Sliding Movement Under Single Triangular Pulse

A	Displacement (mm)			
	t_o = .05 sec		t_o = .10 sec	
	μ = 0.2	μ = 0.3	μ = 0.2	μ = 0.3
.3	0.13	0	0.44	0
.4	0.80	0.02	2.40	0.08
.5	1.32	0.38	5.28	1.52
.6	2.44	0.90	9.76	3.60
.8	5.27	2.54	21.08	10.16

Sliding movements of objects with various friction coefficients for the Koyna earthquake (Long. Comp.) are shown in Table 2. Here also it is seen that as μ approaches A_{max} = 0.63, the extent of movement goes on decreasing and the movement may be difficult to observe in the field for μ > 0.5. Therefore, the determination of minimum value of the upper bound will require very careful observation.

b) **Overturning of Objects:** The overturning movement of a rectangular prism 40 mm x 40 mm x 200 mm high having a/h = 0.2 was computed from Eq. (3) by applying a single triangular pulse having peak acceleration coefficient A and time base t_O. The object would start tilting about one edge of base as soon as the acceleration coefficient exceeded 0.2, but it would fall back to its vertical equilibrium position unless the area under the pulse was large enough to tilt the object to an unstable equilibrium position. For t_o = 0.05 sec, the overturning of the prism occurred when A exceeded 1.15. For to = 0.1 sec, A would be 0.67.

Table 2:- Sliding Displacement Under Koyna Earthquake (Long. Comp.)

μ	Maximum Displacement (mm)		Residual Displacement (mm)	
	Horz. Ground Ground motion	Horz. + Vert. motion	Horz.Ground Ground motion	Horz. + Vert. motion
0.1	23.66	38.14	2.885	25.04
0.2	7.07	24.24	5.66	22.28
0.3	1.25	9.75	1.07	9.57
0.4	0.34	2.13	0.34	2.13
0.5	0.064	0.36	0.064	0.36
0.6	0.0002	0.01	0.0002	0.0005

The overturning oscillations of the same prism were computed by using Eq. (3) and (6) for the Koyna earthquake (Long. and vert. Comps.). The results are plotted in Fig. 5 for υ = 1.0, 0.9 and 0.7. A value of υ = 1.0 means no energy loss in the impact when velocity changes sign. It is seen that the prism overturns in the earthquake with A_{max} = 0.63 in the case when $\upsilon \geq 0.9$ but does not overturn when υ = 0.7. Thus, the fact of overturning gives a definite lower bound but the absence of it may give an upper bound which is too low and the actual upper bound may be much higher.

From the dynamic analysis of sliding and overturning of objects, it turns out that the peak ground accelerations determined by the static approach, Eq. (1) and (2), will definitely represent the minimum that occurred during

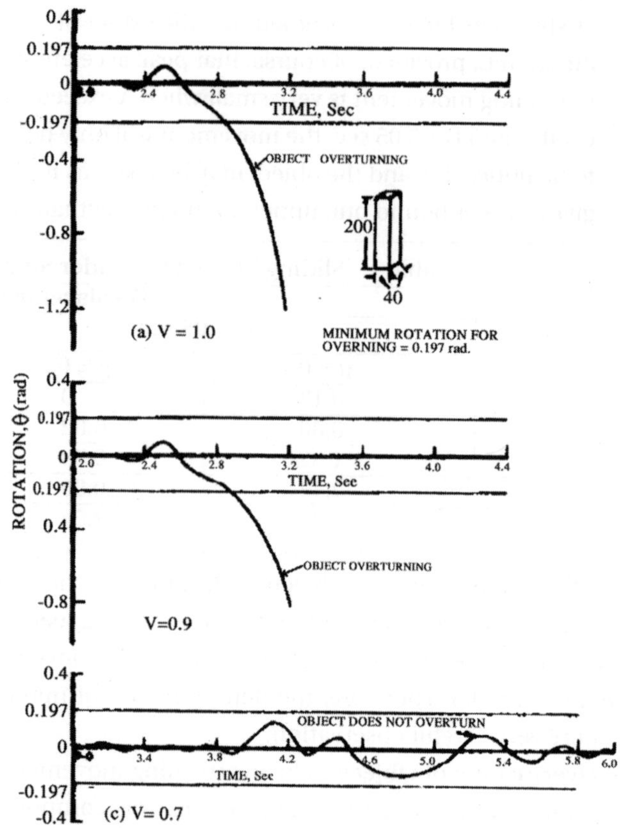

Fig.5: Overturning responses for Koyna earthquake longitudinal & vertical motions combined

the earthquake. For achieving closeness of the computed accelerations to the actual value, as many observations must be taken at a given site as possible.

4.5 Experimental Observations

Solution of Eq. (1) and (2) for estimating $(k_h)_{max}$ requires a knowledge of the properties μ and (a/h). These properties are experimentally determined for the rigid objects. An index map of the Broach region is shown in Fig. 6 and the shapes of some typical objects studied are shown in Fig. 7.

For collecting data during the Broach earthquake, more than 1200 objects were examined in an area of about 20 km radius around the epicentre at Broach and after scanning through their appropriateness, about 900 objects were selected for experimental observations.

A few of these observations giving the narrowest limits on the horizontal ground acceleration are presented in Table 3. The placing of the observed object is also given; ks and ko are coefficients of horizontal ground acceleration corresponding to sliding and overturning respectively. The effect of vertical ground acceleration is neglected.

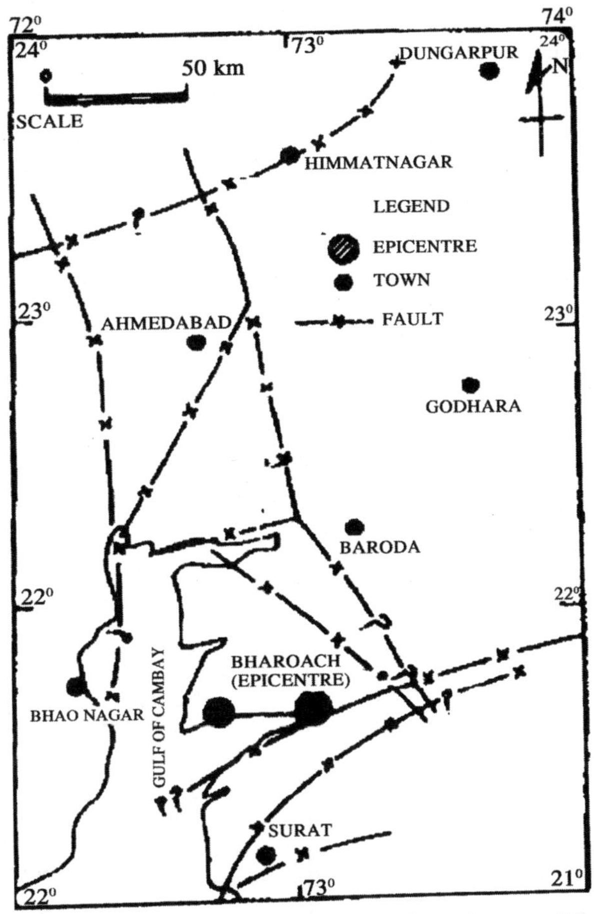

Fig.6: Index Map showing faults & the epicentre of the March 23, 1970 earthquake

4.6 Isoacceleration Map

The maximum ground accelerations at various places obtained from analysis of rigid objects, such as listed in Table 3, have been utilised to prepare an isoacceleration map of the area as explained below:

Since only the limits on the maximum acceleration at any place are obtained, the probable value of maximum ground acceleration cannot be directly predicted. Therefore, a reasonable estimate is obtained on the assumption that the decrease of seismic intensity would be more or less gradual rather than abrupt.

The curves of gradually reducing ground accelerations with distance from the epicentre are drawn passing within the acceleration bounds along various directions which are first marked by straight lines on the plan radiating out from a point near the epicentre (Broach) so that each passes close to as many places of observation as possible (Fig. 8). The observation points are transferred to the nearer radial line by projecting them, as a first approximation, on circular arcs, with Broach as the centre. The limits for the peak ground acceleration at each place are marked on the cross-sections by arrow-heads and a smooth curve is fitted to remove small discrepancies. Two such cross-sections are shown in Figs. 9 and 10 for lines 7 and 8 respectively. Other sections are not included here. From these curves it is easy to measure the distances in the plan corresponding to any desired value of acceleration, and plotting these points on the corresponding radial lines in the plan. Since the correct projecting arc in the plan should be parallel to the acceleration contours, the cross-sections based on earlier projections with circular arcs need adjustment. The final accelerations in the sections and plan are arrived at by trials.

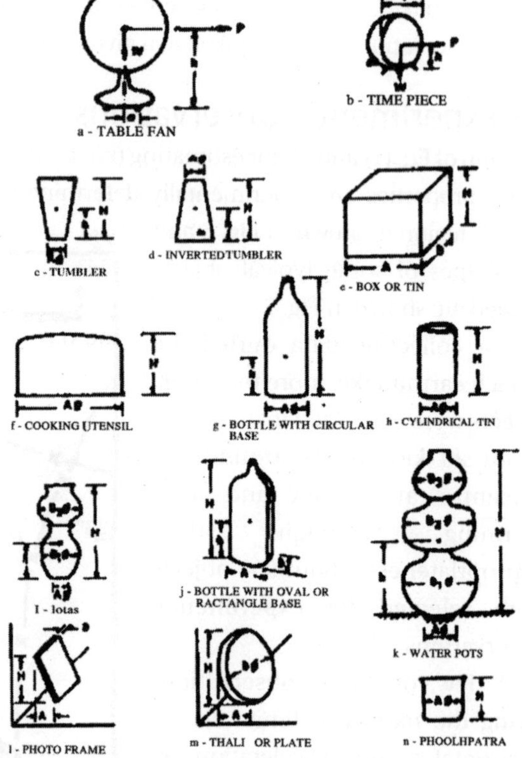

Fig.7:- Objects observed for overturning and sliding motion

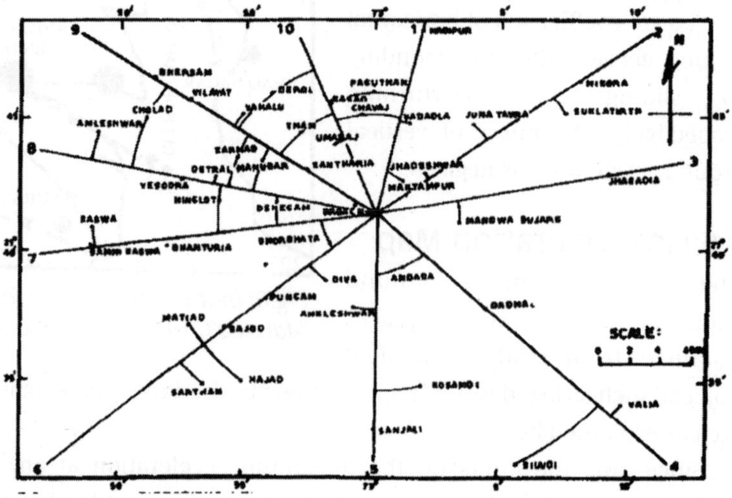

Fig.8: Radial directions 1 to 10 chosen for constructing cross-section of variation of ground acceleration showing projections of observations on them.

Table 3:- Limiting Acceleration Coefficients from Observed Sliding and Overturning at Some of the Places.

S. No.	Place	Distance from Broach (km)	Object	Action	k_s or k_g	Place of Object
1	Dehgam	6	Glass bottle	Overturned	0.18^+	Concrete shelf at 1.85 m
			Earthen water pots	Not overturned	0.21^-	Concrete shelf at 0.2 m
2	Bhorbhata	4	Earthen water pots	Overturned	0.23^+	Concrete shelf at 0.75 m
			Earthen water pots	Not overturned	0.31^-	Floor (Plinth level)
3	Hinglot	10	Earthen water pots	Overturned	0.27^+	Wall 0.7 m high
			Glass bottle	Not overturned	0.29^-	Wood shelf 1.7 m high
4	Dhanturia	13	Earthen water pots	Overturned	0.20^+	Wall 1.0 m high
5	Kangra	18	Metal Plate	Overturned	0.18^+	Wood shelf at 2.07 m
			Earthen water pots	Not overturned	0.23^-	Wall 0.8 m high

Finally, the contour of equal acceleration or the isoacceleration lines are constructed, giving the map shown in Fig. 11. In the map, isoseismals based on intensity observations (5) are also shown for comparison.

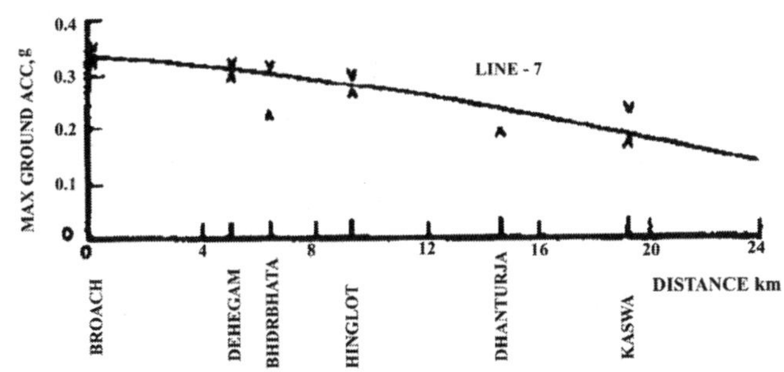

Fig. 9

4.7 Correlation of Results

Isoacceleration map and Seismotectonics: The general trend of the isoacceleration lines indicates their elongation along the course of the river Narmada in the ENE-WSW direction which represents slower attenuation of ground

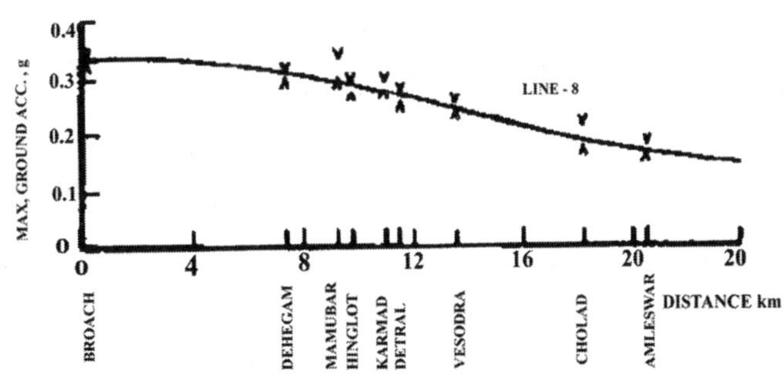

Fig. 10

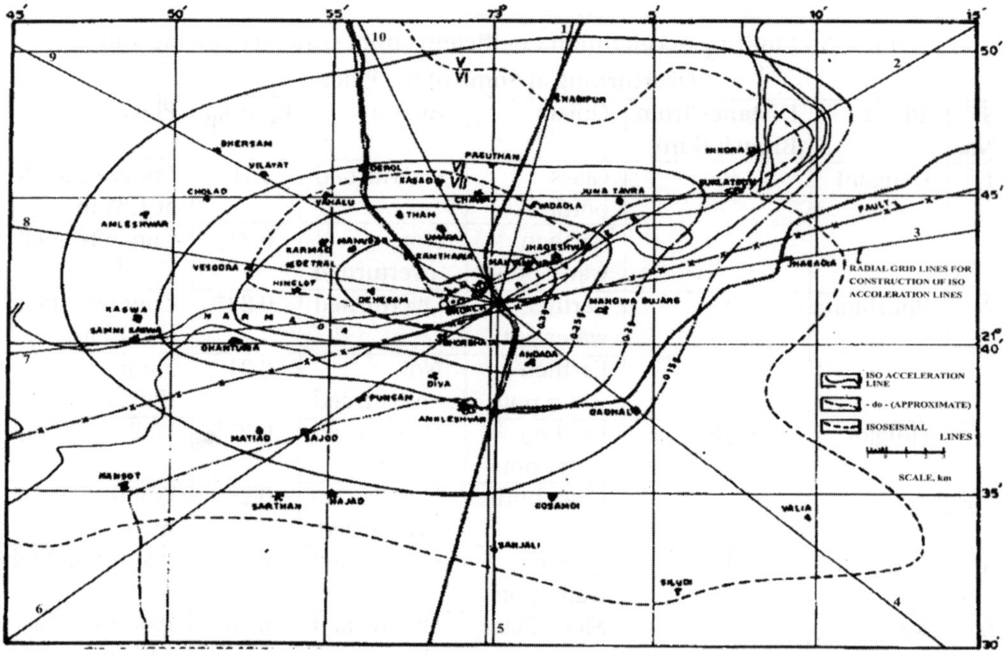

Fig.11: Isoacceleration map for the Broach earthquake of March 23, 1970

acceleration in this direction, a phenomenon indicative of the direction of the fault from which the earthquake originated. South of Broach, the existence of an ENE-WSW fault is confirmed by exploration surveys carried out by the Oil and Natural Gas Commission of India (Figs. 6 and 11). Tectonic movements along the Narmada fault are, therefore, the most probable cause of the Broach earthquake.

The ground acceleration attenuation curves along and transverse to the course of the river Narmada are shown in Fig. 12. For comparison, the attenuations available for three other earthquakes – Koyna (4), Parkfield (6) and San Fernando (7) – are also given. Two empirical relationships (1,8) are also plotted for the data related to the Broach earthquake. It is seen that the attenuation in both the directions is contained within the range of the Koyna and Parkfield earthquakes. Now the Koyna earthquake occurred where the basaltic trap deposits are very dense with wave velocity of 6 km/sec and the absorption of energy will naturally be slow. On the other hand the Parkfield earthquake had a shallow focus and occurred in an area consisting of alluvium of unknown depth, having wave velocities 2.8 to 4.0 km/sec in the upper 4 km depth. Perhaps the attenuation in these earthquakes could be taken as the limits of attenuation curves.

Comparison with Isoseismal map: The isoseismals have an E-V trend roughly fitting with the orientation of the Narmada fault, and like the isoacceleration lines, corroborate with the tectonic features of the region. However, the intersection of an isoseismal with several isoacceleration lines *confirms considerable variation of peak ground acceleration in a given isoseismal.*

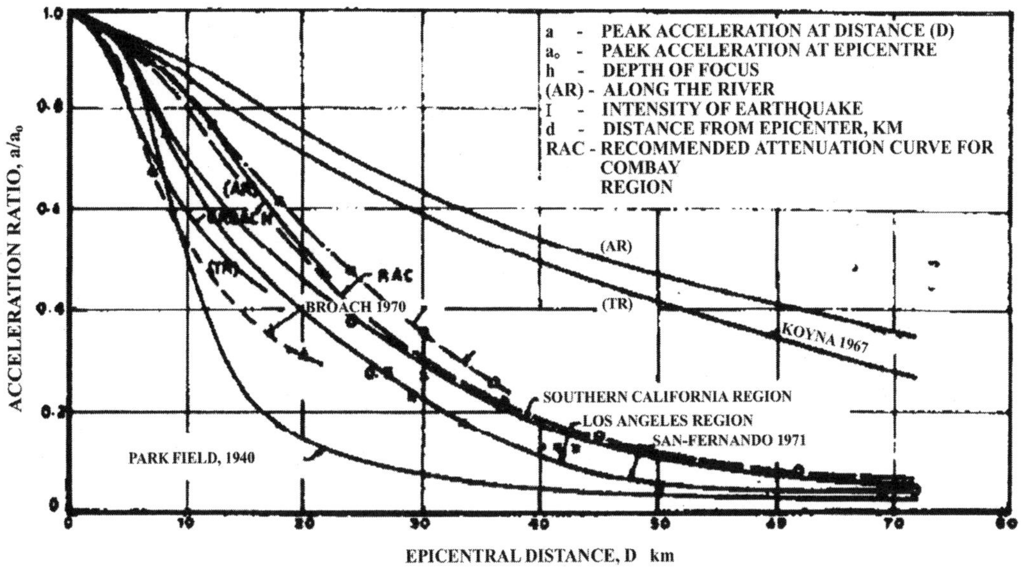

Fig.12: Attenuation of ground acceleration observed in various earthquakes and computed by empirical formulae

4.8 Conclusions

1) The large amount of data available on the sliding and overturning response of house-hold and other objects in an earthquake can be usefully utilised in estimating peak ground accelerations.
2) Reasonably accurate determination of the acceleration attenuation curves in all directions is made possible by this method.
3) The accelerations determined in this manner actually represent the lower bound on maximum ground accelerations and should be looked upon as the probable minimum values.
4) The pattern of the isoacceleration lines is likely to reveal the tectonic features of the site.
5) Attenuation of peak ground accelerations away from the causative fault is obtained at small cost which is not possible to obtain through instrumentation alone. ■

4.9 References

1. Hershberger J., "A Comparison of Earthquake Accelerations with Intensity Ratings", Bull. Seism. Soc. of Am., Vol. 46, No.4, October 1956, pp. 317-320.
2. Krishna, J., Arya A.S. and Kumar K., "Distribution of Maximum Intensity of Force in the Koyna Earthquake of December 11, 1967", Earthquake Engineering Studies, School of Res. & Trg. In Earthquake Engg., University of Roorkee, August 1969.

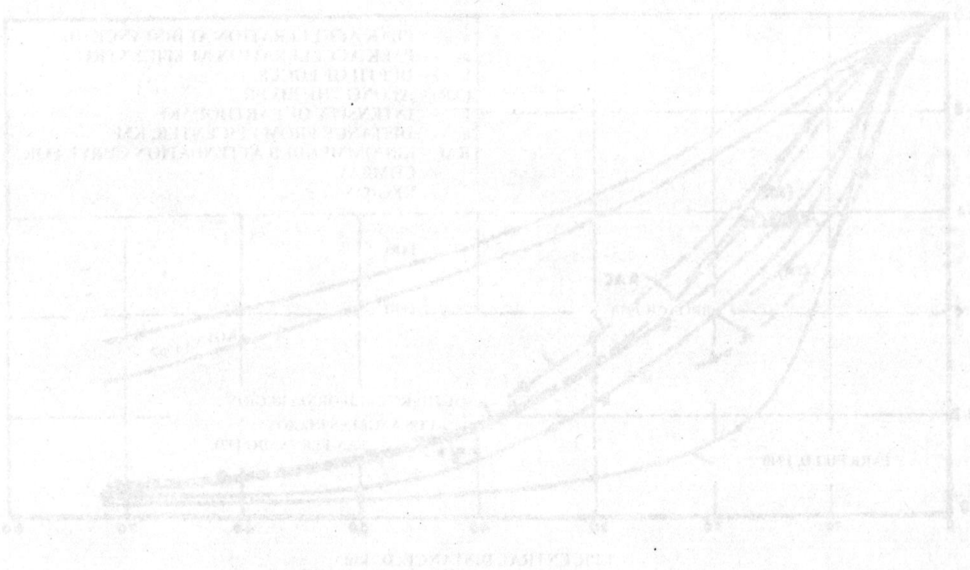

Chapter 5

Concepts and Techniques for Seismic Base-Isolation of Structures[1,2]

5.1 Abstract

This paper briefly reviews the concepts, techniques applicability and benefits of using base-isolation in structures in severe seismic zones. It is a passive way for achieving seismic response control by introducing various types of isolators between the foundation and the superstructure. The system has to perform three functions: horizontal flexibility, energy dissipation and rigidity against normal lateral loads. The rubber-lead bearing appears to be the best isolator so far, performing all three functions efficiently. The performance of isolated buildings during earthquakes in Japan shows clearly the achievement of desired reduction in the seismic response of buildings. Conditions favouring the choice of the base-isolation alternative compared to the conventional elastic or elasto-plastic design are indicated. The base-isolation of masonry buildings through a planned sliding joint is also described.

5.2 Conventional Aseismic Design Approach

As is well known from actual ground acceleration records during earthquakes, obtained near and away from the sources, the ground motion consists of positive and negative peaks of varying amplitude and time intervals between them in any three mutually perpendicular directions chosen for recording. In the whole record, the duration of which could be a few seconds to a few minutes depending on the earthquake magnitude and the type of source, the motion is

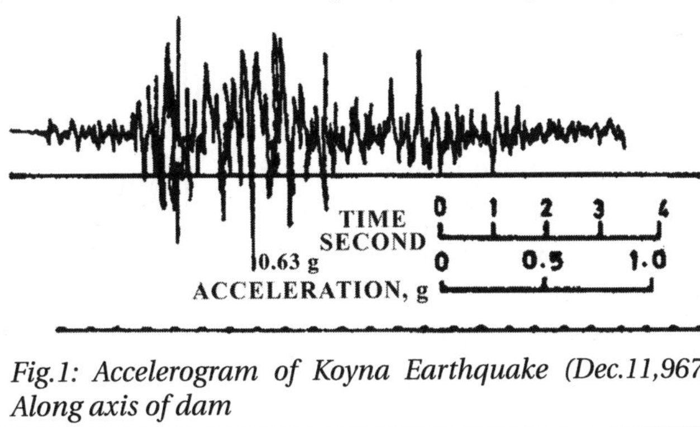

Fig.1: Accelerogram of Koyna Earthquake (Dec.11,967) Along axis of dam

1 *A.S.Arya, Professor & Head, Deptt. Of Earthquake Engineering, University of Roorkee, Roorkee*

2 *Tenth World Conference on Earthquake Engineering, Madrid, Spain, July 1992, pp 6639-6648.*

particularly intense over a short time duration (Fig. 1) which is the main cause of damage of structures and their contents, structural as well as non-structural.

The structures are normally built firmly attached to the ground. Therefore, naturally, they receive all the ground movements at their base. The seismic response of any structural system for a given base motion depends on the mass, stiffness and damping distribution in the system. This response for linear elastic systems to a given earthquake time history is most conveniently represented by acceleration response spectra, the main features of which are the following (Fig. 2):

i) For zero period, absolute acceleration response is equal to the peak ground acceleration.
ii) For low periods (that is the stiff system), the response gets amplified, it is highly peaky and becomes several times the peak ground acceleration. The amplification is significantly affected by damping in the system, being lower for higher damping.
iii) For long period (flexible systems), the acceleration response curve is smooth and lower than the peak ground acceleration, tending towards zero. Here the significance of damping is very much reduced. Unfortunately, most engineering structures and system lie in the low to medium period range, hence, are subjected to very large inertial forces.

Our approach for seismic design at present is to 'confront' the brute forces as generated by the earthquake shaking. That is, the inertial forces are taken into account by designing a *strong* structure. This approach results in increasing the size of structural members and connections and providing additional bracing members and shear walls, etc. Now, since in most structures, *full* catering for these elastic response forces will require very large additional financial inputs over those for normal conditions without earthquake consideration, a strategy of design is employed which is different than that used for usual loads. It may be called the 'fail-safe' design approach. For economical reasons, a 'non-collapse' design is aimed at rather than a 'no-damage' design, that is, the structure is designed for 'lower-than-expected' seismic forces so as to remain in the elastic range and, for the expected maximum seismic motions, it is allowed to go into the plastic stage with sufficiently large deformations or, damage to non-structural and structural elements but ensuring that *collapse* is avoided. Energy absorption through plastic deformations or minor

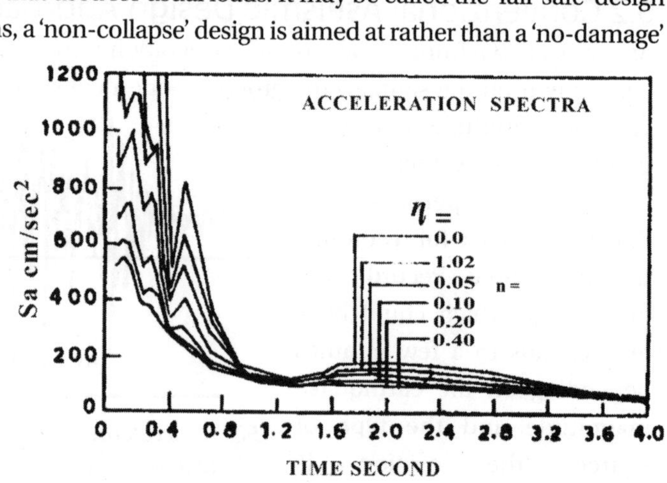

Fig.2: Response spectra for elastic system (Koyna earthquake Dec.11, 1967, component along axis of Koyna dam)

damages is known very well as being capable or reducing the dynamic response effectively. This puts a great responsibility on the designer who should ensure that the structure should possess enough *ductility and energy absorption capacity without strength deterioration* throughout the seismic time-history of future probable earthquake motions. Buildings designed for code-based seismic forces without the requisite detailing for ductility, are likely to collapse in severe ground motions, as happened in the case of the nine-storeyed prefabricated framed R.C. buildings in the Spitak (Armenia) earthquake of Dec. 7, 1988. *Well designed and properly detailed* steel and reinforced concrete structures do exhibit such non-collapse characteristics.

5.3 Response Control Approach

The above-stated approach does ensure safety of lives and properties that would have suffered due to collapse of structures but leaves a number of uncertain questions: how much will be the cost of the damaged non-structural elements, how functional will the damaged/deformed structure remain, what will be the economic losses due to stoppage of transportation, communication, etc. and what economic losses will occur due to the damage of contents of the shaken building. The scenario in this regard will be different between developed and developing countries like Japan and India, between the urban and the rural areas such as Delhi and the area around it, between a zone of very frequency earthquake occurrences and one with infrequent occurrence. In the developed and urban areas, the concerns are increasing for continuity of life-lines besides the basic issue of saving of lives, and keeping the economic losses to the minimum. The need has arisen to consider modifying the design strategy from 'confrontation', to 'accomodation' of the ground motion through 'base isolation'. In this approach, the structure is not attached to the ground firmly but through devices which will transfer the movements of the ground only partially to the structure, more particularly cutting down the intense peaks of the ground motion. The aim, therefore, is to reduce the structural response to such levels that it should remain safe *without* damage, hence, to achieve a 'no-damage' or 'safe' structure. Since in this approach, the intensity of shaking of the structure could be made less than that of the ground, the contents and non-structural elements will also be relieved of intense shaking, hence, should remain undamaged. Thus, the economic losses also will either be prevented altogether or reduced to a minimum.

Besides seismic base isolation, other concepts for 'seismic response control' have also been introduced and developed like the use of energy dissipating devices and resonant mass damper. This paper aims at presenting these concepts in general and the developments that have taken place in the area of seismic base isolation in particular.

5.4 Seismic Response Control Concepts

The concept of seismic response control of structural systems can be understood with reference to the acceleration response spectra given in Fig. 2 from which a number of possibilities emerge to reduce response as follows (2):

First, if flexibility could be introduced in the system for a given mass and damping to elongate its period considerably, the response could be brought down substantially and to the desired level of 'no-damage' situation. This is a very attractive proposition and has given rise to various techniques to achieve the aim.

Second, if damping could be increased without changing the mass or stiffness, the peaks in the response will be eliminated and the response reduced considerably, particularly in the low period range.

Third, the effective vibrating mass could be altered at appropriate instants of time history so as to significantly reduce the amplification of the main structural system.

Fourth, attempts could be made to alter the base motion, hence, the response by an active dynamic actuator system which is actuated by the structural response, continually analysed and monitored through a dedicated computer system.

Fifth, a combination of some of the above approaches may be attempted with a view to evolve an effective, reliable and economical control mechanism to achieve 'no-damage' design for the structure (including its contents) in question.

In order to have a common understanding of the various systems and the terms used for them, the following classification has been recently proposed based upon three points of view (13):

1. Classification based on Basic Principles of Dynamics
   i) A method to control and adjust restoring force characteristics.
   ii) A method to control and adjust damping.
   iii) A method to control and adjust mass.
   iv) A method to adjust input motion (a combination of above methods).
2. Classification based on Realisation Procedure
   i) Passive way
   ii) Active way
3. Classification based on Installed Location
   i) External types (like base isolation)
   ii) Internal types (internal elements)

According to the above classification, the seismic base isolation is an external type, works in passive way and provides a method of seismic response control by adjusting stiffness and damping.

5.5 Base Isolation Fundamentals and Devices

As stated above, the fundamental principle of base isolation is to introduce flexibility at the supports of a structure in the horizontal plane so as to ensure that the time period of the structure is well above the predominant periods of the probable earthquake. Now in this process, the relative displacement amplitude increases, hence, often damping or restraining elements have also to be introduced simultaneously to restrict the extent of relative movement caused by the earthquake. Although this principle is not new, its practical exploitation has occurred only recently in the last about 15 years during which suitable hardware of isolating devices has been developed and actually applied to the constructions of some buildings, bridges and atomic power plants. A short list of references is attached.

There are three basic elements in any practical base isolation system. These are as follows:

1. Decoupling between the superstructure and the base with or without flexible mounting so that the effective period of vibration of the total system is lengthened sufficiently to reduce the force-response.
2. A damper or energy dissipater so that the relative displacements between the structure and its supports can be controlled; and
3. A means of providing rigidity under low in-service load levels such as wind and minor earthquake so that the structure behaves as if fixed at base during normal service loads. These are considered in some detail here below:

5.6 Flexibility

The vibration-isolation principle is frequently used for reducing the transmission of machine vibrations to the buildings by mounting the machine on flexible pads or springs. The isolators used therein are mostly meant for vertical vibration reduction. For seismic protection, introduction of horizontal flexibility is actually necessary, and vertical flexibility is rather undesirable. Therefore, the isolators must maintain vertical rigidity while allowing horizontal flexibility. Steel roller or rubber bearings as used in the bridges for providing longitudinal movements fulfil these objectives. These rubber bearings are constructed in layers by sandwiching steel shims between each layer and bonding together by gluing. The steel shims constrain the lateral deformation of the rubber under vertical load resulting in vertical stiffness several hundred times the lateral stiffness. These bearings are found very practical for use in buildings also. Other possible devices for introducing flexibility include spherical balls, cylinders with spherical ends, cable suspensions, pinned or soft storey columns, sleeved piles and sliding or rocking arrangement, as indicated schematically in Fig. 3.

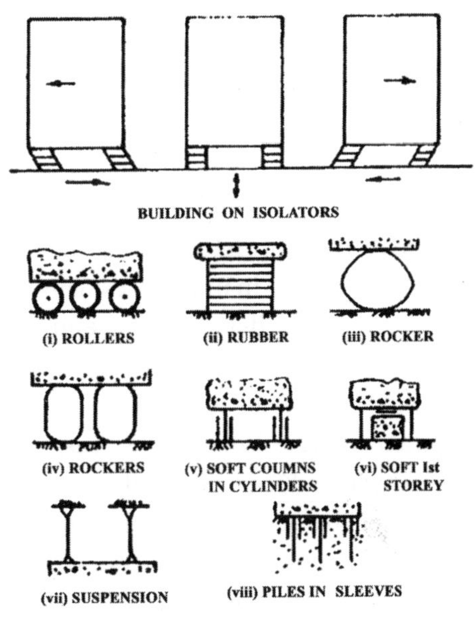

Fig.3: Base isolation devices

The acceleration response spectra of the maximum probable earthquake for a major project in India are shown in Fig. 4. Substantial reductions in the acceleration are clearly seen when the period of vibration of the structure is lengthened from 0.4 sec to 2.0 sec. Such a reduction in force response is primarily dependent on the characteristics of the earthquake ground motion and the original period of the fixed-base structure.

The force reduction effects are most significant where the intense part of earthquake record has predominantly short to medium range of periods say *less than about 0.5 sec* and also the fixed base structure is *stiff type,* so that the isolation system takes the structure to well beyond the peak response. Simultaneously, the additional flexibility of the structure gives rise to large relative displacements across the flexible mount. This would also be clear from Fig. 4 in which an idealised displacement response curve is also shown where displacement are seen to increase with increasing period. The displacement problem has to be overcome by introducing higher damping alongwith the flexible mount or restrainers/stoppers have to be introduced to restrain the displacement reaching undesirable values.

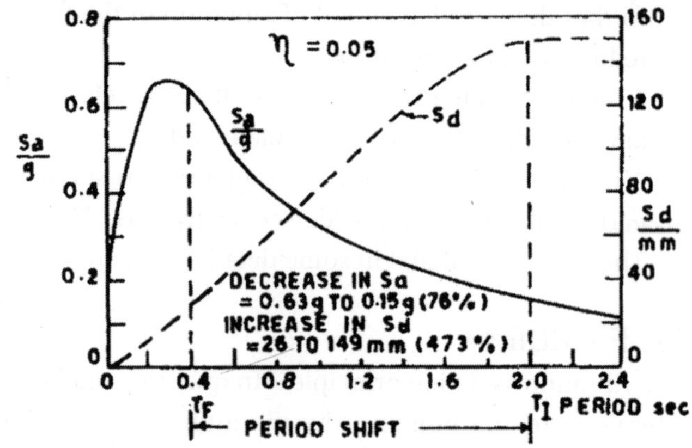

Fig.4: Acceleration and displacement response spectra for design MCE

5.7 Energy Dissipation

The dissipation of kinetic energy during seismic ground motions put into the conventional fixed-base structures takes place by internal damping, friction damping at the supports, and radiation damping through the base and side soils. The damping available in the undamaged small deformation state is rather low, less than 4% of critical in most cases. This value does increase with large plastic deformations and damage. But for the 'no-damage' design approach, such an increase will be undesirable and not to be relied upon. Therefore, where required in base isolated systems, additional damping has to be introduced externally.

One of the effective means of providing a substantial level of damping is through hysteretic energy dissipation. The term hysteretic refers to the cyclic nature of building vibrations in which energy is dissipated as the building

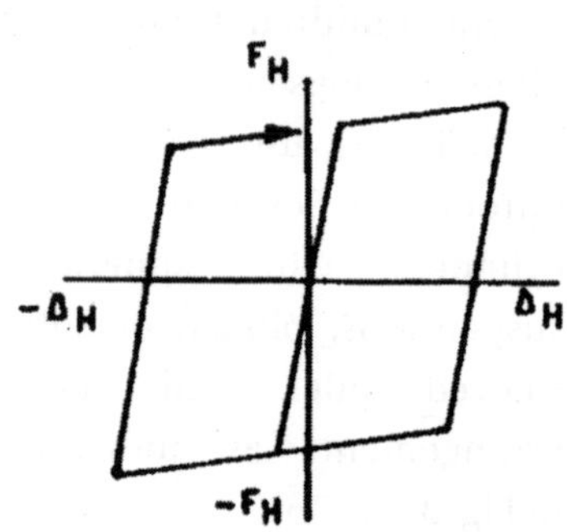

Fig.5: Hysteretic energy dissipation

moves from side to side. Fig. 5 shows an idealised force-displacement loop where the enclosed area is a measure of the energy dissipated during one cycle of motion.

Mechanical devices which use the plastic deformation of either mild steel or lead to achieve this behaviour have initially been developed in New Zealand by the Department of

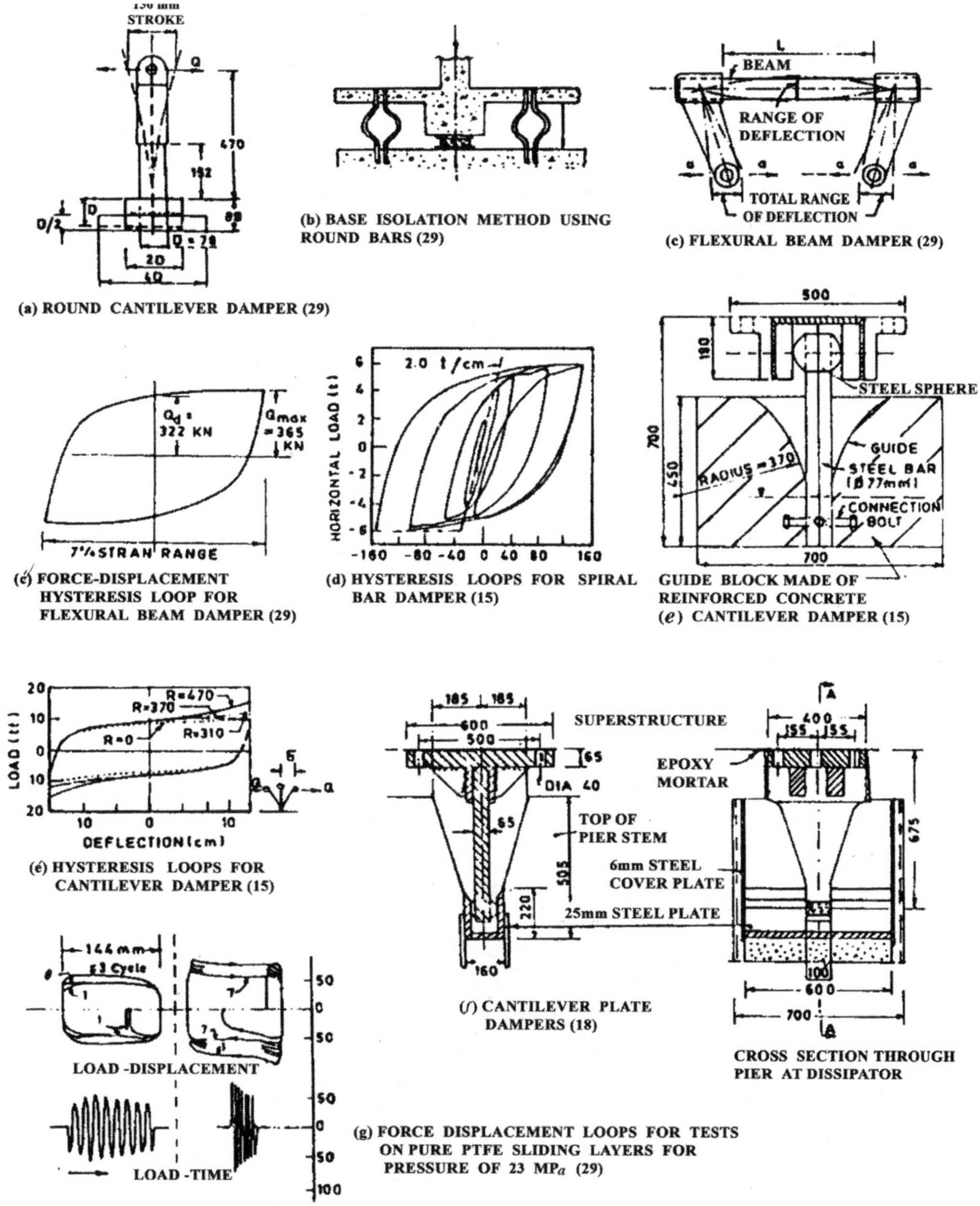

Fig.6: Energy dissipation system using steel plasticity

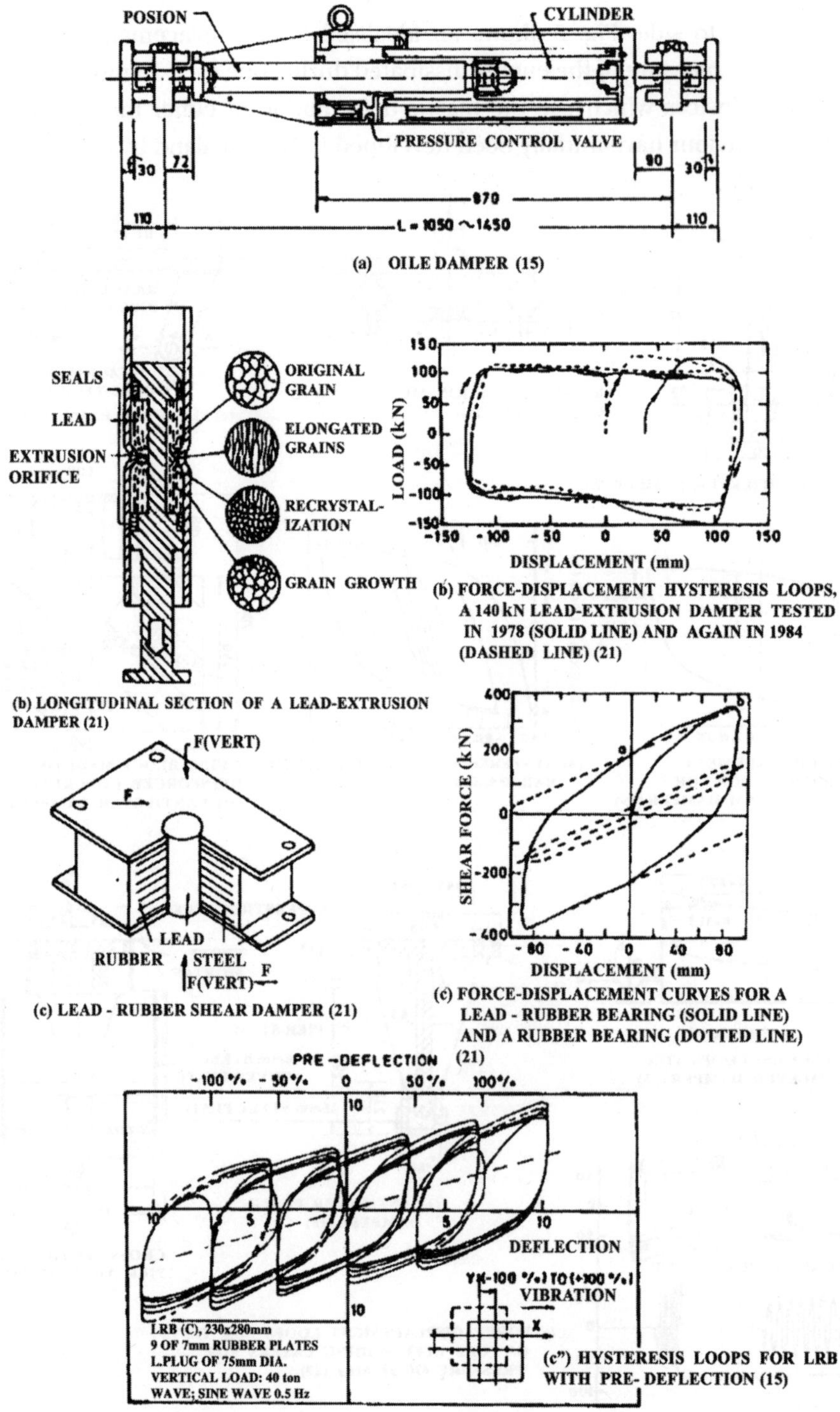

Fig.7: Energy dissipation devices using viscous fluid or lead

Scientific and Industrial Research. Several mechanical energy dissipation devices are shown in Fig. 6. These include round steel bar cantilevers (Fig. 6a, b, e. e'), flexural beam (Fig. 6c,c'), spiral bar (Fig. 6d) and plate cantilevers (Fig. 6f). Another form of energy dissipation is through sliding friction (Fig. 6g). Mild steel bars in torsion and cantilevers in flexural have been tested, refined and are now included in several bridge structures.

Dampers using fluid viscosity and energy dissipation through shearing of lead cylinder are shown in Fig. 7. The lead extraction and lead shear devices have also developed to a high level of sophistication. Lead has a crystalline structure under deformation, but almost instantly regains its original crystal structure when the deformation ceases (See Fig. 7b). For this reason, lead exhibits excellent hysteretic damping properties over many repeated cycles of earthquake motion.

The most highly developed and practical dissipater is the lead-rubber device in which a cylinder of lead is enclosed in an elastometric bearing as shown in Fig. 7(c). It combines in one physical unit, the flexible element and the energy dissipater. In this application, the lead is forced to deform plastically in shear by the steel shim plates. Excellent energy dissipation is possible with this device as shown in Fig. 7(c') and 7(c"). Recent work at the University of California at Berkeley and also the University of Auckland in New Zealand has improved the performance of this device to the point where it can be used in practical applications with the same confidence as with other building materials such as steel or concrete.

5.8 Rigidity under Low Lateral Loads

While lateral flexibility is highly desirable for high seismic loads, it is clearly unacceptable to have a structural system which will vibrate perceptibly under frequently occurring loads such as minor earthquakes or wind loads. Therefore, it is necessary that the flexible mount should have high enough initial stiffness and elastic resistance to balance the specified low lateral loads such as the wind loads and the low earthquake forces specified for elastic design in the applicable standard code of practice in the country such as IS: 875 and IS: 1893 respectively in India.

The lead-rubber bearing and other mechanical energy dissipators provide the desired rigidity against low loads by virtue of the initial elastic stiffness (Fig. 7c'). Some other base isolation system such as rollers will require a separate restraining device for this purpose like a buffer or spring-dashpot system, or sliding friction type device. An oil damper is shown in Fig. 7a which could be used in parallel with rollers, suspension system, etc.

5.9 Some Examples of Base-Isolation Applications

Base-isolation principle has now already been applied to numerous buildings in France, New Zealand, Japan and the USA, many bridges in New Zealand and Japan, and a few other structures like chimney stacks and nuclear power plants. In Japan alone, Kitagawa (13) lists 23 buildings permitted by the Ministry of Construction using base isolation upto 1988. These include 22 RC and one steel building, two upto 10 storeys in height and building area varying from 60 sqm to 8,240 sqm. The isolating and energy dissipating devices used in these structures include the following types:

Isolators : layered rubber bearings

Dampers : round steel bars; lead cylinder in rubber bearings; sliding friction; viscous dashpot, and high-damping rubber

By far the most used device is the rubber-lead bearing (RLB) which performs all the functions of initial high enough stiffness, period elongation of the superstructure and energy absorption through shearing of the lead cylinder. The actual earthquake behaviour of buildings on isolators with seismometers installed in the buildings has been observed in Japan. Two case studies are described in brief below:

1) The Okumura Tsukuba Institute building constructed in 1986 consists of a reinforced concrete four storeyed framed building resting on twelve laminated rubber bearings and provided with eleven sets of spiral steel dampers. The instrumentation consists of nine accelerometers, eight displacement meters, two velocity meters and an anemometer. Free vibrations tests conducted with increasing amplitudes indicated that

- when without dampers, the free vibration period was almost constant at about 1.9 sec, the damping value lying between 2 to 5% of critical.
- but with dampers, the period elongated with amplitude from 1.4 sec to 1.7 sec and damping ratio increased due to hysteretic behaviour of the steel damper from about 4% to 11% of critical.

The building was shaken by two real south-western Ibaraki earthquakes, first on April 10, 1987 with 5.1 Magnitude and second on June 30, 1987 with M = 5.1. The following peak accelerations were recorded (gals):

From the observed accelerations, the isolation effect is clearly seen, the superstructure above the isolators is tending to move as one rigid unit with very little inter-storey deformations and the acceleration level is limited to the lateral load capacity of the isolators.

Table 1:- Observed Behaviour of a Base Isolated Building

Earthquake	Base	Floor 1	Floor 2	Roof
April 10				
Horiz. x	38.6	11.3	6.1	11.1
Vert. z	18.4	22.2	21.4	20.1
June 20				
Horiz. x	203.0	17.0	12.0	20.1

2) The Tahoku University Full Scale Test Buildings are two identical three storeyed buildings, constructed in 1986 at Sendai City in Miyagi Prefecture. One building is conventional, fixed at base, and the other an isolated one using six laminated rubber bearings and twelve oil

Table 2 Observed Behavior of Fixed Base and Isolated Identical Buildings

S.No.	Quantity Observed	Fixed Base	Isolated
1.	Fundamental period, second	0.228	1.37
2.	Magnification of max. resp. acc. (mean of 18 records in Y-Dir.)	3.01	0.91
3.	Peak acc. resp. in Fukushima offing EQ of Feb. 1987 (g-value)		
	X-Dir.	1.50	0.13
	Y-Dir.	0.80	0.16
4.	Peak acc. resp. in Fukushima offing EQ of April (g-value)		
	X-Dir.	0.80	0.10
	Y-Dir.	0.45	0.13

dampers. Free vibration measurement gave the fundamental periods of non-isolated (N.I.) building as 0.228 sec and that of isolated (I) building as 1.37 sec. The acceleration recorded in a number of earthquakes at ground level and the roof of the buildings are summarised in Table 2 from which it is clearly seen that the peak acceleration response in the I-building got limited between 0.1 to 0.16 g when that of the N.I.-building varied from 0.45 to 1.5 g. The mean amplification was 0.91 for I-and 3.01 for N.I.-building. This clearly shows the large reduction in acceleration response, that is, the seismic forces in the isolated building as compared with the conventional fixed base building.

5.10 Base Isolation of Small Masonry Dwellings

The base isolation schemes described above cost a great deal more than the conventional fixed base buildings due to (a) the cost of the isolator and dampers and (b) the cost of additional reinforced concrete beams or flat slab system for supporting the superstructure above the isolators besides the foundation system below the isolators. Against this additional cost, there will be a saving in the design of the superstructure which could now be designed for smaller earthquake inertial forces. However, for small masonry dwellings, the cost of base-isolator application will rather be prohibitive. Observations in India in the 1930 Dhubri earthquake and 1934

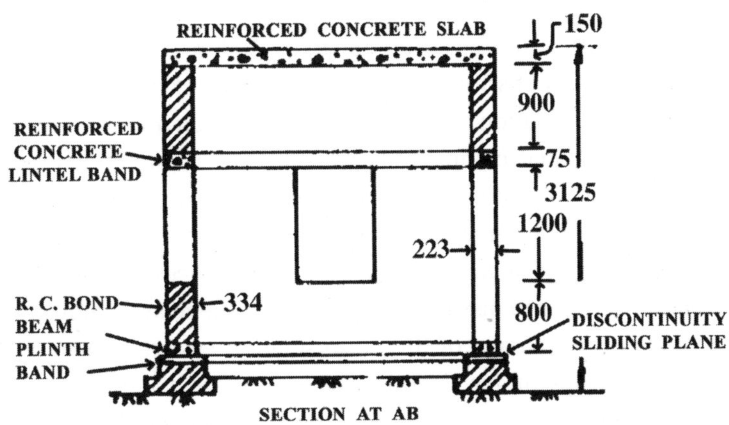

Fig.8: Sliding system for masonry dwellings

Bihar earthquake, and in China during the 1976 Tangshan earthquake showed that wherever sliding of the superstructure was possible either by the method of construction (e.g. not fixing the wooden columns at base) or by a horizontal crack occurring through the masonry joint, the damage to the building was much less. Hence, a simpler methodology was developed and tested analytically as well as experimentally in India (1). The basic concept is to permit sliding of the superstructure masonry at plinth level as shown in Fig. 8. The plinth masonry is plastered smooth and oiled to prevent bonding with the upper structure. A continuous *reinforced concrete* band under all walls is provided above it to ensure integral movement of the superstructure which is well bonded to the band through rich mortar of 1:3 cement: sand mix. Alternatively to oil, black polyethylene sheets could be used. A coefficient of friction of 0.15 to 0.2 is aimed at, which would restrain the building from movement under wind and minor earthquakes but permit sliding under severe shocks. The reduction of response accelerations under the Koyna 1967 and El Centro 1940 earthquake inputs was seen to be as much as by two-thirds and the maximum sliding displacements were upto 20 mm only. A plinth projection of 50 mm on both sides of the superstructure wall should suffice to keep the upper walls on the plinth.

In China, trial has been made (40) using a thin layer of specially screened sand on the sliding surface and constructing the superstructure on this layer. Some prototype buildings have been constructed on this basis in Beijing (40).

5.11 Additional Consideration

Appropriate details are needed in all isolated buildings to provide unbreakable flexible connections of the service pipes (for water, drainage, sewerage, gas, etc.) coming into the building from outside so as to accommodate the relative movements of the building and its base.

5.12 Reduction of Damage Potential by Response Control

The possible consequences of failures of structural systems (buildings, bridges, power plants, and dams, etc.) with regard to loss of lives and economic losses, etc., are much better understood than those caused due to the failure of non-structural components, equipment and contents. A possible qualitative summary of these is given in Ref. 2. The principles of seismic-base isolation and energy dissipation can not only be applied to whole buildings but also to individual components as well. For example, computer installations, spent fuel storage tanks or battery stacks of atomic power plants, control cabinets and switch yard equipments of power stations, individual objects of art in museums, etc., can be supported individually on isolator systems. Frictional devices can be incorporated between partition walls and frames to dissipate energy during the relative movements.

Another advantage of the base-isolation system will be to standardise the plant and equipment design for various sites of different seismicity levels. That is, keeping the superstructure and its components the same, the problem will be to design suitable isolator-dissipator systems for given seismic intensities.

5.13 Conclusion

In conclusion the following initial guidelines are suggested for using the alternative of seismic isolation in place of conventional design and achieving the objectives of desired safety and continuity of functional operations over long periods of time.

First, the following conditions appear favourable for choosing the base isolation alternative:

(a) where the subsoil is moderately stiff to hard; and the building's fundamental period is less than about 0.5 sec, as for example:
   (i) the structure is stiff, say two to ten storeys for moment resistant frames with substantial infill walls, upto 15 storeys for shear walls or braced framed building, and taller for box like rigid building systems; or
   (ii) the structure is of a crate-like plan in masonry with any number of storeys

(b) where horizontal relative motions at the isolators of upto about 150 mm on either side can be permitted.

(c) where the building contains valuable breakable articles, such as hospitals and museum artifacts, and the effective earthquake force acting on building contents must be reduced.

Second, while designing the building with the base isolation system, the following guidelines will be useful :

(a) A full grillage of beams or complete diaphragm plate will be most effective for distributing the vertical and lateral loads to the bearing units.

(b) A clearance greater than the maximum computed bearing displacements for probable maximum earthquakes must necessarily be provided around the building (See Fig. 9).

(c) Appropriate details must be worked out to accommodate the maximum relative displacements across the isolators in service pipes, cladding walls, staircases, elevators, etc.

(d) The isolating bearings should have access for inspection and maintenance and possibility of replacing damageable components, if any.

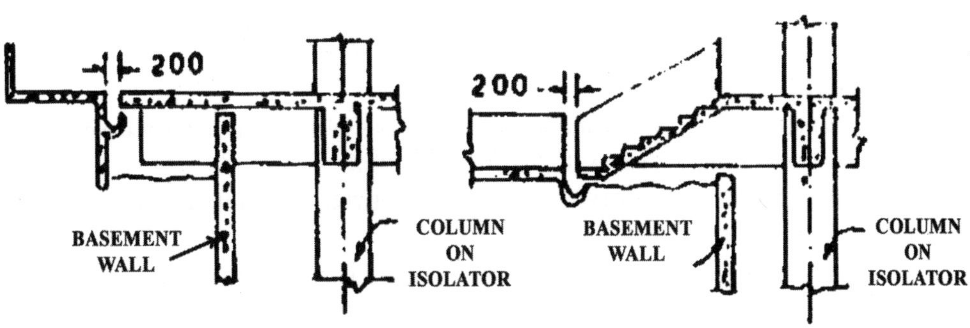

Fig.9: Typical details for clearance (2)

Finally, the following advantages of seismic isolation with energy absorption schemes can be summarised:

(a) The structural deformations going into the inelastic/plastic range and the consequent damage is likely to be completely eliminated; the structure will need designing for much smaller accelerations, hence, should be more economical;

(b) the relative storey displacements (drift) will be reduced, hence, the 'non-structural' damage to cladding, partition walls, etc. will be minimised or eliminated altogether; and

(c) the response accelerations at the higher floor will be much reduced, hence, the damage to equipment, service lines will be minimised.

5.14 Acknowledgement

This study forms part of the work done by the author under the Emeritus Scientist Scheme of Council of Scientific and Industrial Research, New Delhi. ■

5.15 References

1. Arya A.S., "Sliding Concept for Mitigation of Earthquake Disaster to Masonry Buildings", 8th World Conference on Earthquake Engineering, San Francisco, July 1984, Session WA-6.
2. "Base Isolation of Buildings-Architectural Aspects" Dynamic Isolation Systems, Inc., 2855 Telegraph Ave, Suite 410, Berkeley, CA, USA, 1984.
3. Beck, J.L. and Skinner, R.I., "The Seismic Response of a Reinforced Concrete Bridge Pier Designed to Step", International Journal of Earthquake Engineering and Structural Dynamics, Vol. 2, 1974, pp. 343-358.
4. Blakeley, R.W.G., et. Al., "Recommendations for the Design and Construction of Base-Isolated Structures", South Pacific Regional Conference in Earthquake Engineering, Wellington, May 1979.
5. Buckle, I.G., "Factors Affecting the Performance of Lead-Rubber Energy Dissipators", Proc. Bridge Design and Research Seminar, Road Research Unit, Bulletin 73, Wellington, 1984, 157-170.
6. Charleston, A.W., Wright, P.D. and Skinner, R.I., "Wellington Central Police Station: Base-Isolation of an Essential facility" NZ-Japan W/S, Wellington, 1987.
7. "Design of Lead-Rubber Bridge Bearings", Civil Division Publication 818/A, Ministry of Works and Development, Wellington, 1983.
8. Dupont, E.I. de Nemoures & Co. (Inc.), "Suggested Specifications for Filled Teflon TFE Resins Used in Bearing Pads", Delaware, USA.
9. Higashio, Aizawa, hayamizu, "An Experimental Study on a Base-Isolation System Using Laminated Rubber Bearings and Viscous Dampers", 9th SMIRT Report, August 1987.
10. "Highway Bridge Design Brief", Civil Division Publication CDP 701/6 New Zealand Ministry of Works and Development, 1978.
11. Izumi, "Earthquake Response Observation for Base Isolated Buildings", The Building Letter (Japan), August 1987.
12. Jolivet, F. and Richli, M., "A Seismic Foundation System for Nuclear Power Stations", Trans. 4th International Conference Struct. Mech. Reactor Technology, Vol. K(b), August 1977.

13. Kitagawa Y., "Base Isolated Building Structures in Japan", Quarterly Journal of the Federation Internationale de la Precontrainte Notes 1988/3.
14. Kumar K. and Arya A.S., "Retrofitting of Multispan Simply Supported Bridges for Reduction of Earthquake Forces", 8th WCEE, San Francisco, July 1984, Session FA-5.
15. Makiguchi Y., "Present Status of Base-Isolated Building in Japan", Proc. NZ-Japan W/S on Base Isolation of Highway Bridges, Min. of Works and Dev., Wellington, NZ, Nov. 1987, pp. 153-172.
16. Matsuda, Aoyagi, et. al., "Experimental Study on the Dynamic Behaviour of the Base Isolated Building", 9th SMIRT Report, August 1987.
17. Meggett, I.M., "Analysis and Design of Base-Isolated Rinforced Concrete Frame Building," Bulletin of the New Zealand National Society for Earthquake Engineering, Vol. II, No.4, December 1978, pp. 245-254.
18. Park, R. and Blakeley, R.W.G., "Seismic Design of Bridges", Structures Committee Summary, Vol. 3, Road Research Unit, National Roads Board, New Zealand, November 1978.
19. Robison, W.H., 1982, "Test of ANAC's 140 kN Extrusion Damper", Physics and Engineering Laboratory, Report no. 591, lower Hutt.
20. Robinson, W.H. (1982, "Lead-Rubber Hysteretic Bearings Suitable for Protecting Structures During Earthquakes", Earthquake Engineering and Structural Dynamics 10: 593-604.
21. Robinson W.H., Cousins W.J., "Recent Development in Lead Dampers for Base Isolation", Proc. NZ-Japan W/S on Base Isolation of Highway Bridges, Min. of Works and Dev. Wellington, NZ, Nov. 1987, pp. 46-51.
22. Robinson, W.H. and Greenbank, L.R., "An Extrusion Energy Absorber Suitable for the Protection of Structures During an Earthquake", International Journal of Earthquake Engineering and Structural Dynamic, Vol. 4, 1976, pp. 251-259.
23. Robinson, W.H. and Greenbank, I.R., "Properties of an Extrusion Energy Absorber", Bulletin of the New Zealand National Society for Earthquake Engineering, Vol. 8, No. 3, September 1975, pp. 187-191.
24. Robinson, W.H., an Tucker, A.G., "A Lead Rubber Shear Damper", Bulletin of the New Zealand National Society for Earthquake Engineering, Vol. 10, No.3, September 1977, pp. 151-153.
25. Skinner, R.I., Beck, J.I. and Brycroft, G.N., "A Practical System for Isolating Structures from Earthquake Attack," International Journal of Earthquake Engineering and Structural Dynamics, Vol. 3, 1975, pp. 297-309.
26. Skinner, R.I., Heine, A.J. and Tyler, R.G., "Hysteretic Dampers to Provide Structures with Increased Earthquake Resistance", Proceeding Sixth World Conference on Earthquake Engineering, New Delhi, January 1977.
27. Skinner, R.I., Kelly, J.M. and Heine, A.J., "Hysteretic Dampers for Earthquake Resistant Structures", International Journal of Earthquake Engineering and Structural Dynamics, Vol. 3, 1975, pp. 287-296.
28. Skinner, R.I. and McVerry, G.H., "Base Isolation for Increased Earthquake Resistance of Buildings", Bulletin of the New Zealand National Society for Earthquake Engineering, Vol. 8, No. 2, June 1975, pp. 93-101.
29. Skinner, R.I., et al. "Hysteretic Dampers for the Protection of Structures from Earthquakes", Proc. NZ-Japan W/S on Base Isolation of Highway Bridges, Min. of Works and Dev. Wellington, NZ, Nov. 1987, pp. 28-45.
30. Turkington, D.H. et al., "Seismic Design of Bridges on Lead-Rubber Bearings", Pacific Conference on Earthquake Engineering, Wairakei, New Zealand, August 1987, pp. 389-400.

31. Turkington D.H., et al., "Development of Revised Design Procedures for Bridges on Lead-Rubber bearings", Proc. NZ-Japan W/S on base Isolation of Highway Bridges, Min. of Works and Dev. Wellington, NZ, 1987, pp. 132-141.
32. Tyler, R.G., "Damping in Building Structures by Means of PTFE Sliding Joints", Bulletin of the New Zealand National Society for Earthquake Engineering, Vol. No. 10, No. 3, September 1977, pp. 139-142.
33. Tyler, R.G., "Dynamic Tests on PTFE Sliding Layers Under Earthquake Conditions", Bulletin of the New Zealand National Society for Earthquake Engineering, Vol. 10, No.3, September 1977, pp. 129-138.
34. Tyler, R.G., "Dynamic Tests on laminated rubber bearings", Bulletin of the New Zealand National Society for Earthquake Engineering, Vol. 10, No. 3, Sept. 1977, pp. 143-150.
35. Tyler, R.G. and Skinner, R.I., "Testing of Dampers for the Base-Isolation of a Proposed 4-Storey Building Against Earthquake Attack", Proceedings Sixth Australisian Conference on the Mechanics of Structures and Materials, University of Cantebury, New Zealand, August, 1977, pp. 376-382.
36. Tyler, R.G., "A Tenacious Base-Isolation System Using Round Steel Bars", Bulletin of the New Zealand National Society for Earthquake Engineering, Vol. 11, No. 4, December 1978, pp. 273-281.
37. Tyler, R.G., "Tapered Steel Cantilever Energy Absorbers", Bulletin of the New Zealand National Society for Earthquake Engineering, Vol. 11, No. 4, December 1978, pp. 282-294.
38. Tyler, R.G. and Robinson, W.H., "High-Strain Tests on Lead-Rubber Bearings for Earthquake Loadings", Bulletin of the New Zealand National Society for Earthquake Engineering 17 (2), 1984, 90-105.
39. Yasaka, Izuka, "Base Isolation System for Earthquake Protection and Vibration Isolation of Structure", Annual Report of Kajima Institute of Construction Technology, Vol. 34, June 1986.
40. Kie Li Li, "Base Isolation Measure in Aseismic Building in China", 8th World Conference on Earthquake Engineering, San Franscisco, CA, 1984, pp. 791-198.

Chapter 6

Isolator Parameters for Earthquake Response Reduction[1,2]

6.1 Introduction

The conventional approach to earthquake resistant design of structures is to calculate the expected dynamic forces due to an earthquake, superimpose the same on other concurrent loads and then proportion the various elements according to the specified working stresses. An alternate method would be to isolate the main structure from the incoming earthquake forces by mounting the main structure over an isolator system so as to reduce the earthquake forces to the desired extent. The advantages of this approach over the conventional approach are seen to be the following:

a) The main structure mounted over an isolator would be subjected to reduced earthquake forces and, hence, economy in the design can be achieved. If the additional stresses due to the earthquake could be brought down so that they could be absorbed by the permissible increase in working stresses under seismic conditions, the member sections would be determined by the normal dead and live load condition.
b) If the contents of the structure are of special nature, susceptible to shock and vibration hazard, isolation of the main structure would be the ideal solution.

In the present study, an attempt has been made to find the optimum characteristics of an isolator for which the desired reduction in the response of the main structure, for a given earthquake is achieved. The earthquake chosen for the study is the N-S component of the El-Centro earthquake of 1940. The parameters studied are the stiffness and damping of the isolator in relation to those of the main structure. The response is computed in terms of the accelerations of the masses of the structure and the isolator, the absolute displacement of isolator mass and the displacement of main mass relative to that of the isolator.

---

1 *A.S.Arya, Professor, Prakash R., Lecturer, University of Roorkee ,& Gopal R.. Asstt. Engineer, U.P.S.E.B, Lucknow*

2 *VI Symposium on Earthquake Engineering, University of Roorkee, Oct.5-7, 1978,Vol.I., pp. 361-364.*

6.2 Mathematical Model and Basic Equations

A simple mathematical model as shown in Fig. 1(a) has been adopted to represent the structure-isolator system. The system is a two degree freedom system consisting of two single degree structures. The upper structure represents the idealised single degree model of the main structure having mass m_1 spring constant k_1 and coefficient of viscous damping C_1. This system is mounted on the isolator system which is also a linear single degree freedom system having mass m_2, spring constant k_2 and coefficient of viscous damping C_2. The free body diagram with the forces acting on the two masses is shown in Fig. 1(b). Considering the equivalent static equilibrium of two masses, the governing differential equations of motion are obtained as follows:

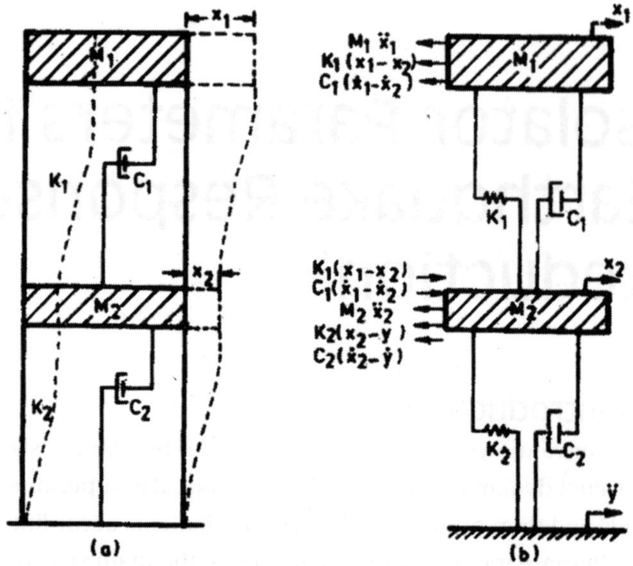

Fig.1: Mathematical model and free body diagram of structure isolator system

$$m_1\ddot{x}_1 + c_1(\dot{x}_1 - \dot{x}_2) + k_1(x_1 - x_2) = 0 \tag{1}$$

$$m_2\ddot{x}_2 + c_2(\dot{x}_2 - \dot{y}) - c_1(\dot{x}_1 - \dot{x}_2) + k_2(x_2 - y) - k_1(x_1 - x_2) = 0 \tag{2}$$

Putting,

$$c_2/m_2 = 2\xi_2 p_2, \quad c_1/m_1 = 2\xi_1 p_1$$

$$c_1/m_2 = 2\xi_1 p_1/\alpha, \quad k_1/m_1 = p_1^2$$

$$k_2/m_2 = p_2^2 \quad k_1/m_2 = p_1^2/\alpha, \quad m_2/m_1 = \alpha$$

in equations (1) and (2) above and manipulating, these equations are brought to the form as given in equations (3) and (4),

$$\ddot{z}_1 + 2\xi_1 p_1(\dot{z}_1 - \dot{z}_2) + p_1^2(z_1 - z_2) = -\ddot{y} \tag{3}$$

$$\ddot{z}_2 + 2\xi_2 p_2 \dot{z}_2 - 2\xi_1 p_1(\dot{z}_1 - \dot{z}_2)/\alpha + p_2^2 - p_1^2(z_2 - z_1)/\alpha = -\ddot{y} \tag{4}$$

where

$(\ddot{x}_1 - \ddot{y}) = \ddot{z}_1, (\ddot{x}_2 - \ddot{y}) = \ddot{z}_2$

$(\dot{x}_1 - \dot{y}) = \dot{z}_1, (\dot{x}_2 - \dot{y}) = \dot{z}_2$

$(x_1 - y) = z_1, (x_2 - y) = z_2$

ξ_1= ratio of damping to critical damping

p_1 = natural frequency of equivalent single degree model of parent structure

Equations (3) and (4) may be written in matrix notations as follows:

$$\{\ddot{z}\} + [D]\{\dot{z}\} + [S]\{z\} = -\{\ddot{y}\} \qquad (5)$$

where, $\{\ddot{z}\}$, $\{\dot{z}\}$ and $\{z\}$ are acceleration, velocity and displacement vectors respectively; $\{\ddot{y}\}$ is ground acceleration vector, [D] is damping matrix and [S] is stiffness matrix.

6.3 Computation of Response

The modified fourth order Range-Kutta method is used for solving equation (5) for getting the time-wise response,(Wang Ping-Chun, 1966). As the objective of the study is to find suitable damping and stiffness characteristics of the isolator for a particular structure, a parametric variation of damping and stiffness is carried out. The time period of a single degree freedom system is a measure of stiffness with respect to mass. As such, the flexibility ratio is indirectly varied by varying time periods of the structure. This is done in view of the convenience in presenting results in the conventional form of response spectra.

6.4 Presentation of Results

The effect of the stiffness and the damping of the isolator in relation to those of the main structure on the acceleration and displacement response of the main structure is studied by plotting two degree acceleration and displacement response spectrum curves. The stiffness ratio is varied by varying the ratio of the time period of the isolator to the main structure. Six values of the ratio of the time periods T_2/T_1 chosen are 1.0, 1.5, 2.0, 2.5, 3.0 and 6.0. The damping of the main structure is kept as 2, 5 and 10% of the critical respectively and the damping ratio of the isolator to the structure ξ_2/ξ_1 is varied from 1 to 4. Thus the effect of damping is studied by plotting 12 response spectrum curves covering a damping range of the main structure from 2 to 10% of the critical and that of the isolator from 2 to 20% of the critical. In what follows, a typical acceleration and displacement response spectrum curve is taken and discussed. Similarly, the effect of the damping on the response of the main structure is discussed. In all the curves, mass of the isolator is kept at 10% of the mass of the main structure.

6.5 Effect of Isolator Stiffness on Acceleration Response

Fig. 2 shows a typical acceleration response spectrum curve showing the effect of the isolator stiffness on the acceleration response of the main structure. In this figure, curve 7 is the acceleration response spectra for the single degree freedom system (without isolator) constructed to show the order of response reduction. Curves 1 to 6 are for varying T_2/T_1 ratios. The damping of the main structure is 5% of the critical and the damping ratio of the isolator to the main structure is 2.0. It is seen from this figure that the resonance occurs with a moderately stiff structure. Further, it can be seen that the peak acceleration response for an unisolated structure corresponds to time period T_1 = 0.5 sec whereas for the structure mounted on the isolator having T_2/T_1 equal to unity, the peak occurs at T_1 = 0.2 sec. Thus, with

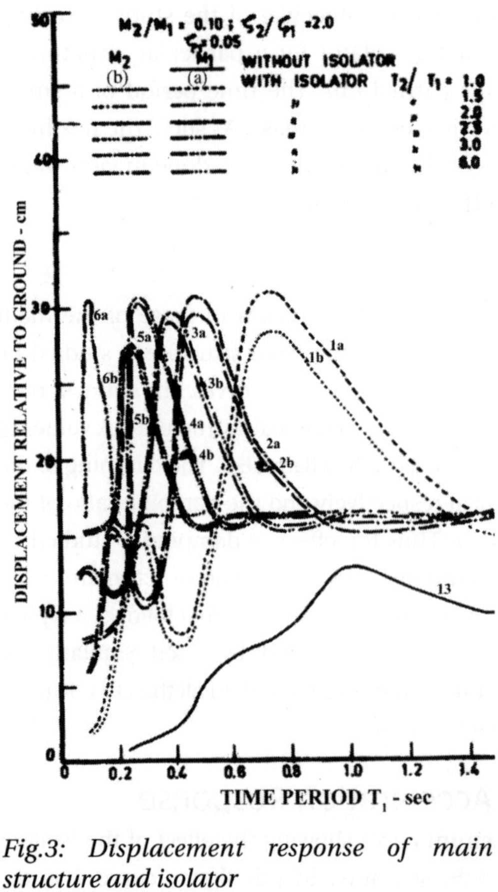

Fig.3: Displacement response of main structure and isolator

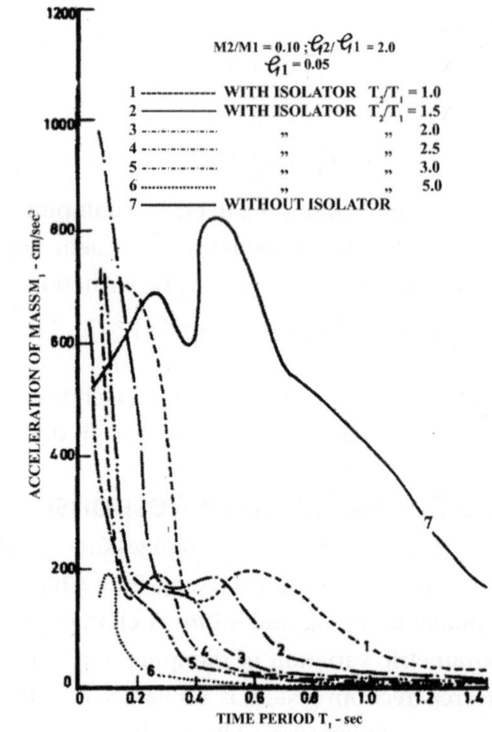

Fig.2: Acceleration response of main structure

increasing ratio of the time periods T_2/T_1, the peak shifts towards the lower value of the time period. This may enable the designer to avoid the resonance peak by mounting the structure over the isolator of suitable stiffness. However, as this study is conducted for a particular earthquake, it is quite possible that for another earthquake, the pattern of the response spectrum curves may be quite different. It is also seen from the figure that an improperly choosen isolator, may increase the peak response instead of decreasing it.

6.6 Effect of Isolator Stiffness on Displacement Response

Fig. 3 shows the total displacement curves for the isolator and for the main structure.

Curve 13 is the displacement spectrum curve for the single degree freedom system (without isolator) plotted to get a comparative idea of the total displacement with and without the isolator. Curves 1a to 6a show the total displacement of the main structure relative to ground whereas the curves 1b to 6b show the total displacement of the isolator. It is seen that the introduction of the isolator, while increasing the total displacement of the main structure, effectively reduces the relative displacement between the structure and the isolator. The forces coming on the columns are, therefore, greatly reduced. Further, it is seen that for increasing ratio of T_2/T_1, the relative displacement reduces and for T_2/T_1 equal to 6, the two curves 6a and 6b overlap, indicating that there is no relative displacement.

6.7 Effect of Damping of Main Structure

It is seen from Fig. 4 that the response of the structure mounted over an isolator remains practically unaffected with increasing value of the damping of the main structure. The reason for this behaviour of the isolated system is that in an isolated structure, most of the deformation occurs in the isolator. The main structure is shifted as a rigid body over the isolator with very small velocity relative to that of the isolator. Consequently, the force exerted by the viscous damper which is the function of relative velocity is very small inspite of the large damping coefficient.

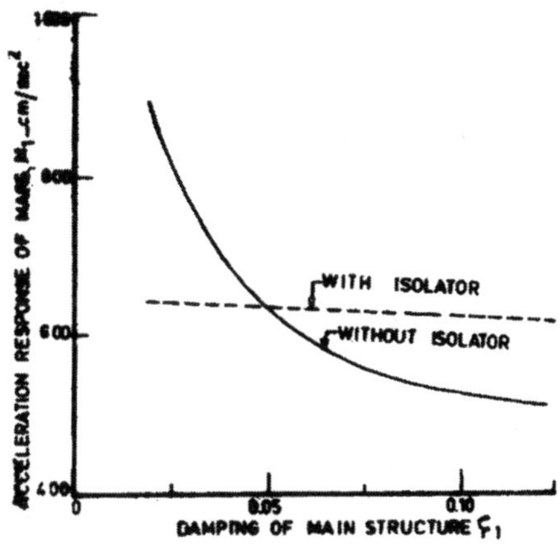

Fig.4: Effect of damping of main structure on its acceleration response

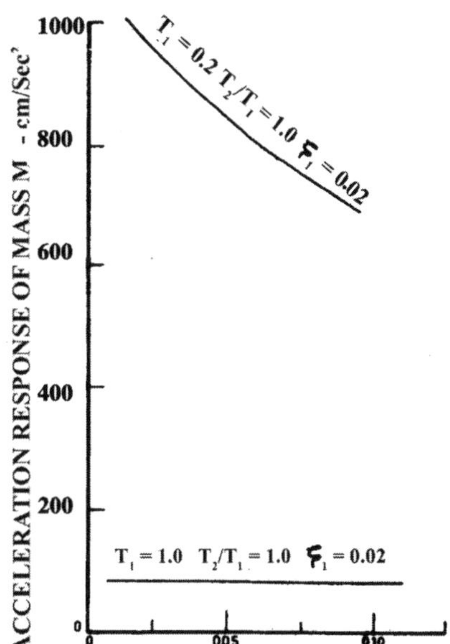

Fig.5: Effect of isolator damping on acceleration response

6.8 Effect of the Damping of the Isolator

In general, with increasing value of isolator damping, the acceleration response of the system decreases, but the reduction in the response is more for stiff structures than for flexible structures. From Fig. 5 it is seen that the isolator damping is very effective for a structure having time period T_1 = 0.2 sec but not so effective for the structure having time period T_1 = 1.0 sec.

6.9 Conclusion

1. For the given ground motion, (N-S component of the El Centro earthquake of 1940) it is seen that the isolation technique is suitable for stiff to moderately stiff structures but is not very useful for structures having fundamental time period of more than 1.0 sec.
2. The ratio of the time period of the isolator to the structure significantly affects the response of the main structure.
3. The introduction of an isolator increases the total displacement in the structure but the relative displacement between the main structure and the isolator is greatly reduced.
4. The damping of the structure mounted over an isolator does not significantly affect the response of the main structure.

The damping of the isolator is effective in reducing the response of the stiff structures but not very effective for structures having fundamental time period of 1.0 sec or more.

6.10 Acknowledgement

The authors are grateful to Dr. A.R. Chandrasekaran, Professor, School of Research and Training in Earthquake Engineering for providing digitised punched data for the N-S component of the El Centro earthquake of 1940. Thanks are also due to Sri Swaraj Singh Saini for typing the manuscript. ■

6.11 References

1. Wang Ping-Chun (1966), "Numerical and Matrix Methods in Structural Mechanics with Application to Computers", Book published by John Wiley and Sons Inc., New York, USA.

Chapter 7

Vulnerability and Risk of Damage to Masonry Buildings in India[1,2]

7.1 Introduction

The term risk is defined as the expected number of lives lost, persons injured, damage to property and disruption of economic activity due to a particular natural phenomenon such as an earthquake, and consequently the product of specific risk to an element and the number of elements at risk. The risk emanating from earthquakes in general is a function of four components, as stated below:

Seismic Risk = f (Hazard, Location, Exposure, Vulnerability).

where the four components are defined below:-

HAZARD- occurrence of an earthquake of sufficient Magnitude (hence: Intensity at the epicentre) capable of causing damage to man-made structures.

LOCATION- (i) How far the 'exposure' is situated from the *Hazard location* the nearer ones being in greater danger than those far away, and (ii) Local site conditions which can modify the *hazard* and/or affect the stability of the exposure, such as topography, soil deposit, water table, etc.

EXPOSURE- Objects and structures built by man which are exposed to the effects of *the 'hazard'*: buildings, bridges, dams, power plant, life-line structure, etc.

VULNERABILITY-Damageability of the *'expcsure'* under the action of the hazard; weaker ones being more vulnerable and 'risky' than the stronger ones.

---

1 *Dr.A.S.Arya, Professor Emeritus, IIT - Roorkee and Chairman, Peer Group for Vulnerability Atlas of India, 2006, Ministry of Urban Affairs, GoI.*

2 *An extract from chapter 1, Vulnerability Atlas of India, 2006.*

In a given *location,* the *seismic hazard* can be assumed as indicated in the Seismic Zoning Map of India (see IS:1893 - 2002) or as determined from specific site dependent studies as carried out for very major projects such as atomic power plants, major dams, major bridges or major industries. The *local soil conditions* have to be considered from the point of view of amplification of the ground motion as well as soil structure interaction. The existing *exposure* in any area is composed of very old, old and new constructions some of which will be well designed and well constructed and others either badly designed or badly constructed. Consequently, their *vulnerabilities* to different earthquake *intensities* will be at different levels, so also the *risk* of their damage and loss could be quite different. Among the above *four factors,* the first three factors are really as they *exist* on which man has hardly any control, but the *fourth* factor of *vulnerability* of the physical assets is where the vulnerabilities can be reduced by engineering interventions of preventive measures such as earthquake resistant design of all new constructions and retrofitting of existing vulnerable constructions, particularly those which may be considered as critical to the safety and well-being of the people.

Reduction of vulnerability of our buildings and other structures and systems, those existing and those being built or to be built is the key to earthquake protection.

This chapter is devoted to the consideration of vulnerability of masonry buildings in India which, as will be seen, comprise the most prevalent construction system in the country.

7.2 House Types

The Census of Housing, 2001 Census of India, gives the following details of houses based on materials of construction for walls and roofs:

a) Type of Roof:
   i) Pitched or sloping including tiles, slate; corrugated iron, zinc or other metal sheets; asbestos cement sheets; plastic polythene, thatch, grass, leaves, bamboo, etc.
   ii) Flat including brick, stone and lime; reinforced brick concrete/reinforced cement concrete.

b) Type of Wall:
   i) Mud, unburnt bricks, stone laid in mud or lime mortar
   ii) Burnt bricks laid in cement, lime or mud mortar
   iii Cement concrete
   iv) Wood or Ekra walling
   v) Corrugated iron, zinc or other metal sheets
   vi) Grass, leaves, reeds or bamboo or thatch, plastic polythene and others

c) Type of Flooring:
   Various types like mud, stone, concrete, wood or bamboo, mosaic floor tiles, etc.

Table 1A: Houses by Material of Wall in India*

	Total census houses	Mud, Unburnt brick		Wood		G.I., Metal, Asbestos, sheets		Burnt Brick		Stone		Concrete		Any other material	
		No. of Houses	%	No. of Houses	%	No. of Houses	%	No. of Houses	%	No. of Houses	%	No. of Houses	%	No. of Houses	%
Total	249,095,869	73,799,162	29.6	3,196,992	1.3	1,998,678	0.8	111,891,629	44.9	25,481,817	10.2	6,540,338	2.6	26,187,253	10.5
Rural	177,537,513	65,807,212	37.1	2,363,200	1.3	876,677	0.5	62,715,919	35.3	20,347,899	11.5	2,253,979	1.3	23,172,627	13.1
Urban	71,558,356	7,991,950	11.2	833,792	1.2	1,122,001	1.6	49,175,710	68.7	5,133,918	7.2	4,286,359	6.0	3,014,626	4.2

* Census of India 2001-Housing Data

Table 1B: Houses by Material of Roof in India*

	Total census houses	Tiles		Slate		G.I., Metal, Asbestos, sheets		Brick		Stone		Concrete		Any other material	
		No. of Houses	%	No. of Houses	%	No. of Houses	%	No. of Houses	%	No. of Houses	%	No. of Houses	%	No. of Houses	%
Rural	177,537,513	62,935,397	35.4	2,364,095	1.3	18,665,296	10.5	9,927,250	5.6	11,906,910	6.7	21,061,294	11.9	50,677,721	28.6
Urban	71,558,356	12,591,573	17.6	444,565	0.6	11,821,919	16.5	4,147,242	5.8	5,246,952	7.3	31,777,933	44.4	5,528,172	7.7
Total	249,095,869	75,526,970	30.3	2,808,660	1.1	30,487,215	12.2	14,074,492	5.7	17,153,862	6.9	52,839,227	21.2	56,205,893	22.6

* Census of India 2001-Housing Data

The distribution of houses based on predominant materials of the wall and roof over the whole of India, according to 2001 Census is shown in Tables-1A & 1B. From Table 1A it is seen that the total number of census houses in 2001 is 249 million out of which the maximum number of houses accounting for 111.9 million (44.9% of the total) belong to the burnt brick wall type. The next in numbers are mud or unburnt brick houses accounting for 73.8 million (29.6% of the total). The stone masonry houses number 25.5 million (10.2% of the total). Thus, these three types which can be jointly *classified as masonry load bearing wall type houses* account for 84.7% of the total. From the past earthquake studies, it is also seen that these types of houses have the worst behaviour in moderate to severe earthquakes and their collapse has led to the maximum number of human and cattle losses. Based on these studies India was the first country to take up the research & development work towards earthquake resisting features to be built into masonry buildings. Table 1B shows that the sloping roofs consisting of tiles, slate, G.I. and asbestos sheet coverings account for about 50% of the houses. These roofs do not provide any diaphragm action at the roof level, introducing a further weakness in the masonry houses.

From the point of view of vulnerability to the earthquake, wind, or flood hazards, it was seen that the type of flooring had hardly any significance, hence, omitted from consideration. Now the roof types and wall types cannot be grouped together since, separately listed, unlike what could be done in the 1991 Census. An appropriate grouping for whole of India for the walls and roofing will be categorised as follows:

7.2.1 Wall Types

Type A1 - built using mud or unburnt brick or block

A2 - built using stone in any mortar

B - built using burnt brick using any mortar

C1- built using concrete wall or concrete frame

C2 - constructed using wood frame and panelling

X - those not following under the above types, such as grass, thatch, bamboo, plastic, polythene; G.I., metal, asbestos sheets, any other material (all collected under one type X here)

The building categories defined under MSK Intensity Scales combine Types A1 and A2 under A, Type B under B abd Types C1 & C2 under C. There is no category defining X types but this is retained in the Vulnerability Atlas as X. Thus, building categories will be taken as below:

Category - A : Buildings in field-stone, rural structures, unburnt brick houses, clay houses.

Category - B: Ordinary brick building; buildings of the large block and prefabricated type, half-timbered structures, building in natural hewn stone.

Category - C : Reinforced building, well built wooden structures.

Category - X : Other materials not covered in A,B and C. These are generally light structures.

7.2.2 Roof Types

Category - R1: Light Weight (grass, thatch, bamboo, wood, mud, plastic, polythene, GI metal, asbestos sheets, other materials)

Category - R2: Heavy Weight (tiles, slate)

Category - R3: Flat Roof (brick, stone, concrete)

With these groupings, the vulnerability of each subgroup could be defined separately for any given Intensity of earthquake. The risk levels of the various categories of houses for the earthquake hazards are shown in Table-3, and explained in the following sections.

7.3 Risk of Damage to House Types

The damage risk to various house types is based on their average performance observed during past occurrences of damaging events. In view of numerous variations in the architectural planning, structural detailing, quality of construction and care taken in maintenance, the performance of each category of houses in a given event could vary substantially from the average observed. The Intensity scales (Modified Mercalli or MSK) represent average observations. The provisions in the MSK Intensity scales are appropriately presented in Table 2 where the extent of damage is correlated with the type of house construction and the earthquake Intensity.

Table 2: Seismic Intensity vs Damage to Buildings

Building Type	MSK Intensity VII	MSK Intensity VIII	MSK Intensity IX
A. Mud and adobe houses, random-stone partial collapse constructions	*Most* have large deep cracks *Few* suffer partial collapse	*Most* suffer partial collapse	*Most* suffer complete collapse
B. Ordinary brick buildings, building of large blocks and prefab type, poor half timbered houses	*Many* have small cracks in walls	*Most* have large and deep cracks	*Many* show partial collapse *Few* complete collapse
C. Reinforced buildings well built wooden buildings	*Many* have fine plaster cracks	*Most* may have small cracks in walls. *Few* may have large deep cracks	Many may have large deep cracks Few may have partial collapses

Note:
In MSK Intensity VI area, *Many* buildings of Type A sustain slight damage and *Few* may even have moderate damage like small cracks in walls and fall of fairly large pieces of plaster. Buildings of Types B and C remain almost free of damage except for occasional fine cracks.

* *Most* = about 75%, *Many* = about 50%, *Few* = about 5%

Source : Arya A.S., Bulletin ISET, September, 1990.

Under seismic occurrence, the following observations have been made in many cases:

(a) All Masonry Houses (Types A and B)

- Quality of construction comes out as a major factor in the seismic performance particularly under intensities MSK VII to IX. Good quality constructions perform much better than poor quality constructions in any type. Appropriate maintenance increases durability and maintains original strength.
- Number of storeys in the house and the storey height are other factors. The higher the storey and more the number of storeys, the greater is the observed damage.
- Size, location and number of door and window openings in the walls also determine seismic performance, since the openings have a weakening effect on the walls. Smaller and fewer openings, located more centrally in the walls are better from the seismic performance viewpoint.
- Architectural layout, particularly in large buildings, that is, the shape of the building in the plan and elevation, presence of offsets and extended wings, also play an important role in initiation of damage at certain points and its propagation as well. More symmetrical plans and elevations reduce damage and unsymmetrical ones lead to greater damage.
- Where clay/mud mortar is used in wall construction, its wetness at the time of the earthquake is a very important factor in the seismic performance since the strength of fully saturated mortar can become as low as 15% of its dry strength.

(b) Wooden Houses

- Quality of construction, that is, seasoning of wood and the joinery are important in seismic and cyclonic wind performance. The better the quality, the better the performance.
- Wood decays with time due to dry rot, insect and rodent attack, etc., therefore, the joints tend to become loose and weak. The state of maintenance of the wooden building will determine its performance during earthquake, high wind, as well as flooding.
- In houses with sloped roofs, a shallow angle for the roof, extended eaves prcjection, and reentrant corners lead to higher damage.
- In light roofs, pressures often exceed the dead weight, leading to blowing-off of roofs.

(c) Reinforced Concrete Houses

- Multi-storey RC frame buildings resting on soft soils and having soft first storey unconnected wall panels and floating columns in the superstructure collapsed even in Seismic Zone III.
- Besides bad quality of configuration planning and structural design, poor quality of construction led to total collapse of 5 to 10 storeyed RC frame buildings.
- In reinforced concrete construction, good structural design and detailing and good quality construction only would ensure excellent performance. Carelessness in any of these can lead to poor behaviour under both earthquakes and cyclones.

The average risk levels to various categories of houses for various hazards and their intensities are defined here below for use in the house vulnerability tables.

7.4 Damage Risk Levels for Earthquakes

The damage risk to various house types is defined under various seismic intensities on the MSK Intensity scale. The following damage risks are defined based on this Intensity scale.

Very High Damage Risk (VH)

Total collapse of buildings

High Damage Risk (H)

Gaps in walls; parts of buildings may collapse; separate parts of the building lose their cohesion; and inner walls collapse.

Moderate Damage Risk (M)

Large and deep cracks in walls fall of chimneys on roofs.

Low Damage Risk (L)

Small cracks in walls; fall of fairly large pieces of plaster, pantiles slip off; cracks in chimneys, part may fall down.

Very Low Damage Risk (VL)

Fine cracks in plaster; fall of small pieces of plaster.

Based on the data in Table 2 and the above damage risk levels, the risk table for various building types in various seismic zones is constructed in Table 3

Table 3: Damage Risk to Housing under Various Earthquake Zone Intensities

Type of Wall	Level of Risk in Seismic Zone			
	V	IV	III	II
A1 - Mud and Unburnt Brick	VH	H	M	L
A2 - Stone Wall	VH	H	M	L
B - Burnt Bricks Wall	H	M	L	VL
C1 - Concrete Wall	M	L	VL	VL
C2 - Wood wall	M	L	VL	VL
X - Other Materials	M	VL	VL	VL
Type of Roof				
R1 - Light Weight Sloping Roof	M	M	L	VL
R2 - Heavy Weight Sloping Roof	H	M	L	VL
R3 - Flat Roof	Damage Risk as per that for the Wall supporting it			

Building Category : (By Wall Material)

Category - A : Buildings in field-stone, rural structures, unburnt brick houses, clay houses

Category- B : Ordinary brick building; buildings of the large block and prefabricated type, half-timbered structures, building in natural hewn stone

Category - C : Reinforced concrete building, well built wooden structures

Category - X : Other materials not covered in A,B,C. like light sheets and biomass materials

Note : Damage Risk is indicated assuming heavy flat roof in cases A, B and C (Reinforced Concrete) building

Building Category : (By Roof Material)

Category - R1 - Light Weight Sloping Roof (Grass, Thatch, Bamboo, Wood, Mud, Plastic, Polythene, GI Metal, Asbestos Sheets, etc)

Category - R2 - Heavy Weight Sloping Roof (Tiles, Slate)

Category - R3 - Flat Roof (Brick, Stone, Concrete)

7.5 The Housing Risk Tables

By correlating the house types, the hazard intensities on the maps and the damage risk levels, the housing damage risk tables have been generated and presented in the Vulnerability Atlas of India, 2006. For the country as a whole, for each state and Union Territory also, an overall risk table has been developed. Each table also gives at the top of each column of hazard intensities, the percent of total area of the country, state or district covered by the Table, lying under the various hazard intensities. Such tables are then prepared for each of the districts and collected state-wise.

Three such risk tables are included here as examples: Table 4 gives the risk to housing for India, Table 5 shows the risk to housing for the State of Bihar and Table 6 shows the risk to housing for the District of Madhubani in Bihar.

According to Table 4: 99.3 million housing units belong to the most risky Category A type, out of which 10.8% are under very high risk of collapse, 17.5% are under risk of heavy damage and 30.8% have risk of moderate damage consisting of wide cracks and partial collapses. On the other hand, the burnt brick wall houses (Category B type) number 111.9 million i.e 44.9% of the total number of houses in the country. These are liable to high risk of damage i.e, destruction and partial collapse in 11% of the area of the country.

Considering the State of Bihar in Table 5, it is seen that16.5% area of the State is in Seismic Zone V, Category A buildings account for 23.2% and Category B buildings account for 42.2%. Another 62.4% of the State area lies in Seismic Zone IV, thus, 78.9% of the total State area lies in Seismic Zones IV & V together under which the Category A buildings can totally collapse, as had happened in the 1934 earthquake in that area. The risk to life is very high.

Now considering the Madhubani District of Bihar in Table 6, it is seen that the whole district is classified in Zone V in which 18.7% are of Category A type i.e. 140,000 houses are liable to total collapse in a severe earthquake in that area and the Category B buildings amounting to 234,000 houses can have very heavy damage, including partial collapses. It shows that the risk to life and property is very high.

Thus, from the Housing Vulnerability Tables, the concerned administrative and disaster management authorities in the states and districts can visualise the extent of damage risk under earthquakes at any time in the future.

7.6 Risk of Earthquake Damage in Urban Areas

The risk of earthquake damage has been growing because of increasing urban development in seismically active areas and the vulnerability of older buildings, which may not have been built or upgraded to current building codes. Given below is an example of the damage that happened in the city of Kobe during the earthquake of January, 1995

Table 4: Distribution of Houses by Predominant Materials of Roof and Wall and Level of Damage Risk(INDIA)

Wall / Roof		Census Houses		Level of Risk under EQ Zone			
		No. of Houses	%	V	IV	III	II
				Area in %			
INDIA				10.8	17.5	30.8	40.9
WALL							
A1 - Mud & Unburnt Brick Wall	Rural	65,807,212	26.4				
	Urban	7,991,950	3.2				
	Total	***73,799,162***	***29.6***	*VH*	*H*	*M*	*L*
A2 - Stone Wall	Rural	20,347,899	8.2				
	Urban	5,133,918	2.1				
	Total	***25,481,817***	***10.3***	*VH*	*H*	*M*	*L*
Total - Category - A		***99,280,979***	***39.9***				
B - Burnt Bricks Wall	Rural	62,715,919	25.2				
	Urban	49,175,710	19.7				
	Total	***111,891,629***	***44.9***	*H*	*M*	*L*	*VL*
Total - Category - B		***111,891,629***	***44.9***				
C1 - Concrete Wall	Rural	2,253,979	0.9				
	Urban	4,286,359	1.7				
	Total	***6,540,338***	***2.6***	*M*	*L*	*VL*	*VL*
C2 - Wood wall	Rural	2,363,200	0.9				
	Urban	833,792	0.3				
	Total	***3,196,992***	***1.2***	*M*	*L*	*VL*	*VL*
Total - Category - C		***9,737,330***	***3.9***				
X - Other Materials	Rural	24,049,304	9.7				
	Urban	4,136,627	1.7				
	Total	***28,185,931***	***11.4***	*M*	*VL*	*VL*	*VL*
Total - Category - X		***28,185,931***	***11.3***				
TOTAL BUILDINGS		249,095,869					
ROOF							
R1 - Light Weight Sloping Roof	*Rural*	*69,342,56*	*27.8*				
	Urban	*17,350,091*	*7.0*				
	Total	***86,692,658***	***34.8***	*M*	*M*	*L*	*VL*
R2 - Heavy Weight Sloping Roof	*Rural*	*65,299,492*	*26.2*				
	Urban	*13,036,138*	*5.2*				
	Total	***78,335,630***	***31.4***	*H*	*M*	*L*	*VL*
R3 - Flat Roof	*Rural*	*42,895,454*	*17.2*				
	Urban	*41,172,127*	*16.5*				
	Total	***84,067,581***	***33.7***	*Damage Risk as per that for the Wall supporting it*			
TOTAL BUILDINGS		249,095,869					

Categories of building types including wall and roof are explained under Table 3 as well as the text

Table 5: Distribution of Houses by Predominant Materials of Roof and Wall and Level of Damage Risk BIHAR

Wall / Roof		Census Houses		Level of Risk under EQ Zone			
		No. of Houses	%	V	IV	III	II
				Area in %			
BIHR				16.5	62.4	21.1	
WALL							
A1 - Mud & Unburnt Brick Wall	Rural	3,555,951	21.8				
	Urban	202,255	1.2				
	Total	***3,758,206***	***23.0***	*VH*	*H*	*M*	
A2 - Stone Wall	Rural	13,950	0.1				
	Urban	5, 543	-				
	Total	***19,473***	***0.1***	*VH*	*H*	*M*	
Total - Category - A		***3,777,679***	***23.2***				
B - Burnt Bricks Wall	Rural	5,576,724	34.2				
	Urban	1,311,233	8.0				
	Total	***6,887,957***	***42.2***	*H*	*M*	*L*	
Total - Category - B		***6,887,957***	***42.2***				
C1 - Concrete Wall	Rural	20,199	0.1				
	Urban	15,945	0.1				
	Total	***36,144***	***0.2***	*M*	*L*	*VL*	
C2 - Wood wall	Rural	81,917	0.5				
	Urban	13,965	0.1				
	Total	***95,882***	***0.6***	*M*	*L*	*VL*	
Total - Category - C		***132,026***	***0.8***				
X - Other Materials	Rural	5,327,185	32.6				
	Urban	191,680	1.2				
	Total	***5,518,865***	***33.8***	*M*	*VL*	*VL*	
Total - Category - X		***5,518,865***	***33.8***				
TOTAL BUILDINGS		16,316,527					
ROOF							
R1 - Light Weight Sloping Roof	Rural	6,095,954	37.4				
	Urban	280,474	1.7				
	Total	***6,376,428***	***39.1***	*M*	*M*	*L*	
R2 - Heavy Weight Sloping Roof	Rural	5,171,169	31.7				
	Urban	380,658	2.3				
	Total	***5,551,827***	***34.0***	*H*	*M*	*L*	
R3 - Flat Roof	Rural	3,308,783	20.3				
	Urban	1,079,489	6.6				
	Total	***4,388,272***	***26.9***	*Damage Risk as per that for the Wall supporting it*			
TOTAL BUILDINGS		16,316,527					

Categories of building types including wall and roof are explained under Table 3 as well as the text

Table 6: Distribution of Houses by Predominant Materials of Roof and Wall and Level of Damage Risk(Dist. MADHUBANI, BIHAR)

Wall / Roof		Census Houses		Level of Risk under EQ Zone			
		No. of Houses	%	V	IV	III	II
				Area in %			
MADHUBANI				100	0	0	0
WALL							
A1 - Mud & Unburnt Brick Wall	Rural	138,256	18.4				
	Urban	1,695	0.2				
	Total	*139,951*	*18.6*	*VH*			
A2 - Stone Wall	Rural	285	-				
	Urban	5	-				
	Total	*290*	-	*VH*			
Total - Category - A		*140,241*	*18.7*				
B - Burnt Bricks Wall	Rural	214,360	28.6				
	Urban	19,293	2.6				
	Total	*233,653*	*31.2*	*H*			
Total - Category - B		233,653	31.2				
C1 - Concrete Wall	Rural	834	0.1				
	Urban	20	-				
	Total	*854*	*0.1*	*M*			
C2 - Wood wall	Rural	3,462	0.5				
	Urban	89	-				
	Total	*3,551*	*0.5*	*M*			
Total - Category - C		*4,405*	*0.6*				
X - Other Materials	Rural	363,977	48.6				
	Urban	7,097	0.9				
	Total	*371,074*	*49.5*	*M*			
Total - Category - X		*371,074*	*49.5*				
TOTAL BUILDINGS		749,373					
ROOF							
R1 - Light Weight Sloping Roof	Rural	475,451	63.4				
	Urban	10,548	1.4				
	Total	*485,999*	*64.8*	*M*			
R2 - Heavy Weight Sloping Roof	Rural	131,414	17.5				
	Urban	3,852	0.5				
	Total	*135,266*	*18.0*	*H*			
R3 - Flat Roof	Rural	114,309	15.3				
	Urban	13,799	1.8				
	Total	*128,108*	*17.1*	*Damage Risk as per that for the Wall supporting it*			
TOTAL BUILDINGS		749,373					

Categories of building types including wall and roof are explained under Table 3 as well as the text

Earthquake Disaster in Kobe, Japan, January 1995

- Kobe city area = 544 sq.km., Osaka area = 221 sq.km
- Kobe city population 1.47 million, Density 270/ha
- Osaka city population 2.50 million, Density 1132/ha
- Earthquake Magnitude M 7.2
- Epicentre: Close to the city of Kobe (Not like this in the last 1500 years)
- Maximum Intensity JMA 7, MM ~ XI
- Coastal Areas Liquified; Fires 631
- Killed 5,420; Missing 2
- Injured 43,6654
- Houses (total damaged 198,403)

Fully destroyed	91,285	(46.0%)	Old type
Seriously damaged	731,132	(36.9%)	New type
Partially damaged	33,986	(17.1%)	

- Public buildings destroyed 525
- Other Buildings 3,080 old type R.C. before 1980 code damaged more
- Roads damage points 9369
- Water supply suspended in 1.2 million households (for 5 days)
- Gas supply suspended in 857,000 households (resumed in 367,000 users in 4 weeks)
- Rail and road damaged and traffic suspended at many sections for weeks to months.

Similarly, the infrastructure facilities provided in any city are liable to get damaged depending upon the type and quality of construction. Given below are some of the sectors which can get affected and pose a great threat to life and property

7.6.1 Inner (old) city areas

The core area of a city is one of the most vulnerable areas as far as a natural hazard like earthquake is concerned. This is primarily due to:

- presence of narrow streets (which will be a major constraint in conducting rescue and relief operations),
- existence of old buildings and structures which were not designed/constructed keeping in mind earthquake safety,
- poor infrastructure facilities like exposed electric poles with hanging electric wires,
- choked sewers, drains, etc.

From the past earthquakes, we have seen that the narrow streets get blocked as the houses on both sides of the road collapse there killing the people moving (More than *300 children and teachers were crushed to death in Anjar during the Kachchh earthquake of 2001 due to the building collapsing on them from both sides. In Kobe, city fire tenders could not go on to streets as the streets were filled with fallen facades, display boards, etc.)*

7.6.2 Water supply and sanitation.

The disruption of water and sewer lifelines could seriously affect emergency facilities, fire suppression systems, telecommunication systems, and drinking water supplies.

- *Emergency facilities:* Water could be unavailable to serve emergency facilities.
- *Telecommunication systems:* The cutoff of water supplies, with the resulting shutdown of cooling systems, could render computer-dependent telecommunication systems inoperable.
- *Water supplies:* If distribution systems were damaged, drinking water would have to be trucked in until the water system could be restored. Businesses would have to remain closed until the water service was restored because of health and fire hazards.

(Due to the earthquake in 2001, the drinking water supply was adversely affected in 1,340 villages of the 5 earthquake affected districts of Gujarat. The most severely damaged facilities were: two dams, two water treatment plants, over 1,500 km of pipelines. The RCC underground tanks and overhead tanks could withstand the impact, whereas masonry structures collapsed. Other damage included rupture and dislocation of transmission pipelines, resulting in leakages. The total amount of damage was estimated to be Rs.233 crores. The estimated cost of reconstruction, including upgrading for hazard resistant construction, was around Rs.451 crores).

7.6.3 Sewerage and solid waste management.

- *Sewers:* As a result of loss of power and damage to pump station buildings and equipment, sewage (in reduced volumes because of water system dysfunction) would overflow from manholes into city streets, back up into basements, and run into drainage conduits. Sewers in areas that suffer significant permanent ground deformation would be destroyed completely. Sewers in less affected areas might sustain damage that would go unnoticed, presenting a hazard to health.
- *Sewage treatment plants:* In an earthquake, sewage treatment plants not designed for seismic resistance would sustain damage. Raw or inadequately treated sewage that had reached the sewage treatment plant would be discharged into the receiving water body.

7.6.4 Urban transportation including roads, highways, expressways, MRTS, and metro projects.

The consequences of failure in a transportation lifeline due to a natural hazard can involve:

- Direct loss of life due to collapse or structural failure of the lifeline.
- Indirect loss of life due to inability to respond to secondary catastrophes, such as fires, and/or provide emergency medical aid.

- Delayed recovery operations.
- Release of hazardous products (e.g., losses from tank cars derailed by track failure, gas leaks from ruptured utility lines).
- Direct loss of property and utility service (e.g., the collapse of a bridge carrying utilities)
- Losses due to interruption of access.

Historically, bridges have proved to be vulnerable to earthquakes of MSK Intensity VIII or higher, that is, in seismic Zones IV & V, sustaining damage to substructures and foundations and, in some cases, being totally destroyed as substructures fail or superstructures are unseated from their supporting elements. Pavements are also vulnerable to earthquake damage, due principally to ground failure such as liquefaction.

Although railroad systems suffer damage similar to that suffered by highways, their operation is much more sensitive to permanent ground deformation than highway operation. A survey of damage to railroad components during past earthquakes shows damage to bridges, embankment failures, vertical and horizontal track misalignments, tunnel misalignments, failure of tunnel linings, structural damage to railroad buildings, and overturned rail cars and locomotives.

Ports and waterways are, by their nature, constructed on soft, saturated sites that are susceptible to site amplification effects and/or soil failure. Historically, damage due to earthquakes can lead to a tsunami, massive flow and flooding due to liquefaction, and structural damage to wharves and container cranes. Even relatively minor damage can close a port for an extended period of time, and loss of export revenue can have a crippling effect on some economies.

Failures of airport runway pavements have occurred in the past because of ground deformation and/or liquefaction effects. The potential for severe structural damage appears to be lower for airports than for other transportation systems, principally because the basic components are pavements and buildings. However, even minor structural damage can cause closure of a facility and severely impede recovery efforts. Control towers at airports are particularly vulnerable to contents damage and consequential loss of operation. Similarly, loss of electric power, telecommunication, or radar equipment due to relatively minor structural damage can have a major impact on both local and regional air traffic operations.

(Due to the earthquake in 2001, in the road sector, maximum damage was caused to bridges and culverts. The old Surajbadi bridge, which had been constructed in the 1960s, suffered significant damage during the earthquake. The bridge was closed for traffic for the first two days. It was temporarily restored for slow single-lane traffic. The railway infrastructure suffered relatively minor destruction. Most of the damage was to buildings - stations and staff quarters.

With regard to ports, there was significant damage to the Kandla port. Five of the ten dry cargo jetties developed major cracks. The oil jetty and a small wharf were damaged. Twelve of the forty ports managed by the Gujarat Maritime Board were damaged. Most of the marine structures, cargo handling equipment, storage facilities, residential/office buildings, roads, bridges, etc., were destroyed.

The damage to the transport sector was assessed at about Rs.233 crores for roads; Rs.79 crores for railways; Rs.198 crores for ports; and Rs.19 crores for airports).

7.7 Conclusion

The foregoing vulnerability and risk assessment clearly indicates that the major parts of the country are liable to high to severe damage during a future earthquake which had been considered probable to occur in Seismic Zones III, IV & V. The existing building stock of 249 million housing units as per the census 2001 consists mostly of non-engineered buildings without having desirable earthquake resistant features. 85% of these buildings consist of walls of mud and unburnt brick, or stone, or burnt brick, constructed using any mortar varying from clay mud, lime cinder or Surkhi or lean cement mortar. Unfortunately, in the absence of proper techno-legal and techno-financial regime, the number of unsafe buildings is increasing at the rate of about 2.5% of the existing stock every year. The risk is increasing even more rapidly in the urban areas not only to buildings but also to the urban infrastructure. A concerted effort of all stake holders is called for towards earthquake risk reduction by constructing all new assets conforming to the earthquake resistant building codes as well as a national effort towards retrofitting of existing critical lifeline structures. ■

7.8 References.

1. *Vulnerability Atlas of India*, 2006, prepared through Peer Group appointed by Ministry of Urban Affairs and Poverty Alleviation, Government of India [to be published by Building Materials and Technology Promotion Council (BMTPC)].
2. Anand S. Arya, "Guideline on Provision of Hazard Safety Aspects in Projects Eligible for Jawaharlal Nehru National Urban Renewal Mission (JNNURM) Assistance", 2006.

Chapter 8

Behaviour of Buildings and Foundations During Earthquakes[1,2]

8.1 Introduction

The problem of earthquake resistant construction of masonry buildings has attracted the attention of several research workers during the last few years and a number of papers have been published on the subject (1-8). The reports regarding damage to structures during the past earthquakes have brought out the weaknesses in construction and suggested improvements for future construction such as those contained in the reports (9,10) of the Bihar earthquake of 1934 and Koyna earthquake of December 11, 1967. The aim of this paper is to review briefly the available information and summarise the main principles of earthquake resistant construction.

8.2 Behaviour of Different Constructions In Past Earthquakes

The random vibrations which are associated with earthquake motions and propagated in all directions at speeds of about 5 km. per second actually subject the structures to large scale field tests. As a result, the poor constructions collapse, the weak ones suffer damage to a large extent, the strong ones get away with minor damage, and the exceptionally sound constructions remain intact. Since all parts of a structure are shaken, the weak links cannot escape damage. They give in first and, in turn, damage the stronger parts as well.

Earthquakes have occurred in most parts of the world. Therefore, all types of constructions have been put to this type of severe test. For instance, in India, the buildings constructed from brick, stone, mud, timber or a combination thereof have been impacted in the seismic regions (9, 10, 11). Occasionally, some reinforced concrete buildings have also been present in earthquake affected areas and their behaviour has presented a striking contrast to that of masonry constructions. The old Japanese earthquakes presented the behaviour of wood framed buildings, with or without brick panel filling (12). Reinforced concrete-block constructions and

---

1 *A.S.Arya, Professor & Assistant Director, School of Research and Training in Earthquake Engineering, University of Roorkee, Roorkee*

2 *Indian Builder, Vol.18, No.14, January 1970, pp 26-39.*

reinforced concrete framed constructions are more recent examples (13,. 23). The Alaskan earthquake of 1964 presented full size tests of framed constructions as well as that of prefabricated constructions in reinforced and pre-stressed concrete (14). The Caracas earthquake of 1967 tested multi-storeyed reinforced concrete buildings with hollow brick panel fillings (15). These are the kinds which are being used in Yugoslavia and in India too, but they are using the solid bricks instead. At many places the the shaking of the structures has been the reason for its damage or collapse but in others, the foundation has been the villain. Its settlement led to the straining of the structure. Thus the behavior of most types of constructions can be studied separately. The Chilean earthquake of May 22, 1960 offers at one place a comparative study of various constructions which are particularly used in buildings which are a few storeys high (16). During this earthquake, about 45,000 dwellings of various types were damaged of which about 10% were damaged beyond repair. Table 1 presents a comparison of their earthquake behavior. The percentages given in the Table refer to the total number of buildings of a particular type. The order of usefulness with respect to loss of life, as indicated therein, has been worked out on the basis of percentages of 'dangerous' and 'destroyed' buildings combined. Similar behavior has been seen in the earthquakes in India where frequently random rubble masonry and composite (Kutcha-pucca) constructions of mud and burnt bricks are often encountered (9,11). These types may at best be graded slightly higher than an unreinforced adobe but below unreinforced brickwork. The diagonally braced timber frames, which are often used in old houses in Kashmir valley (17), with or without brick nogging are highly resistant to earthquakes and may be classed with the first two types in Table 1.

From the numerous observations of damage and non-damage during earthquakes, such as the Chilean earthquake described above, the constructions can be divided into the following four categories indicating their suitability.

(i) *Highly suitable* constructions are steel or reinforced concrete rigid frames and diagonally braced timber buildings. Such buildings have minimum weight, high strength to lateral forces and high ductility or deformation capacity which are the most desirable qualities for resisting the applied forces and absorbing the kinetic energy fed into the structure by the shaking ground.

(ii) *Moderately suitable* constructions are reinforced block or reinforced brick masonry and timber frames with brick noggin or sufficient brick walling acting with it. These buildings have moderate weight, high lateral load resistance and moderate ductility.

(iii) *Slightly suitable* constructions are unreinforced brick, block or stone masonry buildings with horizontal runners of timber, reinforced concrete or reinforced brick work at plinth, floor and roof levels having proper connections at corners; or the same type of buildings without the runners but constructed in good cement or lime mortars and having flat roofs like reinforced concrete slabs. These buildings have large weight, some amount of lateral strength and little ductility. They can be much improved by introducing vertical steel bars at corners and junctions of longitudinal and cross walls, and by introducing a

Table 1:- Comparative Steady of Damage in Different Constructions
(Total dwellings considered 6,125)

Sl. No.	Type of construction	Total number of dwellings	Repair-ably damaged %	Dangerous must be dis-mantled %	Destr-oyed %	Total of dam-aged ones %	Order of Use fulness with respect to loss of life
1.	Reinforced brickwork	1,781	8.3	1.4	0.8	10.5	II
2.	Reinforced concrete blockwork	5	20.0	0.0	0.0	20.0	I
3.	Unreinforced brickwork	1,149	37.6	33.6	11.6	82.8	VI
4.	Unreinforced concrete blockwork	6	16.7	33.3	16.7	66.7	VII
5.	Combined reinforced and unreinforced brickwork	1,334	37.7	20.8	3.5	52.0	V
6.	Wood frame	1,516	24.8	8.0	1.9	34.7	III
7.	Wood frame with masonry	147	53.0	19.7	3.5	76.2	IV
8.	Unreinforced adobe	187	17.0	52.5	23.0	92.5	VIII

reinforced concrete band at lintel levels of all storeys as recommended in IS: 4326-1967 Code of Practice for Earthquake Resistance Construction of Buildings (18). These provisions have been found to cost about 4 to 8% of the cost of buildings in Seismic Zone IV (19). With such strengthening measures, the buildings can be brought to almost the same level of suitability as reinforced block or reinforced brick masonry.

(iv) *Unsuitable* constructions are unreinforced brick or stone masonry construction in mud or weak mortar, composite (kutcha-pucca) constructions, adobe and mud huts. Such buildings have large weight, little or no lateral strength and almost no ductility. The lateral strength of brick or block constructions can be improved by constructing the jambs of openings and a few courses at plinth and floor levels in cement sand mortar and using reinforced concrete or reinforced brick lintels over openings with sufficient length of bearings, say 20 to 25 cm, over the jambs. Jack or flat arches for covering openings must be avoided, otherwise the tie rods must be used to keep them intact.

Besides the factors of weight, strength and ductility, the other important factor is *quality of workmanship.* Damage is found to be less in well constructed buildings which follow the standard specifications than in the poorly constructed buildings. Incidentally, the quality of construction also generally drops down with the order of suitability mentioned above, because of the nature of materials involved and the skill required to do the job. For example, a reinforced brick construction will usually have better workmanship than an unreinforced one.

8.3 Effect of Foundation Soil Upon Structural Behaviour

Softness of soil has been observed to have pronounced effect on the structural behaviour of buildings during an earthquake as was well evident in the Bihar earthquake of 1934(9) and Kern Country (USA) earthquake of 1952 (20). In the former, houses founded on rock out-crop suffered much less damage than similar buildings in the valleys resting on alluvial deposits. Table 2 shows the behaviour of different types of construction resting on different types of foundation materials as observed in the Kern Country earthquake.

The general trend of damage observed in most earthquakes is similar to that shown in Table 2, that is, the damage increases with the softness of the ground. But the reverse also

Table 2:-Effect of Foundation Material on Damage

S.No.	Type of Building	Foundation Material	Damage
1.	Steel or Reinforced Concrete	Thick alluvium	None
2.	Framed Building	-do-	Little or no damage
3.	Reinforced Block	-do-	Little or no damage except for cracking of unsupported façade
4.	Unreinforced Brick or Block	-do-	Extensive damage or collapse
5.	Stone masonry	(i) Thick alluvium (ii) Rock	Moderate damage to collapse Little or no damage
6.	Adobe	(i) Thick alluvium (ii) About 3m deposit (iii) Rock	Collapse Extensive cracking Slight cracking or no damage

happens sometimes, as in the case of Long Beach (USA) earthquake of 1933 (21) where the damage to buildings on soft soil on the beach was somewhat less than to those on more firm ground. It appears that the short period structures suffered more damage in that earthquake than long period ones due to the short period characteristics of the earthquake.

Some soils like poorly graded sands and sand-gravel mixtures, classified as SP in IS: 1498-1959, are found to lose their structure when vibrated in a dry condition. This causes a large amount of settlement. They liquefy and lose their shear strength if saturated with water and subjected to vibrations. In this condition, the buildings sink into the ground. The same type of behaviour is seen with the water-bearing, soft alluvial soils. Large areas liquefied during the Bihar earthquake of 1934 and Dhubri (Assam) earthquake of 1930 (22). Most striking examples of liquefaction of soil and sinking of buildings occurred in the Niigatta (Japan) earthquake of 1964 (23). The contrast in the behaviour of structures founded differently also provided the remedy against such failures. The buildings which were founded on point bearing piles remained standing vertical, although the soil slumped down at the surface. Those having shallow footing, however, sank, tilted or overturned. Therefore, point bearing piles must be used where loose soils having *standard penetration value* N less than 10 are encountered. Driven piles are preferred since the vibrations and compaction caused by them will improve the soil through which they pass. Friction piles may be used in the case of soft clays.

It follows, therefore, that from the point of view of behaviour during an earthquake, buildings should be founded on rock, where available. Otherwise, the following types of foundations may be adopted in the decreasing order of preference depending upon the height, size and importance of the building:

(1) Bearing piles in cohesionless material resting on stiff soil having a high N value.
(2) Friction piles in cohesive material.
(3) Solid raft under the whole buildings.
(4) Continuous reinforced concrete strip footings running in both directions, interesting and monolithic with each other.
(5) Individual reinforced concrete footings connected together by plinth beams.
(6) Continuous unreinforced strip footings with plinth level band (reinforced concrete runner).
(7) Unconnected individual footings or unreinforced strip footings.

Sand piles may be used for compacting, draining and consolidating loose soft fills.

8.4 General Planning and Details

In addition to the main factors or type and quality of construction of structure and its foundation, there are other more or less important factors influencing the behaviour of buildings during earthquakes. There are briefly stated in the following:

---

* *For current provisions, refer IS:4326 - 1993 and IS: 13828 - 1993*

(a) *Plan and elevations* - Buildings irregular in plan or elevation are found to develop torsional stresses. Therefore, those having symmetry in plan and elevation are better. Compact plans are seismically better than extended plans with several projections. E, U, T or L shaped plans must be provided with separation sections, (2,18) so as to reduce them to an assemblage of rectangular units.

(b) *Roofs and Floors* - Roofing and flooring units, when used instead of monolithic slabs, are to be tied together and fixed to the supporting members to prevent their dislodging due to shaking. Therefore, corrugated iron or asbestos sheets are better than earthen tiles, slates, etc. Joists of timber or reinforced concrete, if used anywhere for supporting flooring units, should be blocked and tied together so as not to allow any relative displacement between them. Jack arched roofs are to be avoided unless ties are used in every span.

(c) *Load Bearing Walls** - Damage is found to increase with height. The collapse of a multi-storeyed building then, is much more disastrous in terms of loss of life and property. Therefore, height of masonry buildings may be restricted to about three storeys.

Studies carried out on the effect of openings on the strength of walls indicate that they should be small and more centrally located (1, 24). IS: 4326-1967 provides the following restrictions on the size and position of openings:

(i) The openings shall preferably be located away from the corner by a clear distance equal to at least 1/4 of the height of opening.

(ii) The length of all openings shall not be more than half the length of the wall between consecutive/end cross walls.

(iii) The horizontal distance (pier width) between two openings shall not be less than 1/2 of the height of the shorter opening.

(iv) Where the openings do not comply with the above requirements, they should either be boxed in reinforced concrete or reinforcing bars provided all around them through the masonry.

(d) *Projecting Parts* - Overhanging parts such as projecting cornices, balconies, parapets and chimneys are the first to fall during an earthquake. Not only is there damage to the building, but such parts, when they do fall, injure the people who are running out of the houses or moving on the streets. Such projecting and over hanging parts should be avoided as far as possible or enough care should be taken to reinforce them and anchor them to the main structure adequately.

(e) *Suspended Ceilings* - Suspended ceilings which are often used for aesthetic reasons, are usually brittle, weak and incapable of resisting horizontal forces. The result is that during an earthquake, they crumble and fall down. Thus, special care is required in the design of suspended ceilings – if they cannot be avoided. They should be strong and rigidly tied to the roof, or be ductile enough to withstand the strains during ground motion.

Similarly, the plaster on the ceiling frequently falls down (11). The thickness of such plaster should be kept to a minimum.

(f) *Damage to Non-Structural Parts* - During past earthquakes, it has sometimes happened that whereas the structural frame was strong enough to resist the earthquake forces, the non-structural elements, like brick filling in a timber frame, which is not supposed to carry any other loads besides its own weight, have fallen out of the frame (11). Therefore, it is necessary that the non-structural parts be well tied to the structural framing. To avoid damage to window frames or glazing, the drift in buildings should also be limited to about 1.5 cm per storey height.

From the behaviour of buildings during past earthquakes, it may be concluded that the most desirable qualities for earthquake resistance are *light weight, high lateral load* resistance, large ductility and *a non-yielding foundation.* The building should also have simple, regular plan and elevations, well integrated construction of all units, with as few openings in walls and as few projecting parts as possible.■

8.5 References

1 J. Krishna and A.S. Arya, "Building Construction in Seismic Zones of India", Second Symposium on Earthquake Engineering, University of Roorkee, November 1962.

2. J. Krishna and A.S. Arya, "Earthquake Resistant Design of Buildings", Journal of the Institution of Engineers (India), Vol. XLV, No. 7, March 1965.

3. J. Krishna and B. Chandra, "Protection Against Earthquake Damage", Publication and Information Directorate, Council of Scientific and Industrial Research, New Delhi, 1966.

4. A.S. Arya, "Design and Construction of Masonry Buildings in Seismic Areas", Bulletin of the Indian Society of Earthquake Technology, Vol. 4, No. 2, April 1967.

5. A.S. Arya, "Construction of Small Buildings in Seismic Areas", Bulletin of the Indian Society of Earthquake Technology, Vol. 5, Nos. 3 and Sep.-December 1968.

6. J. Krishna and B. Chandra, "Strengthening of Brick Buildings in Seismic Zones", Proceedings, Fourth World Conference on Earthquake Engineering, Santiago, Chile, January 1969.

7. J. Monge, "Seismic Behaviour and Design of Small Buildings in Chile" Proceedings, Fourth World Conference on Earthquake Engineering, Santiago, Chile, 1969.

8. A.I. Churayan and S.A. Djabua, "On One Method of Increasing the Seismic Stability of Brick Buildings", Proceedings, Fourth World Conference on Earthquake Engieering, Santiago, Chile, 1969.

9. "The Bihar-Nepal Earthquake of 1934" Memories of the Geological Survey of India, Vol. 73, 1939.

10. "Koyna Earthquake, December 11, 1967" Report of Committee of Experts Vol. I, New Delhi, April 1968.

11. N. Gosain and A.S. Arya, "A Report of Anantnag Earthquake of February 20, 1967", Bulletin of the Indian Society of Earthquake Technology, Sept. 1967.

12. "Tottori Earthquake of 1943, and Fukui Earthquake of 1948," personal communication with Prof. T. Hisada, Director Building Research Institute, Tokyo, Japan.

13. Miyagi Earthquake of 1962, personal communication with Prof. T. Hisada, Director Building Research Institute, Tokyo, Japan.

14. F.J. Wood, "The Prince William Sound, Alaska, Earthquake of 1964 and After shocks," Vol. II, ESSA, Coast and Geodetic Survey, 1967.
15. M.A. Sozen, "The Caracas Earthquake of July 29, 1967" Journal of the American Concrete Institute, Vol. 65, No. 5, May 1968, pp. 394-401.
16. K.V. Steinbrugger and R. Flores, "The Chilean Earthquakes of May, 1960: A Structural Engineering View Point", Bulletin of the Seismological Society of America, Vol. 53, No. 2, pp. 225-307, February 1963.
17. L.S. Srivastava etc., "Badgam Earthquake of September 2, 1963", Bulletin of the Indian Society of Earthquake Technology, Vol. I, No. 1, January 1964, p. 83.
18. IS: 4326-1967 "Code of Practice for Earthquake Resistant Building Construction", Indian Standards Institution, New Delhi, January 1967.
19. IS: 1893-1966, "Criteria for Earthquake Resistant Design of Structures", Indian Standards Institution, New Delhi, November. 1967.
20. Steinburgge and Moran, "An Engineering Steady of the Southern California Earthquakes of July 21, 1952 and After shocks", Bulletin Seismological Society of America, 1953.
21. R.R. Marrel, "Earthquake Damage to Type III Buildings in Long Beach, 1933", in "Earthquake Investigations in the Western United States, 1931-1964", U.S. Department of Commerce publication No. 41-2, p. 221.
22. R.R. Martel, "Earthquake Damage to Type III Buildings in Long Bech, 1933", in "Earthquake Investigations in the Western United States, 1931-1964", U.S. Department of Commerce publication No. 41-2, p. 221.
23. E.R. Gee, "The Dhubri Earthquake of the 3rd July, 1930", Memories of the Geological Survey of India, 1931.
24. B.H. Falconer, "Niigata Earthquake, Japan, 1.02 PM, 16 January 1964", International Institute of Seismology and Earthquake Engineering, Tokyo, Japan, 1964.
25. V.K. Agnihotri, "Strength of Single Storey Brick Shear Walls Against Earthquake Forces", M.E.Thesis University of Roorkee, 1962.

Chapter 9

Building Construction Before and After Earthquake[1,2]

9.1 Introduction

An earthquake constitutes a natural hazard which acts all of a sudden, and though lasting only for a few seconds, causes widespread destruction of various man-made structures. Among various natural hazards, an earthquake has no parallel so far as the destructive power is concerned for demolishing all types of structures ranging from small buildings to gigantic dams. The history of deaths caused by various disasters like cyclones, volcanic eruptions, floods, etc. shows that maximum loss of life in a single event has also occurred in an earthquake in China. And most deaths have been caused by the collapse of small buildings used as dwellings by the people. Therefore, if the state of design and construction of small buildings could be improved by incorporating earthquake resisting features, much loss of life and limb could be prevented in future shocks. The aim of this paper is to present the state of art of such building constructions as are prevalent today, their weak and good features, and the desirable modifications for achieving adequate seismic resistance. The scope is limited to the consideration of non-engineered constructions, that is, those which are traditionally constructed by the people without qualified engineer's assistance.

9.2 Pre-Earthquake Construction

The idea of pre-earthquake constructions may be taken as referring to those building systems as are constructed by the people without specific consideration of the seismic aspects. Although most buildings of this type have little earthquake resistance, in some areas like Kashmir and Assam in India, certain forms of construction have been developed through intuition over ages, like Dhajji Diwari, Brick Nogged wood frame, Stud wall construction using Ikra or bamboo mat walling, which are inherently earthquake resistant. Therefore, traditional constructions in different parts of India and the world, have widely varying

---

1 *A.S.Arya, Professor & Head, Deptt. Of Earthquake Engineering, University of Roorkee, Roorkee*

2 *Proc. National Meet on Earthquake Mechanism and Mitigation, Deptt. Of Science and Technology, GoI, New Delhi, August 1986, pp. 103 - 120*

performance results during earthquakes and, consequently, need various degrees of strengthening measures. The usual defects and weaknesses of these buildings are fairly well known now and are brought out in this paper for such construction types.

Another study related with existing buildings is to determine the *minimum* retrofitting features necessary for improving their seismic resistance so that in a probable earthquake intensity, the chance of collapse is minimised. The main problem here is to see if a proposed improvement technique is feasible at all for an existing building and at what cost. It is, of course, certain that introduction of any such feature in an existing building will be much more costly and inconvenient than incorporating the same in a new building. Therefore, earthquake resisting features must be provided in all new buildings for efficiency and economy.

9.3 Post-Earthquake Reconstruction

The post-earthquake period requires a number of activities, namely, immediate rescue and relief operations, provision of temporary shelters, restoration of services, and restarting of economic activities. Then come the reconstruction activities, including repair, restoration and seismic strengthening of damaged buildings; planning, design and construction of new dwellings; supply of materials of construction for self-help projects; and training and education of engineers, builders and artisans in earthquake engineering principles, ideas, and features, etc. The reconstruction period provides an unique opportunity for introducing appropriate technology ideas, since the people are most receptive and eager to learn as to how to protect their dwellings. Most earthquake resistant features could easily be built into the new designs and guidelines, and the additional costs involved will be minimal. Advantage can be taken of this situation for propagating the hazard resistance ideas and features in other earthquake prone areas of the country also.

9.4 Common Building Systems in India

During the 1971 Census of India, data was collected about the various house types in the country. An abstract of the percent of various construction types is given in Table 1. It is seen from this table that the most common types of constructions in India are as follows:

a) Earthen Houses
b) Burnt-Brick Buildings
c) Stone Masonry Buildings
d) Wooden houses
e) Miscellaneous types of huts

Note: Cement Concrete Buildings which are only 0.33% of all houses, are not considered here.

The prevalent structural systems for the above construction types, their earthquake behaviour, defective and weakness factors, good features as well as desirable precautionary and strengthening suggestion are described in the following sections:

9.5 Earthen Houses

9.5.1 Construction Types:

The basic material for their construction is clayey earth either in mud lump form, or rammed earth in wood forms or rectangular shaped bricks, called Adobe. Usually, the naturally available soil is used without treatment except for mixing of some fibrous material like straw, hay, human and horse's hair, etc. For one storeyed houses, the walls are kept 30-35 cm thick for Adobe, and 40-50 cm thick for other types. The thickness is increased in the lower storey for 2 or 3 storeyed hosues. Sometimes, the thickness of walls is tapered upwards. In some

Table 1 - Predominant Materials Used In Houses* In India

(1971 Census, Number of Houses Rural 98.5 million, Urban 24.4. million)

Wall Material	Percent of Rural Houses	Percent of Urban Houses	Percent of all Houses	Roofing Material	Percent of Rural Houses	Percent of Urban Houses	Percent of all Houses
Clay Mud	48.67	15.10	41.99	Tile, slate, shingle	37.12	33.11	36.32
Unburnt Bricks	8.35	3.26	7.34	Corrugated iron other metal sheets	4.68	12.02	6.30
Burnt Bricks	15.55	59.08	24.21	A.C. sheets	0.40	2.47	0.81
Stone	14.17	12.00	13.74	Brick & lime concrete	1.52	4.09	2.03
Cement Concrete	0.14	1.11	0.33	Stone slabs	2.65	6.13	3.34
Wood	1.32	2.28	1.51	Cement concrete or R.B. slabs	2.93	23.10	6.92
Metal Sheets	0.14	1.76	0.46	Grass leaves, reeds, thatch, wood, mud, bamboo	50.42	18.92	44.15
Grass Leaves, reeds, bamboo	11.57	5.26	10.31				
Others	0.09	0.14	0.10	Others	0.08	0.26	0.11

*The term house here includes: residences 72.98%; shop-cum-residences 9.86%; residence-cum-household industry 1.41%; hotel, sarai, dharamashlas, inspection and tourist homes 0.14%; shops, including eating places and offices 2.93%; factories, worksheds 1.3%; places of gathering and worship 1.23%; others and vacant 19.1%.

variations, wooden elements in the form of sawn lumber or original tree-branch form are inserted between courses, particularly at the corner and T-junctions of walls.

The room enclosures in earthen houses are usually of small dimensions, 3 m x 4 m or so in plan, particularly when pitched roofs are used. But in case of flat roofs consisting of wooden log beams or sawn lumber, the length of the room may even be 10 m long, keeping the clear span upto 3 m.

In two or three storeyed houses, which are used mostly in dry climates, the intermediate floor consist of wood beams and planks in some cases, while in some others they are heavy, consisting of wooden beams, reeds or tree branch decking and earth cover.

In 'Dhajji Diwari' construction in Kashmir, pillars of adobe 60 cm x 60 cm in size are constructed every 1.5 m to 2 m on centres, the intervening space being filled with bricks or stones. The floors are of wood joists and planks. The joists rest on wooden wall plates and also sometimes have wall plates on top of them. These wooden wall plates form runners all around the house. The roofs are usually pitched.

Another variation of earthen houses is where the pitched roof of thatch or tiles is supported independently on wooden posts just on the outside of the clay enclosures.

A better quality construction than the pure adobe construction is the composite wall construction in which the outer face of the wall is built in burnt bricks laid in mud mortar, the inner is built from unburnt bricks and the two are bonded together by header bricks. The outer wythe protects the walls from erosion during rains and helps in carrying the vertical loads of the floors and roof better.

9.5.2 Earthquake Behaviour of Earthen Buildings:

The performance of earthen houses during earthquakes of MM VII or more has been generally very poor, consisting of wide cracks in walls and separations of walls at corners and complete collapse of walls, roofs and floors, leading to death and injury to the residents. Due to heavy mass of debris, rescue work of buried people also has been found to be difficult and time consuming. Even more so if the streets get blocked by fallen debris. Single storeyed adobe or rammed earth houses with flat heavy roofs have shown fair behaviour during the Dhamar, Yemen earthquake of December 1982 even in MM VIII area in that they did not collapse and cracking damage was minimum, but most two and three storey houses collapsed completely. The main weaknesses and defects in earthen houses are as follows:

- Poor strength of material in tension and shear and not high enough compressive strength for two or three storey constructions against earthquake stresses.
- Poor bond between walls meeting at right angles.
- Too large openings and too close to the corners.
- Small bearing length of lintels across openings.
- Eroded/deteriorated unrepaired state at the time of earthquake.
- More than one storey height and large storey heights.

- Heavy roofs and floors.
- Poor erosion resistance against rain and flooding.

The following may be considered as good features of such houses and some of the variations used in their construction:

- Cheap in initial cost, particularly if constructed through self-help or community voluntary activity.
- Good thermal insulation against heat and cold.
- Resistant to fire and bullets.
- Fibrous admixture like straw and cowdung make the adobe and rammed earthstrong.
- Tapering walls providing better stability against lateral forces.
- Large savings of energy used in construction.
- Use of burnt brick outside façade is a good feature against water and wind erosion.
- Application of mud-cowdung mix plaster on outside faces and its yearly repair are good for maintenance of the house.
- Wooden wall plates in continuous runner form provide integrity to the enclosures as a box against lateral forces.

9.5.3 Precautionary And Strengthening Measures* :

It is rightly said, what a clay house needs is a shoe at the base and hat at the top. Water is the most detrimental to it. To avoid trouble due to water, the following are recommended:

- Use foundation with burnt brick or stone in lime or cement mortar upto plinth.
- Use plinth level above the storm water flow level.
- Keep the roof with good projection, say half a metre beyond the walls to drain the rain water away from the walls.
- Make the roof waterproof by embedding a layer of black polythene sheet in the roof clay cover.
- Use a water proof plaster on the outside of the house.

From the earthquake view point, the following may be done:

- Restrict length of rooms to about 4 m maximum and storey height to about 2.8 m.
- Use symmetrical rectangular house plan, arrange openings symmetrically.
- Restrict height to one storey only.
- Restrict openings width to one-third the wall length and place these away from the corners by about three times the wall thickness.
- Use clayey soil with about 20% clay and adequate fibrous admixtures, for both adobe units and for mortar.

** For current provisions, refer IS:13827 - 1993*

- Use adequate bond beams on all outer and inner walls with full continuity at junctions. Bond beam could be of wood or bamboo. Reinforced concrete bands have been used in USA for retrofitting of Adobe Heritage Building.
- Select roof of minimum weight as far as possible.

In the above recommendations, the bond beam is the only component requiring additional cost. If more investment can be made, it will be desirable to,

- Use water proofing admixture, like cement, lime or asphalt in making adobe and mortar.
- Use compacted adobe with machines like the Cinva-Ram or RRE compacting machine.

In the case of seismic intensity of MM IX or more, vertical reinforcing of adobe walls by bamboo or cane, pretreated against rats and white ants, will be desirable for greater safety against collapse.

9.6 Fired-Brick Buildings

9.6.1 Construction Types:

The basic material for these constructions is fired bricks, prepared in different sizes in various countries, by hand-molding or machine molding, and fired in different types of clamps and kilns. The standard nominal size including mortar joints in India is 23.8 cm x 11.4 cmx 7.6 cm (9" x 4 1/2" x 3"). Modular bricks having nominal size 20 cm x 10 cm x 10 cm have also been suggested but have not yet been much adopted in practice. Thus, the following wall thicknesses are commonly adopted:

Partition walls, non-load bearing	11.4 cm (4 1/2" or 1/2 brick)
Load bearing walls, one storey in mud mortar	23.8 cm (9" or one brick)
Load bearing walls, two storeys in cement or lime mortar	23.8 cm (-do-)
Load bearing walls, first of 3 storey in cement or lime mortar	35.2 cm (13 1/2" or 1 1/2 brick)

The room sizes depend on the types of roof. Where flat roofs are used consisting of wood joists, the room width is restricted to 3 to 3.4 m and length can be any thing from 4 m to 8 m or even more. Sometimes, wider rooms are obtained by using an intermediate girder of steel or timber. The newer constructions in town and cities are making increasing use of reinforced concrete or reinforced-brick slabs and room widths may be upto 3.5 m, using slabs, and wider using T-beams.

The wall construction in India mostly uses the English bond, that is, 'stretchers' and 'headers' in alternate courses. Use of queen-closer near the ends permits breaking of vertical joints. With standard bricks, the overlap is 5.7 cm (2 1/4"). Normally, a 'toothed' joint is made between the walls, meeting at right angles, although the joints in practical constructions are rarely fully filled.

Foundations normally consist of stepped type strip footings below the walls, starting from rammed earth and brick bats. The bricks used in foundations are often second class, either slightly under-burnt or over-burnt and the plinth masonry is kept half-brick thicker than the first storey walls. In first class constructions, the base may be made using lime concrete with brick aggregate or lean cement concrete of 1:4:8 mix.

A variation in this form of construction is the use of brick piers 1 1/2 Bk x 1 1/2 Bk spaced at about 3 m spacing with unburnt brick infill panels. The piers have independent footings projecting by about half brick in each direction beyond the pier. It is not unusual in these cases to use wooden ring beams on top of walls to receive the hipped roof.

Brick masonry is not only used for small dwellings but also for schools, dispensaries and other community buildings. In these cases, larger enclosures are used such as 5 m x 7 m and roofs consist of RC T-beams and slabs. In such buildings, thicker walls, better mortar and qualified engineering supervision is used. Although the National Building Code of India gives the procedures for 'calculated' masonry, wall thicknesses are rarely calculated. For earthquake safety, these buildings need special design and cannot rationally be covered by the general rules as applied to houses due to the longer spans of walls and larger heights.

9.6.2 Earthquake Behaviour of Brick Buildings:

The performance of brick buildings during earthquakes has been much related with the type of roof, the mortar used, and the quality of construction. Performance has been poor with pitched roofs having no binding effect on walls, poorer with mud or weak mortars and still poorer with poor construction quality which is seen to improve with the quality of mortar since supervision becomes stricter.

The behaviour of brick buildings having arched roofs without ties has been catastrophic, as also of those having wooden joist-clay type flat roofs over long rooms.

Buildings with rigid slab roofs have generally behaved much better than others due to their binding effects on walls as well as the diaphragm action by which lateral load is transferred to shear walls available in the direction of the earthquake motion.

Cracking is frequently observed in diagonal or cross form in the masonry piers between the openings, vertical cracks near the corners leading to separation of perpendicular walls through the toothed joints, and horizontal bending cracks in the walls which are at right angles to the predominant direction of the earthquake.

The following may be considered as defects and weaknesses in brick buildings:

- Poor strength of material in tension and shear, particularly where mud mortar or lime-sand mortar weaker than 1:3 or cement mortar weaker than 1:6 is used.
- Toothed joint causing a vertical plane of weakness between perpendicular walls.

- Large openings and their placement too close to the corners.
- Eroded mud mortar due to lack of protection or maintenance.
- Very long rooms having long walls unsupported by cross walls.
- Unsymmetrical plan of building, or with too many projections.
- Poor quality of construction.
- Use of heavy roofs having flexibility in plan.
- Use of light roofs with little binding effects on walls.

The following are good features of brick buildings:

- Durable construction lasting hundreds of years with minor maintenance.
- Traditionally constructed through local artisans.
- Comfortable interior, reasonable insulation against heat and cold when one brick thick walls are used.
- Resistant to rain and flooding if erosion from the surface of mud mortar is prevented with cement or lime pointing or plaster, or these mortars are used for masonry.
- Resistant to fire.
- Burnt brick buildings are considered of status in the society.
- Well constructed and integrated brick wall enclosures provide good stability against vertical as well as lateral loads, permitting use of even thinner walls than one-brick for load bearing constructions.
- Use of rigid slab floors and roof is a good feature for improving stability of brick wall enclosures.

9.6.3 Earthquake Protection Measures*:

For earthquake resistance, the following measures have been found effective not only in preventing collapse but controlling the propagation and widening of cracks:

- Symmetry and rectangularity of the building in the plan.
- Symmetry in the location of openings.
- Simplicity in elevations, that is, avoidance of ornamentations, large cornices, etc.
- Intersecting internal walls in good number so as to divide the total plan in squarish enclosures not more than about 6 m length.
- Total width of openings in a wall not more than 50% of wall length in one and two storeyed houses and not more than 40% in three storeyed buildings.
- Width of piers between openings or from opening to wall corner not less than half the height of opening.
- Use of steel or wooden dowels going into walls meeting at corners or T-junctions to provide effective bonding between them, say, every fourth course.

** For current provisions, refer IS:4326 - 1993*

- Use of bond beam or band of reinforced concrete at lintel levels of openings and serving as lintel too. This is one single feature which is most effective in ensuring the integrity of enclosures like a rigid box.
- For buildings in MM IX or higher zones, vertical reinforcement at critical sections of the building, viz. corners and junctions of walls and jambs of openings, going from foundation into the floors and roof slab, so as to provide strength and ductility to the shear walls.

Seismic analysis of 3 to 5 storeyed residential flat type buildings constructed in brickwork in cement or lime mortar situated in MM VII and VIII zones indicates that tensile and shear stresses increase rapidly in the shear wall piers as the building height increases. It will, therefore, be advisable to restrict the building height to three storeys in MM IX Zone and to four storeys in MM VIII Zone.

9.7 Stone Masonry Buildings

9.7.1 Construction Types:

The basic unit of stone masonry is natural stone which is used in masonry work in *random rubble, half-dressed or fully dressed* state, the mortars being none (that is, dry masonry), mud, cement, or lime. Random rubble in mud being the cheapest construction, where stone is locally available, is the most common of the three types in hilly rural areas. Fully dressed stone construction is very costly in terms of labour and is, therefore, sparingly used. In half-dressed stone masonry, the outer facia is constructed with stones, with the outer face dressed to rectangular shape but the 'inside *wythe*, consists of field stones'. It thus forms two, more or less independent *wythes* with mortar and chip filling in between them. The wall thickness varies from 45 cm to 75 cm and could even be 90 cm thick in 3 to 4 storeyed buildings. When used with wood beams, clay floors and roof,the building becomes extremely heavy, easily 5 t/m^2 of covered area. Foundations are also constructed in stone but where exposed rock is available on the surface, the construction is started from the rock itself. A few variations are stated below:

- Single storey; pitched roof sloping both says. Walls about 2 m high with gables in stone masonry; rooms 3 m x 4 m.
- Single storey with flat roof consisting of wood-logs used as beams, with tree branches and bushes making the deck with clay topping; rooms 3 m x 4.5 m, but sometimes long rooms upto 8 m are also used.
- Single storey laterite stone units, walls 25 to 30 cm thick, with hipped roof. Rooms may be wide and long 6 m x 9 m using inner row of wood posts.
- Single storey stone houses with RC slab roof, rooms 3.5 m x 4 m or so; height typical 2.4 m.
- Two storey with attic floor extra, intermediate floors of wooden joists and boards with clay-cowdung topping. Rooms upto 3 m x 7 m, stone gables.

- Three to four storeyed, 10 m x 12 m in plan with heavy clay floors and roof having some rooms 3 m wide and 10 to 12 m long.

9.7.2 Performance During Earthquake:

The performance of stone masonry buildings during earthquakes has been most disastrous except where fully dressed stone with cement mortar is used. There are innumerable instances of complete collapses of such buildings in all seismic countries of the world leading to death or serious injury to the inmates. The following are the defects and weaknesses in stone masonry buildings of random rubble or half dressed types:

- Weak in tension and shear and unstable configuration of stones when shaken from the initially constructed position, causing the wall to collapse due to heavy vertical loads.
- Very weak bond between walls at right angles to each other leads to very easy separation.
- Delamination of wall into separate outer and inner wythes due to the absence of 'through' or 'bond' stones.
- Easy shattering and collapse of stone gables.
- Poor quality of construction.
- Poor state of maintenance.
- Very heavy mass.
- Flexible floors and roof, little binding effect on walls, no diaphragm action.

The following are the good features of stone masonry buildings:

- Durable construction in non-seismic areas.
- Traditionally constructed through local labour and local materials.
- Good insulation from heat and cold due to large wall thicknesses.
- Resistant to fire.
- Stone buildings with half-dressed stone facia and pointing considered of status in the society.

9.7.3 Protection Measures*:

For stone masonry buildings, all the protective measures as for brick buildings are found useful. For random rubble and half dressed stone masonry, the following additional measures are absolutely necessary:

- Provision of 'through' stones or 'bonding' elements along the wall thickness at regular internals of about one element per 0.7 sq.m of wall area.
- Restriction of the wall thickness to not more than 45 cm since larger thickness encourages undesirable filling material inside, adding to mass but reducing strength.

** For current provisions, refer IS:4326 - 1993*

9.8 Wooden Buildings

9.8.1 Construction Types:

Wooden buildings include those in which the main load bearing structure consists of wooden columns and frames, although the cladding may be of brick or stone. Wood will include logs, sawn lumber of various wood species as well as bamboo. The systems of construction can broadly be classified in two main forms as described below:

Stud wall construction consists of wooden sill plates laid at the plinth level into which are framed vertical wooden posts at short distances of about 0.8 - 1.2 m apart. The studs carry wood plates at their top into which they are framed and the connections between these plates at corners and junctions of walls stabilise the framing. The sill plates are fixed to the masonry plinth by means of iron bolts or wire. Horizontal members of lighter cross section are inserted between the studs at regular intervals. These are also either notched into the studs or spiked to them. The roof is pitched or hipped type of wood rafters and purlins to which the sheeting is nailed or bolted. The wall cladding consists of sheets, boards etc. Sometimes, the panels are diagonally braced also.

The *brick nogged timber frame* consists of heavy corner columns, sills, intermediate verticals at about 1 m on centres, wall plates, horizontal noggin members, and diagonal braces in typically alternate panels, framed into each other. The space between framing members is filled with tight fitting burnt or unburnt brick masonry in stretcher bond, called brick nogging. Flat stone (slate, laterite) filling is sometimes used. The roof is invariably hipped or pitched, with sheeting or tile covering. Thatch is also sometimes used. Most houses of this type are two storeyed, having an intermediate and an attic floor. Sometimes, more than two storeys are also used.

The foundation of both types of wood houses is usually random stone masonry in mud mortar. Cement or lime mortar is used for official buildings.

There are other types also having extremely low cost, like log houses, wattle and daub, bamboo and thatch type, huts used by very poor people. The pattern varies very widely from place to place. These are usually safe during earthquakes. They would be critical if constructed in cyclone affected areas. There, usually the people load them with heavy things like stones so that they are not blown off.

9.8.2 Performance During Earthquakes and Wind Storms:

The earthquake performance of wooden buildings has generally been good, particularly that of the wooden frame, and also where the cladding consists of sheeting, boarding, Ikra walling, bamboo matting, etc. But the brick or stone infills have frequently shown movement out of the plane of frames. The most dangerous aspect of wooden building has been their *poor fire resistance* and the danger of fire, during earthquakes, due to kitchen fires as well as due to short circuiting of electric wiring. The sheeted roofs as well as the light framework are vulnerable to cyclonic winds and tornados. Unreinforced masonry chimneys frequently fall down.

The following can be listed as the poor features or defects in wood houses of common construction:

- Short life of untreated wood due to rot and white ant attack.
- Loss of strength of wood due to weathering action, particularly when it is exposed to the action of sun and rain.
- Loosening of joints due to drying with age and weathering action, etc.
- Prone to fire hazard.
- Falling of brick noggin during earthquake shaking.
- Tearing apart of roof and frame during high winds.
- Use of brick or stone chimney flues.

The following are the good properties and features of wood construction for seismic forces:

- Light weight.
- Good tensile and excellent impact or shock resistance.
- Braced as well as nogged frames have high lateral load resistance.
- Joints with steel straps stay tight for long and maintain the frame in excellent stiff condition.

9.8.3 Precautionary and Strengthening Measures*:

To make wood houses durable and weather resistant, the following actions are found effective:

- Use of anti-termite treatment of the house site particularly inside the foundation masonry and over a metre or so outside.
- Use of rodent shield just above plinth.
- Use of seasoned and chemically treated wood.
- Use of steel straps at the splices and joints with screwed or bolted connections, or nails with clenching across the grain on the opposite face.
- Use of waterproof paint on all exposed surfaces.
- Use of fire-resistant paint on inside surfaces.
- Adequate, say 50 cm, projection of roof beyond wall frame for protection against rain.

For achieving resistance against lateral loads of wind and earthquakes*, the following are necessary:

- Installation of diagonal braces in all vertical planes of the wood frames or stud walls.
- Bolting down of the sill plates to the plinth masonry
- Continuous top plate all around as a 'band', connecting all wall frames.
- Installation of horizontal diagonal braces at the corners and junctions of walls for providing resistance against torsional effects.

** For current guidelines, refer IS:4326 - 1993*

- Holding down of sheeting to purlins, purlins to rafters, and so on, with adequate resistance against wind uplift.
- Use of asbestos pipe chimney or reinforced brickwork for flue gases.
- Fixing of brick noggin to wooden members through shear connectors or hold-fasts.

9.9 Conclusion

The study of prevalent construction systems indicates clearly that the main items for improving the seismic resistance of existing traditional buildings are very well understood from a qualitative stand point. The quantitative contributions of the various strengthening features still need detailed shake table tests. All the same, the foremost among these features are the following:

(a) Provision of a bond beam or band continuously on all external and internal walls in all types of masonry walls and wooden frameworks. It will help even the existing buildings or damaged ones, if 'banding' or belting is provided all around the external walls and cross-ties across the buildings so as to tie the bands together.

(b) Achievement of integrity of stone masonry walls by providing 'through' or 'bonding' elements and that of wood frames by providing steel straps at all joints. These may be installed in existing buildings too.

(c) Provision of vertical tensile elements in higher than VIII Intensity zones; though it is difficult to provide these in existing buildings but it will help a great deal if the wall surfaces are steel-meshed and gunited with cement.

Achieving structural box like integrity of the walls of the existing or post-quake buildings will surely safeguard them from total collapse during a future earthquake.

9.10 References

1. Arya A.S., "Design and Construction of Masonry Buildings in Seismic Areas", Bulletin of Indian Society of Earthquake Technology, Vol. 4 No. 2, pp. 25-37, April 1967.
2. Arya A.S., "Construction of Small Buildings in Seismic Areas", Bulletin of Indian Society of Earthquake Technology, Vol. 5, No.s 3 and 4, September-December, 1968.
3. Arya, et al., "Basic Concepts of Seismic Codes, Vol. I, Part II" Non-Engineered Construction, Pub., by International Association for Earthquake Engineering, 1980 and reprinted by India Society of Earthquake Technology, 1981.
4. Arya A.S., B. Chandra and P. Rani, "Seismic Risk to School Buildings in Some Asian Countries", Bulletin of Indian Society of Earthquake Technology, Vol. 18, No.1, March 1981, pp. 1-44.
5. Arya A.S., Srivastava L.S., Gupta S.P., "Report of Indian Earthquake Engineering Mission to Yemen Arab Republic on Dhamar Earthquake of December 13, 1982", Roorkee, March 1983.
6. Arya A.S. and T. Boen, "Earthquake Resistant Construction of Earthen Housing", International Workshop on Earthen Buildings in Seismic Areas, Albuquerque, N.M., U.S.A., May 24-28, 1981.
7. Census of India 1961, Vol. I, Part IV-4 (iii) Report on HOUSE TYPES and VILLAGE STTLEMENT PATTERNS

in India, pub. by A Mitra, Registrar General and ex-officio census commissioner for India, New Delhi.

8. *Census of India 1971, Series I, Part IV-B, HOUSING TABLES*, Pub. by A. Chandrasekhar, Registrar General and Census Commissioner of India, New Delhi.
9. Chandrasekaran, A.R., Srivastava L.S. and Arya A.S., "Behaviour of Structures in Koyna Earthquakes of December 11, 1967", Indian Concrete Journal, Part I in Vol. 43, No. 12, December 1969 and Part II in Vol. 44, No. 1, 1970.
10. G.S.I. Officers "The Bihar-Nepal Earthquake of 1934", Memoirs of the Geological Survey of India, Vol. 73, Calcutta, 1939.
11. IS: 4326-1976, "Code of Practice for Earthquake Resistant Construction of Buildings", Indian Standards Institution, New Delhi, 1977.
12. Krishna J. and Arya A.S., "Damage in Kashmir due to June 6, 1966 Hindukush Earthquake", III Symposium of Earthquake Engg., University of Roorkee, November., 1966.
13. Middlemiss C.S., "The Kangra Earthquake of 4th April 1905", Memoirs of the Geological Survey of India, Vol. 38, Calcutta, 1910.

Chapter 10

Design and Construction of Masonry Buildings in Seismic Areas[1,2]

10.1 Introduction

In the past severe earthquakes all over the world, the masonry buildings have generally been damaged the most because of their heavy weight, small or no tensile strength, small shearing resistance, lack of proper bonding between longitudinal and cross walls and poor workmanship. Yet such buildings are being put up in the conventional way for reasons of climatic suitability, low cost and local availability of materials, widespread knowledge of methods of construction, etc. It is quite apparent that it will not be possible to do away with this kind of construction in the seismic areas, particularly in the developing countries. Therefore, finding effective methods of strengthening such buildings is of paramount importance so that the danger to life and property during future earthquakes is minimised. An important requirement of such methods is that they should be cheap so that the additional cost of strengthening is not more than what one would like to spend on insurance.

For closer examination, a building is dissected into various components to establish where each component lacks strength against earthquake forces and suggest means to strengthen them. The discussion here is based on brick buildings in India not taller than four storeys. The conclusions can easily be interpreted for similar construction in other countries. It is assumed that the bricks are not weaker than the building mortar in tensile, shear and compressive strengths.

10.2 Structural Action

During ground motion, an inertia force acts on the mass of the building due to which the building tends to remain stationary while the ground moves. If the building is rigidly fixed into ground, the inertia forces will cause horizontal shears in the building, the magnitude of which will be a function of the ground motion and stiffness and damping characteristics of

---

1 *A.S.Arya, Professor & Assistant Director, School of Research & Training in Earthquake Engineering, University of Roorkee, Roorkee*

2 *Bulletin, Indian Society of Earthquake Technology (ISET) ,Roorkee, Vol.4, No.24, April 1967, pp 25-37.*

the building. However, if relative motion is possible between the ground and the building, the building may slide to and fro and the forces generated, during the earthquake, which would tend to shear the building, would be small. This type of action may be visualised if the ground under a railway wagon is imagined to shake along the direction of rails. It has clearly been demonstrated in the 1930 Dhubri (Assam) and the 1934 Bihar-Nepal earthquake that where a relative displacement was possible between the ground and the superstructure, the building suffered less than other similar buildings fixed into the ground.

As a result of the shaking, the roof tends to separate from the supports, the roof covering tends to get dislodged, the walls tend to tear apart and if unable to do so, the walls tend to shear off diagonally in the direction of motion. If filler walls are used within steel, concrete or timber framing, they may fall out of the frame bodily unless properly tied to the framing members. Thus, on the whole, it is indicated that *a necessary condition for earthquake resistance is that the various parts of the building should be adequately tied together.*

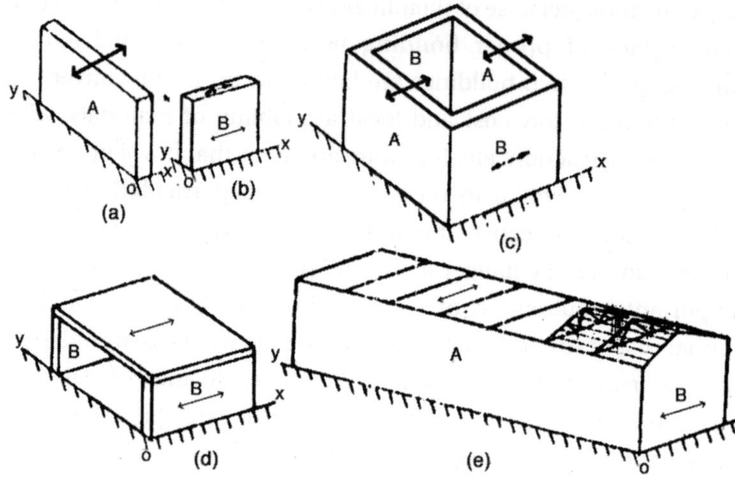

Fig. 1: Behaviour of various wall under lateral force

Consider the structural elements and structures shown in Fig. 1. In (a), wall A is free standing and the ground motion is acting transverse to it. The force acting on the mass of the wall tends to overturn it. The resistance of the wall is obviously very small. In (b), the free standing wall B is subjected to ground motion in its own plane. It is clear that in this case, because of its large depth in the plane of bending, the wall will offer great resistance. Such a wall is termed as a shear wall. It may be noted that for ground motion parallel to y-axis, wall A will act as a shear wall and wall B will topple over. Now consider the combination of walls A and B as shown in (c). For the x-direction of motion as shown, walls B act as shear walls and besides themselves offering resistance against the collapse of walls A as well. Walls A now act as vertical slabs supported on two vertical sides and bottom and subjected to the inertia force on their own mass. Near the vertical edges, the wall will carry bending moments in the horizontal plane for which the masonry has little strength. Consequently, cracking and separation of the walls may occur. If, however, a horizontal bending member (like a timber beam or reinforced concrete or reinforced brick runner) were inserted at a suitable level in wall A and continued in wall B, this would take care of the bending tensions in the horizontal plane. So far as vertical bending is concerned, the wall gets pre-compression due to self weight and can be made to take care of bending tensions. The situation will be the same for walls B for ground motion along y-axis. Thus, the bending member will be required in walls B also. Therefore, a horizontal runner, strong in bending, is required around the enclosure. Such a runner is termed as A seismic band and depending upon which level it is used at, it is called a *roof band, lintel band, gable band or plinth band.*

In Fig. 1(d), a roof slab is resting on walls B and the earthquake is along the x-axis. Assuming that there is enough adhesion between the slab and the walls, the slab transfers its inertia force at the top of walls B, causing shearing and overturning actions in them. To be able to transfer its force on to the walls, the slab must have enough strength in bending in the horizontal plane. Whereas R.C. or R.B. slabs possess such strength inherently, other types of roofs or floors such as timber joists with timber plank or brick tile covering will have to be connected together and fixed to the walls suitably so that they are able to transfer their inertia force to the walls. This point is discussed in greater detail later. After this load transfer, the walls B must have enough strength as shear walls to withstand the applied inertia force and their own inertia force. It is quite clear that the structure shown at (d), when subjected to ground motion along y-axis, will collapse very easily because walls B have very little bending resistance in the y-direction. Hence, bracing arrangement, either by shear walls or by diagonal braces, must be provided in the y-direction.

Lastly, consider the barrack type structure carrying roof trusses as shown at (e). The trusses rest on walls A and walls B are gabled to receive the purlins of the end bays. The ground motion is along the x-axis. The inertia forces will be transmitted from sheeting to purlins, trusses and from trusses to walls A. The end purlins will transmit some force directly to the gable ends. In order that the structure does not collapse, the following arrangements must be made:

(i) the trusses must be anchored into the walls by holding down bolts,

(ii) walls A, which do not get much support from walls B in this case, must be made strong enough in vertical bending as cantilevers, or

(iii) some suitable arrangement must be made to transmit the force horizontally to end walls B which are well suited to resist the force by shear wall action.

Strengthening of walls A in the vertical direction may be done by providing adequate pilasters under the roof trusses. For horizontal transfer of forces, a *band* may be provided at the top of the wall to which the roof trusses are connected, or *diagonal bracing* may be provided in the plan at the level of main ties of the trusses which should extend from one gable end to the other and be suitably connected to the end walls B. The inertia force of the top half height of walls A may be assumed to be resisted by such a band or bracing system.

Now if the ground motion is along the y-direction, walls A will be in a position to act as shear walls and all forces may be transmitted to them. In this case, the purlins act as ties and struts and transfer the inertia force of roof to the gable ends. A sloping band is required at the top of gable ends to transfer this inertia force from gable ends to walls A and where this band joins the roof band at the top of walls A.

The above discussion leads to the following requirements for structural safely of masonry buildings during ground motions:

1. A free standing wall must be designed as a vertical cantilever.
2. A shear wall must be capable of resisting all horizontal forces due to its own mass and those transmitted to it.
3. The roof or floor elements must be tied together and be capable of transferring their inertia force to shear walls by bending in horizontal plane.
4. The trusses must be anchored to supporting walls and have an arrangement of transferring their inertia force to end walls.
5. The shear walls must be present along both the axes of the building.
6. The walls must be effectively tied together to avoid separation at vertical joints due to shaking.

Horizontal bands may be provided for tying the walls together and transferring the inertia load horizontally to the shear walls.

10.3 Design of Seismic Bands

As stated above, the seismic bands are required for tying the walls together and imparting horizontal bending strength to them. The forces, which the bands are subjected to, are usually indeterminate. In many cases, arbitrary dimensions are adopted. An approximate method is suggested here for arriving at minimum dimensions of the band at any level. This is explained by the following numerical example.

10.3.1 Example 1*

Fig. 2 shows a single room building with its roof removed. The walls are 20 cm thick in modular bricks built in 1:6 cement sand mortar. The roof, which is other than R.C. or R.B. slab, weighs 600 kg/m^2. The bands are required for earthquake coefficient of 0.16.

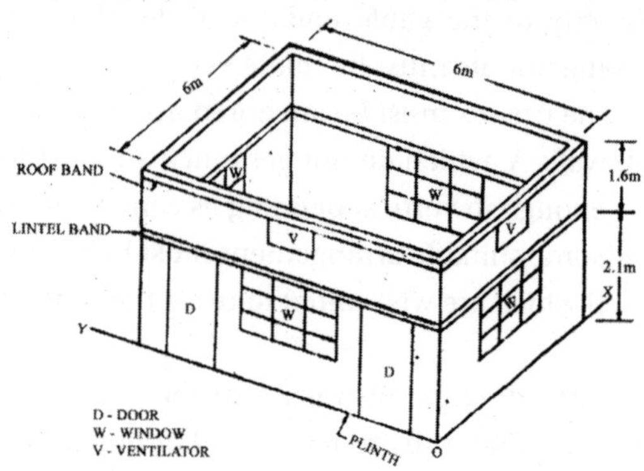

Fig. 2: A single room, seismic band design

* *In these examples, the working stress design method using mild steel reinforcement and concrete of grade M150 (28 days cube crushing strength of 150 kg/cm²) was used. The reader should now use the design as per IS:456-2000.*

Considering the door and window openings, two bands will be suitable, one at lintel and the other at roof level. A band may be necessary at plinth level also under certain conditions. Let the earthquake force act along x-axis. The lintel band divides the long wall in two portions, the lower portion spanning vertically between the plinth and lintel band and the upper portion spanning between the lintel band and roof band.

Taking the weight of masonry 1920 kg/m³, weight of 20 cm wall is 384 kg/m².

Checking Vertical Bending of Wall

Vertical span is 2.1 m with *flat* ends.

Moment = 0.16 x 384 x 2.1^2/12 = 22.6 kgm/m

Bending Stress = 22.60 x 100/[100 x (20^2 /6)] = + 0.338 kg/cm²

Weight of wall and roof per m approximately

= 384 x 1.4 + 600 x 6.4 x 4.4/[2(6.2 + 4.2)]

= 1350 kg.

Stress = 1350/(100 x 20) = 0.675 kg/cm²

Combined stresses = 0.675 $\pm$ 0.338 = 1.013, 0.337 kg/cm² (compressive)

Therefore, the wall is safe.

Lintel Band

Neglecting openings, horizontal load on lintel band

q_h = 0.16 x (2.1 + 1.4) 2 x 384 = 107 kg/m

Assuming continuity of band at corners, bending moment in band

M = 107 x 6.2^2/10 = 411 kgm

F = 107 x6/2 = 321 kg

Taking the following stresses for design:-

Allowable bending compression in concrete	50 kg/cm²
Allowable shearing stress in concrete	5 kg/cm²
Allowable tensile stress in steel	1400 kg/cm²
Modular ratio	18

Taking permissible increase in stress due to earthquake 33 1/3 % and taking the compression steel into account (because reinforcement is required on both faces due to reversible nature of force) or using steel beam theory, the area of steel on each face

A_t = 41100/[1.333 x 1400 x 15] =1.471 cm²

The thickness of the band may be determined from the consideration of diagonal tension. Limiting the shear stress to the allowable value, thickness

b = 321/[1.333 x 5 x 0.867 x 17.5] = 3.17 cm

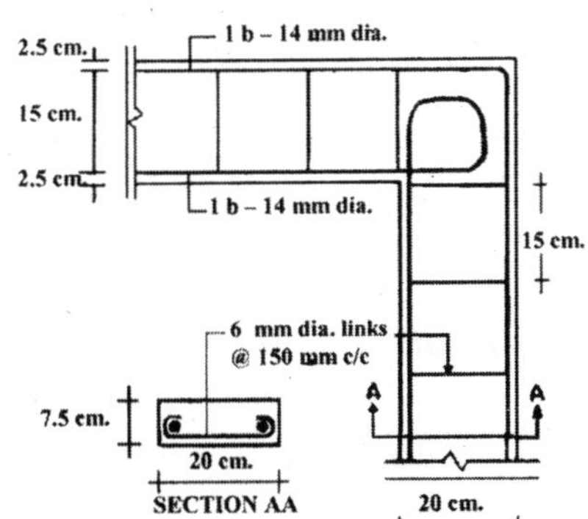

Fig. 3: Reinforcement detail

A minimum thickness of 7.5 cm may be adopted. One bar 14 mm dia may be used on each face. Links consisting of 6 mm dia with hooks at both ends may be used @ 15 cm/cc to keep longitudinal bars in position. Fig. 3 show the reinforcement detail near a corner.

As an alternative to the reinforced concrete, reinforced brick work may be used. In the present case, the steel requirement is the same. Two 10 mm dia bars may be located in two consecutive motar courses on each face. The motar must be 1:3 cement sand mortar, its thickness being such as to provide at least 6 mm cover to the bar, say 25 mm in the present case. Fig. 4 shows the cross section and the arrangement of bending in plan.

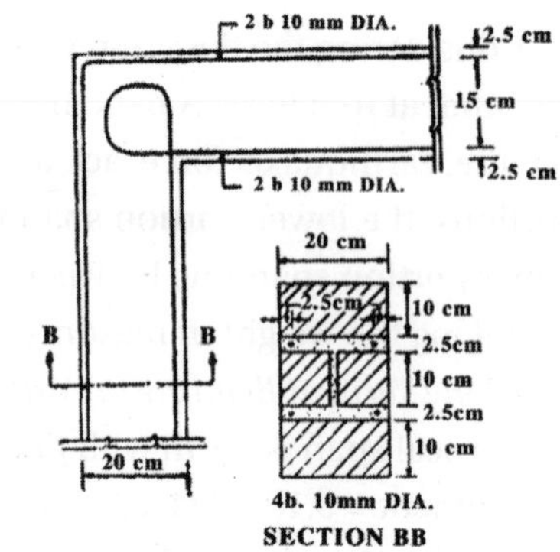

Fig. 4: R. B. Band

Roof Band

The inertia force due to the load as coming on the roof band depends on the arrangement of roofing elements. Let us assume that the roofing elements are connected with the long walls. Inertia load on roof band

q_h = 0.16 (600 x 4.4/2 + 384 x 1.4/2) = 254 kg/m

As for the lintel band, the section of 7.5 cm x 20 cm may still be maintained. In this case, two bars 22 mm dia, one on each face, will be found to be sufficient.

10.3.2 Example 2

A barrack, 15 m long between end walls and 6 m wide carries AC sheet roofing over timber trusses. The roof weighs 100 kg/m^2 of horizontal area. The walls are 30 cm thick and 4 m tall above plinth, the top level of door and window openings being 2.5 m. The strengthening is required against an earthquake coefficient of 0.107.

Anchorage

Holding down bolts as required for wind shall suffice.

Roof Band

Span	= 15+0.30	= 15.3 m
Load from roof	= 0.107 x 100 x 6.6/2	= 35.4 kg/m
Load from wall	= 0.107 x 1920 x 0.30 (4.0-2.5)/2	= 46.1 kg/m
Total load		= 81.5 kg/m

M	$= 81.5 \times 15.3^2 / 10$	=	1905 kgm
F	$= 81.5 \times 15/2$	=	611 kg

Proceeding as in Example 1, a 15 cm x 30 cm band in M 150 concrete is sufficient when reinforced with 4 bars 16 mm dia, one at each corner and 6 mm diameter stirrups 25 mm apart.

Horizontal Bracing

Alternative to the roof band, horizontal bracing system may be used. Let the trusses be 3 m apart and the main tie be divided into 3 panel lengths of 2.1 m each. Let a horizontal truss be provided as shown in Fig. 5, which will be able to take the horizontal force from roof as well as both the longitudinal walls. In large halls, such trusses may be provided along both the long walls.

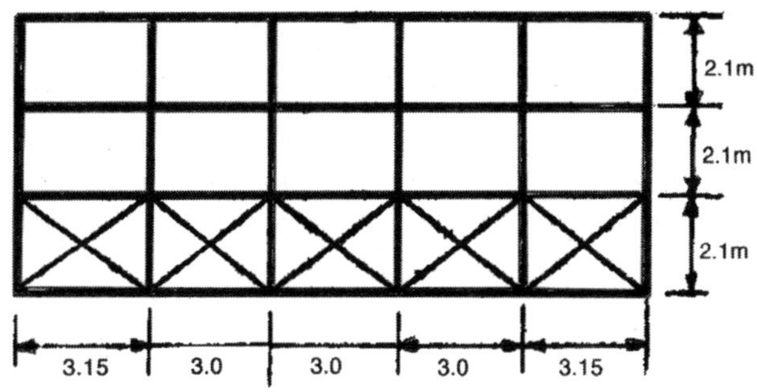

Fig. 5: Horizontal bracing

Panel point load $= (35.4 + 2 \times 46.1) \times 3 = 383$ kg.

Maximum shear in end panel $= 383 \times 2 = 766$ kg.

Maximum force in tension diagonal
assuming the compression diagonal to buckle $= (766/2.1) \times \sqrt{(3.15)^2 + (2.1)^2} = 1380$ kg

Use may be made of 12 mm dia. rods.

Max. force in chord member $= (1/2.1)\,(766 \times 6.15 - 383 \times 3.0) = 1700$ kg

For timber trusses, the members may be of wood and for steel trusses, of steel, designed for a force of 1700 kg, as usual.

Maximum force in end members $= 766 + 383/4$
$= 862$ kg.

The members at end should normally be R.C. or R.B. so as to continue around the walls as a binding member in the shear walls.

Lintel band may be designed as in Example 1.

10.4 Free Standing Walls

Free standing walls, like compound walls, may be designed in two ways. If the mortar has doubtful or no tensile strength, the design must be made for no-overturning. It may be shown that with a factor of safety of 1.5, the minimum thickness of such a wall b is given by

$$b = 1.5\ C.h \qquad (1)$$

where C is the seismic coefficient and h is the height of the wall.

But when the mortar used is believed to have a known tensile strength, the minimum thickness of wall may be arrived at by using a factor of safety of 2.0. For brick work in 1:6 cement mortar, tensile stress of 1.0 kg/cm^2 may be permitted for design against earthquake forces. It can be derived that in this case minimum value of b will be given by

$$b = 3Cw\,h^2 / (f + w\,h) \qquad (2)$$

where w is the unit weight of masonry and f is the allowable tension.

10.4.1 Example 3

Given h = 160 cm, C = 0.12, w = 0.00192 kg/cm^3, f = zero or 1.0 kg/cm^2, the thickness of free standing wall is required.

For no tensile strength, using Eq. (1)
b = 1.5 x 0.12 x 160 = 28.8 cm, say 30 cm.

For given tensile strength, using Eq. (2)
$b = 3 \times 0.12 \times 0.00192 \times (160)^2/[1.0 + 0.00192 \times 160]$
b = 18 cm, say 20 cm.

10.5 Partition Walls

Non-bearing partitions should also be checked for stability against earthquakes because their collapse may result in damage to property and life. They should in general be built into the bearing walls and be stayed against deflection at top as far as possible. If the top has to be kept free, horizontal steel bars may be inserted below the top course, one on each face and anchored into the bearing walls at ends. This would act as a sort of band. The stability of wall may then be checked in bending as a vertical slab.*

An effective way of achieving stability of infill walls against out of plane overturning will be to provide a seismic band of 75 mm thickness with 2 longitudinal 8 mm dia. HSD (TOR) bars at the general door lintel level anchored into the RC columns at the ends.

10.6 Roofs and Floors

The more common types of roofs and floors used in conjunction with masonry are the following:

i) Roof trusses,

ii) Jack arches,

iii) Slab or slab and beams (reinforced concrete or reinforced brickwork of cast in situ construction)

iv) Joists (timber, reinforced concrete, prestressed concrete, precast slab units, stone slabs)

A general distinction is to be made among them. Those which are flat and are bonded or tied to the masonry have a binding effect on the walls and do not require extra tying arrangement in the form of a band for the walls at roof or floor level. The examples are slabor slab and beam construction directly cast over the walls, or jack arch floors or roofs provided with horizontal ties and laid over the masonry walls through good mortar like 1:6 cement sand mortar or equivalent in tensile strength. Others which simply rest on the masonry walls will offer resistance to relative motion through friction only which may not be relied upon. In such cases, other positive means are to be taken to tie the roofing and flooring elements together, to fix them to the walls and also tying the walls together.

To illustrate the point, let us consider the case of a timber joists floor of conventional construction in which the joists about 7.5 cm wide by 10 cm deep are placed across spans of about 3 m at a centre to centre spacing of 25 to 30 cm. These are then covered by timber planks nailed to the joints or by brick tiles placed directly over the joists. Clayee earth is then laid for water proofing and roof or floor may finally be finished as desired. When timber planks are used, they act as effective ties to bind the joists together. But brick tiles have no such binding effect. In fact, during an earthquake, a small relative displacement of the joists will be enough to bring down the tiles, damaging life and property in the room. In such cases, the joists must be tied together at fixed spacing apart so that the tiles above them not dislodged. This may be achieved by *blocking* the space between the joists by means of timber blocks having the same depth as the joists, fixing to them by nailing. Besides blocking at the ends, cross-bridging planks may be nailed to the joists at mid-span.

In order to tie the so formed joist-grill to the walls, a timber runner may be used on the wall under the grill to which the joists may be spiked. The timber runners will also be capable of tying the walls together if they are sufficiently large in section and full length pieces are used in all walls, and jointed together firmly at their junctions. In that case, the timer runners may be built into the walls. If a separate tying arrangement is desired for the walls, an R.B. or R.C. band may be provided between the wall and roof or floor. The design of such a band has been illustrated in Example 1 earlier. *

** For appropriate details regarding prefab roofing elements, reference may be made to IS:4326-1993, cl.9.*

In the case of other kinds of elements, suitable details may be worked out for connecting the units together and fixing them to the band. For example, for precast R.C. planks iron straps and bolts, welding the protruding reinforcement, laying cast-in-situ concrete, etc. are some of the devices that may be used for the purpose.

10.7 Design of Shear Walls

As has been stated earlier, shear walls are the main members for transferring all earthquake forces to the foundations. Tests have indicated that the factors determining their strength are too many and varied. The greatest source of error and uncertainty is the workmanship. Any effort to use the theory of elasticity for analysing the stresses in brick or stone masonry, therefore, proves futile. Hence, the simplest available approach for their analysis should generally suffice. One such procedure for shear walls is explained below:

Let masonry wall have openings as shown in Fig. 6 and carry horizontal inertia forces due to the weight of roof and wall as indicated. For analysing the stresses in such walls, it is assumed that the rotational deformations of the portions above and below the openings are much smaller than those of the piers between the openings and are neglected. Points of contraflexure are assumed at the mid-points of piers and shear is assumed to be shared among the piers such that their tops deflect by the same amount.

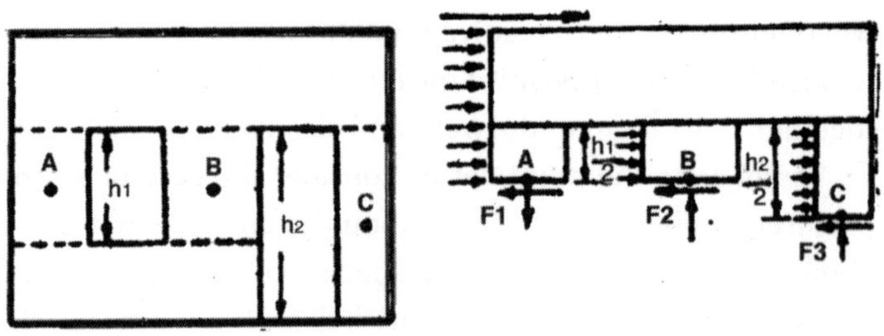

Fig. 6: Assumed contraflexure points in piers

For calculating the deflections, the piers are assumed to be retained against rotation at the ends but as free to deflect; and shear deformation are also considered. Thus if the horizontal shear in a pier is H, the deflection at its top is given by

$$\Delta = (Hh^3/12EI) + (1.2Hh/GA) \qquad (3)$$

where h is the height of the pier, I its moment of inertia about the axis of bending, E the modulus of elasticity, G the modulus of rigidity and 1.2 is a factor required in the calculation of shear deflection of a member having a rectangular section. The shear stiffness of a pier may be defined '*as the load required to deflect its top by unit deflection*'. Hence the shear stiffness is given by

$$S = H/\Delta = 1/[(Hh^3/12EI) + (1.2Hh/GA)]$$
$$= (12\,EI/h^3)/(1 + 14.4\,EI/GAh^2) \qquad (4)$$

If the thickness of the wall is b and the width of pier is d, then expressing I and A in their terms and taking G = E/2 for masonry,

$$S = (12\ EI/h^3) \times 1/[1 + 2.4(d/h)^2] \qquad (5)$$

The total horizontal shear in the piers may be distributed between them in the ratios of their shear stiffnesses. Let the shears produced be F1, F2, etc. Then the piers carry shears F, and bending moments Fh/2 at their top and bottom sections. The resulting values of maximum shear stress, q, and bending stress, Pb in any pier will be given by

$$q = 1.5\ F/bd \qquad (6)$$

$$p_b = 3\ Fh/bd^2 \qquad (7)$$

The total weight of the structure above the top of piers W may be assumed to produce uniform direct stress po in them. Thus,

$$p_o = W/\Sigma A \qquad (8)$$

where ΣA represents the sum of areas of all the piers in the plan of the building at any level.

The overturning forces are determined as follows:

Let the areas of the piers be A1, A2 and A3 (Fig. 7) and the combined centre of gravity of the three areas be at G. Assuming the stress in the piers to be proportional to the distance from G, the forces in the three piers are:

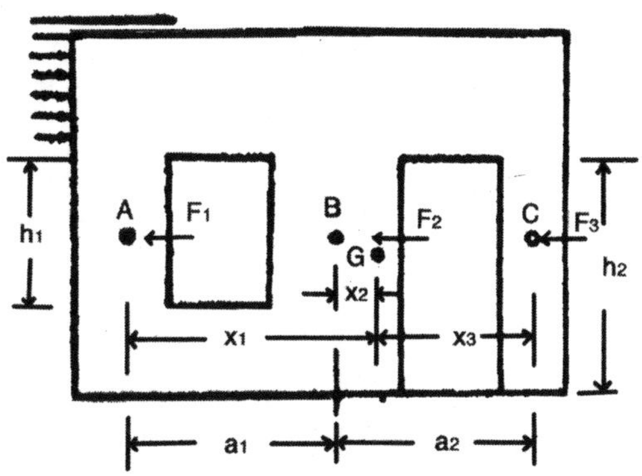

Fig. 7

$- K x_1 A_1, - K x_2 A_2$ and $K x_3 A_3$,

where K is the coefficient of proportionality, and x_1, x_2 and x_3 are the distances of the points considered (in the three piers) from G. If M is the moment of the horizontal forces about A, we get,

$$K = [M + F_3 (h_2 - h_1)/2] / (A_1x_1^2 + A_2x_2^2 + A_3x_3^2) \qquad (9)$$

Knowing the constant of proportionality K, the stresses due to overturning, pt, may be found as follows:

- K x_1 in pier A
- K x_2 in pier B
+ Kx_3 in pier C (10)

Combining all the stresses, the maximum and minimum stresses on the cross-section of each pier can be determined. That is, the resultant total stress p is given by

$$p = p_b + p_o + p_t \qquad (11)$$

which may be positive or negative depending upon the relative magnitudes and signs of the component stresses.

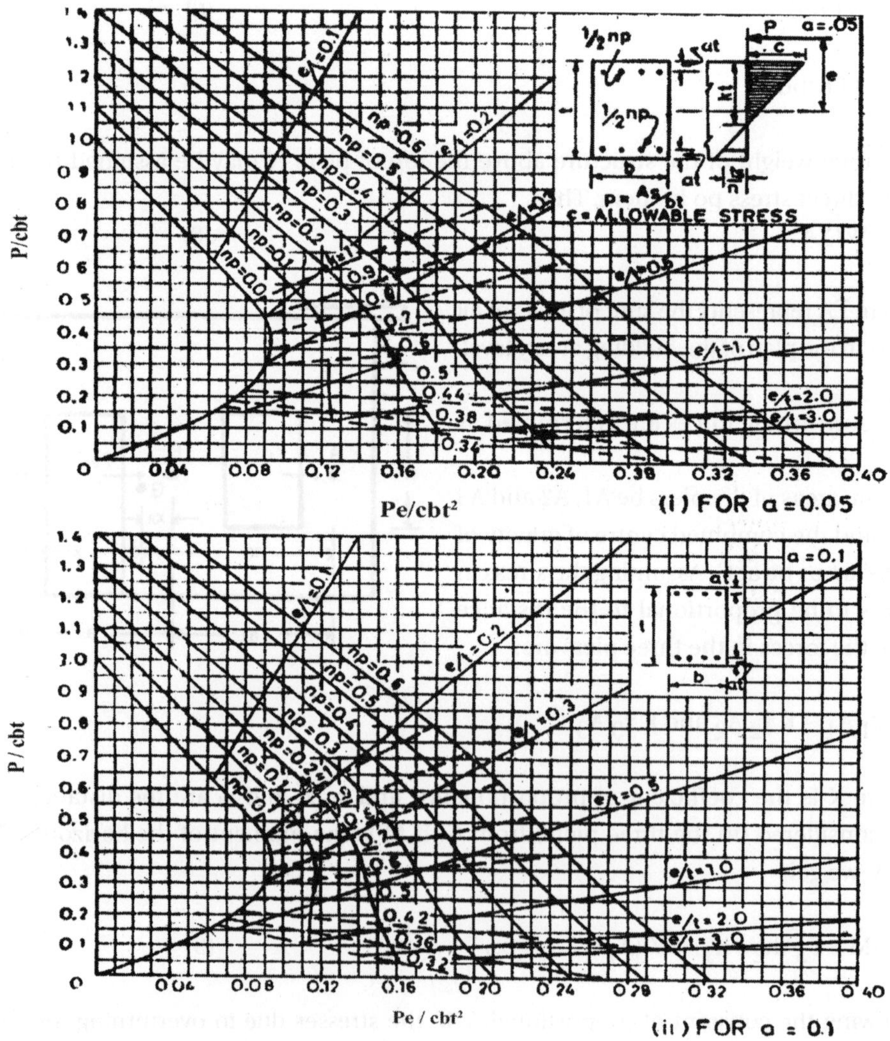

Fig. 8: Direct and bending stress design of piers

Such calculations are to be made along each axis of the building for reversible earthquake force. If the stress becomes tensile, vertical steel reinforcement may be designed according to the usual theory of reinforced concrete for combined direct and bending stresses. In any case, the compressive stress must remain within the safe compressive stress of the type of brickwork. For values of allowable stresses and other specifications reference may be made to IS: 1905 1961. For designing the sections under direct and bending stresses combined, the charts given in Fig. 8 may directly be used. The modular ratio of steel to brickwork may be assumed 40 for this purpose.

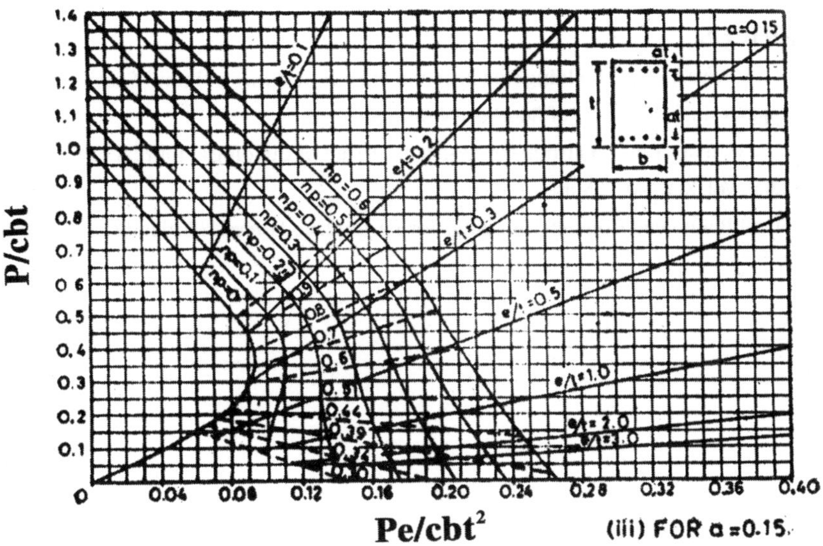

(iii) FOR a =0.15.

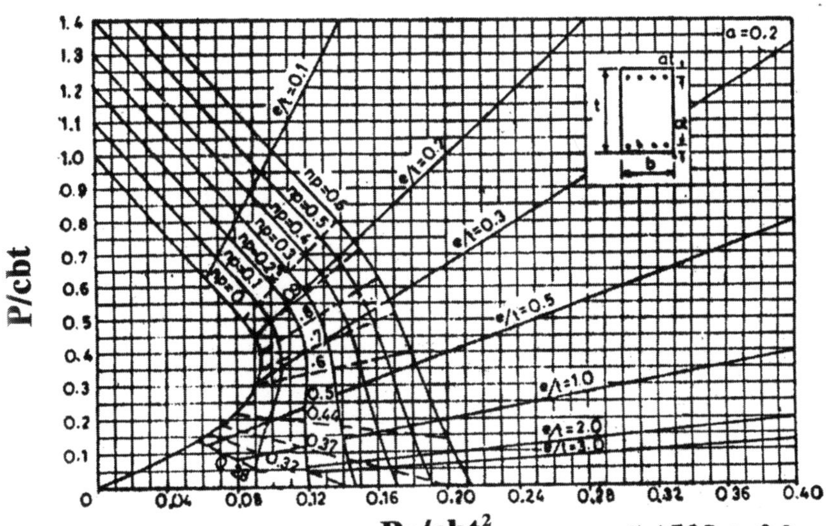

(iv) FOR a=0.2

Fig. 8

Fig.8: continued

10.8 Conclusion

It may be concluded that if the tying and reinforcing arrangements as explained and illustrated in the text are adopted, a masonry building will act as one integral box/crate like unit for resisting inertia forces caused by the ground motion and have sufficient strength to withstand them. The provision of reinforcement is expected to impart some ductility to the otherwise brittle structure which will further help it to absorb energy and delay its collapse.■

10.9 References

1. Krishna J. and A.S. Arya, (1962) "Building Construction in Seismic Zones of India", Second Symposium on Earthquake Engineering, University of Roorkee, November 1962.
2. Krishna J. and A.S. Arya, (1965) "Earthquake Resistant Design of Buildings", Journal of the Institution of Engineers (India), Vol. XLV, No. 7, March 1965.
3. "Analysis of Small Reinforced Concrete Buildings for Earthquake Forces", Portland Cement Association, Chicago, 1955.

Chapter 11

Construction of Small Buildings in Seismic Areas[1,2]

11.1 Abstract

The general principles of earthquake resistant construction as evidenced by studies of damage and non-damage in various past earthquakes are first described. Then recommendations are given for constructing buildings in masonry and timber or tubular frame work. Finally a few designs of small houses are presented which could be adopted for construction of houses in a seismic area.

11.2 Introduction

The Koyna earthquake of December 11, 1967 rocked the supposedly stable area of Maharashtra causing irrepairable damage to about 40,000 houses, rendering lakhs of people homeless. The author visited the area along with two of his colleagues and prepared a report on the damages observed to the buildings and the prevalent construction in the area and, as a first step, suggested an earthquake resistant design with timber or pipe supports for rebuilding the houses in Koynanager, utilising the plinths, roof trusses and sheets of damaged buildings. However, the necessity was felt to bring out detailed design of small earthquake resistant houses which could be constructed on a large scale out of readily available materials.

This paper presents the general principles of earthquake resistant construction, detailed specifications to be adopted for load bearing wall buildings and those using timber or pipe supports and bracing members. A few designs of small houses are also included which could be adopted for construction of houses in a seismic area.

11.3 General Principles Regarding Constructional Features

A study of the damage to buildings during past earthquake indicates that the following principles may be observed in the design and construction of buildings to make them earthquake resistant.

---

1 *A.S.Arya, Professor & Assistant Director Earthquake Engineering, University of Roorkee, Roorkee*

2 *Bulletin of the Indian Society of Earthquake Technology, Vol.5, Nos. 3 & 4, September-December, 1968, pp 89-105*

Materials: Since the earthquake force is mainly a function of mass besides the ground acceleration and the stiffness of the structure, a building should be constructed as light as possible, consistent with the structural safety and functional requirements. In particular, the roofs and upper storeys of the building should be built of light materials or of light construction as far as possible.

Continuity of Construction: It has been observed that a building having elements which bind the walls, columns and other structural components together has stood the earthquake shocks better. It would, therefore, be of advantage to have the floor slabs continuous as far as possible. If precast beams and concrete slab units are used, they should be connected together by positive means like in situ concrete, welding of protruding reinforcement or bolting, etc.

Overhanging Parts: Overhanging parts such as projecting cornices, balconies, parapets and chimneys are the first to fall during an earthquake. Not only is there damage to the building but such parts, when they fall, injure the people who are running out of the houses or moving on the streets. Such projecting and overhanging parts should be avoided as far as possible or enough care should be taken to reinforce them and anchor them to the main structure adequately.

Suspended Ceilings: Suspended ceilings often used for aesthetic reasons, are usually brittle and weak and incapable of resisting horizontal forces with the result that during an earthquake they crumble and fall down. Thus, special care is required in the design of suspended ceilings if they cannot be avoided. They should be strong and rigidly tied to the roof or ductile enough to withstand the strains during ground motion.

Similarly, the plaster on the ceiling frequently falls down. The thickness of such plaster should be kept to a minimum.

Separation of Adjoining Structures: Adequate separation of adjoining structures (or the parts of the same structure) having differential heights or rigidities is necessary to avoid damage due to out of phase vibrations during an earthquake resulting in hammering action between them.

Workmanship: All reports regarding damage to structures during earthquakes have repeatedly pointed out that structures built with the same material and in the same manner but with different quality of workmanship have behaved differently. Those having better workmanship stood the earthquake shaking but those of poor workmanship were shattered to pieces. Therefore, the greatest emphasis must be laid on good quality of workmanship in all construction.

Damage to Non-Structural Parts: During the past earthquakes, it has sometimes happened that whereas the structural frame was strong enough to resist the earthquake forces, the non-structural elements like brick filling in a timber frame, which is not supposed to carry any other loads besides its own weight, have fallen out of the frame. Therefore, it is necessary that the non-structural parts should be well tied to the structural framing. To avoid damage to window frames or grazing, the drift in buildings should also be limited.

Fire Safety: In the past, damage to life and property has occurred due to fires, caused by short circuiting of electric wires or kitchen fires during an earthquake. Mention may be made of the San Francisco (1906) and Tokyo (1923) fires after the earthquakes. Therefore, buildings should be constructed to be fire resistant and other safety precautions against fire should be taken.

Shape of Building: It has been observed that a building having a simple squarish plan and symmetry both with respect to mass and stiffness fares much better during an earthquake than long narrow or irregularly shaped buildings. Buildings having E, U or L shape plans tend to develop cracks at junctions of various limbs. In such cases, the building should be sectionalised into parts by providing separation sections so that each part becomes rectangular, its length not exceeding about three times the width. At such sections, a complete separation of the parts should be made except below plinth level. The plinth beams or foundation beams and the footings may be continuous.

If the two separated parts are dissimilar in their height or weight and rigidity, they may collide with each other during an earthquake and get damaged. Therefore, a minimum gap of about 1.5 cm per storey should be provided between them. Such a gaps may be suitably covered by fragile or collapsible cover in the form of asbestos or metallic sheets, if so desired.

Foundation: In several earthquakes in India and abroad, liquefaction of soil, occurrence of sand fountains or settlement of soil has been observed leading to catastrophic failure of structures. Therefore, soft alluvium and loose sands and low lying water logged ground should be avoided and foundations should be supported on hard and dense strata. Where poor situations are unavoidable, adequately tied and capped pile foundations should be employed for major buildings. For small buildings, continuous reinforced concrete strip footings may be used.

Roofs and Floors: Light roofing materials have obvious advantage in reducing the inertia force at top of building. To avoid the dislodging of the roofing units, they should be tied to the supporting members. For roof coverings, corrugated iron, aluminium and asbestos sheets are preferable to earthen tiles, slates, etc.

In order to check relative movement due to shaking, joists of timber, reinforced concrete or steel, where used to support individual flooring units like brick tiles and prefabricated doubly curved units, should be restrained by blocking the space between them at their ends and by fixing bridging members at their midspan or third-span points. Likewise, steel beams supporting jack or flat arches should be tied together in each bay to restrain their lateral movement and retain the arch uncracked.

In this respect, continuous reinforced concrete or brick slabs are better. They have the further advantage of providing affective binding effect on the walls and columns and may therefore, be preferred.

11.4 Recommendations for Small Masonry Buildings

Seismic Zones: For the purpose of these recommendations, the area is assumed to be divided into the following seismic zones:

(a) Highly seismic having M.M. Intensities IX and over, referred to as Zone A subsequently. Such intensities were not observed in the Koyna earthquake.

(b) Moderately seismic having maximum M.M. Intensity of VIII, referred to as Zone B.

(c) Lightly seismic having maximum M.M. Intensity of VI and VII, referred to as Zone C.

(d) Non-Seismic Design having maximum M.M. Intensity of I to V referred to as Zone D.

No special precautions are required in zone D except good workmanship. Great care is of course, needed in Zones A and B.

Materials and Construction: Well burnt bricks, solid or hollow concrete blocks and squared stones of adequate strength should be used for masonry work. It has repeatedly been observed (Kapkote 1958, Anjar 1956, and Agadir 1960) that random rubble masonry is extremely weak against shaking and should generally be avoided. If economic and other conditions compels its use, it may be used in Zone C with the specifications that it is brought to courses at not more than 60 cm intervals and 'through' bond stones are provided at a maximum horizontal spacing of 1.2 m in every levelling course and the height is not more than one storey. Its use should, however, be avoided in Zones A and B. The mortars should be no leaner than those given in Table 1.

Table 1:- Mortar Specification

S.No.	Cement	Lime	Sand	Surkhi*	
1	1	1	6	-	In Zone A
2	1	2	9	-	
3	1	-	6	-	In Zones
4	-	1	3	-	B & C
5	-	1	-	3	

*Brick powder or cinder

Wall: The height of load bearing masonry walls should preferably be restricted to about a maximum of three storeys. The minimum thickness of load bearing walls in different storeys may be kept as traditionally adopted. Suggested thicknesses are; one unit thick or (20 cm minimum) upto two storeys in height or for the upper two storeys of three storeyed building and 1 1/2 unit thick (or 30 cm minimum) for the bottom storey of three storeyed building. These minimum thicknesses may be used when walls are connected to cross walls not more than 6 m apart. For larger spacing, thickness should be increased.

* *For current zones and design details for small building, refer IS:1893 (Part - 1) - 2002 and IS:4326 - 1993*

Openings in Bearing Walls: Studies carried out on the effect of openings on the strength of walls indicate that they should be small and more centrally located. IS: 4326-1967 provides the following restrictions on the size and position of openings.

(a) The openings shall preferably be located away from the corner by a clear distance equal to at least 1/4 of the height of opening.

(b) The length of openings shall not be more than half the length of the wall between consecutive cross walls.

(c) The horizontal distance (pier width) between two openings shall not be less than½ of the height of the shorter opening.

(d) The vertical distance from an opening to an opening directly above it shall not be less than 60 cm nor less than 1/2 of the width of the smaller opening.

(e) Where the openings do not comply with the requirements (a) to (d), they shall either be boxed in reinforced concrete around or the reinforcing bars provided around them through the masonry. See Fig. 1.

Observations in past earthquakes show that use of arches to span across the openings is a source of weakness and shall be avoided unless steel ties are provided.

Bond Beams, Runners or Bands: For tying the various walls of a building so that the building acts as one unit at the time of an earthquake, runners or bands of reinforced concrete or reinforced brick work should be provided in all the load bearing walls at different levels as specified for each zone and type of construction. Such runner shall be called the *plinth band* when used at the plinth level; *lintel band* when used at the lintel level; *roof band* when used just below the roof or floor and *gable band* when used at the top of gable end below the purlins.

The plinth band may be made in M150 (1:2:4) or richer concrete, having a minimum thickness of 7.5 cm (sufficient for damp proofing), width equal to the thickness of the wall above it and reinforced with two bars 10 mm dia, one near each face of the wall held together with 6 mm dia links provided every 20 cm. Such a band need not be provided where the soil is stiff having a bearing capacity 20 t/m^2 or larger.

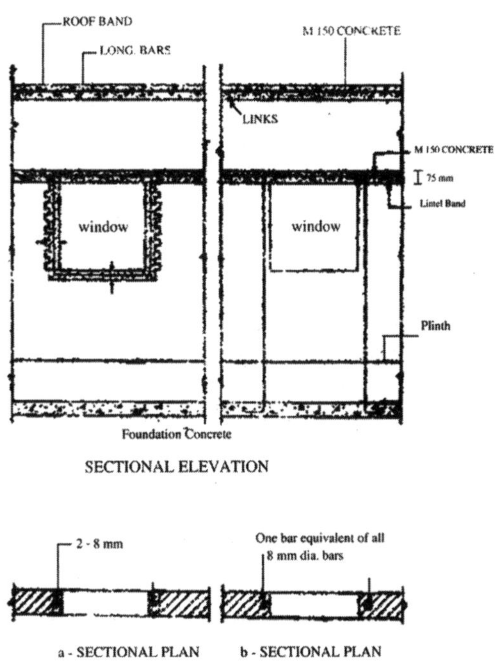

Fig.1: Strengthening around openings

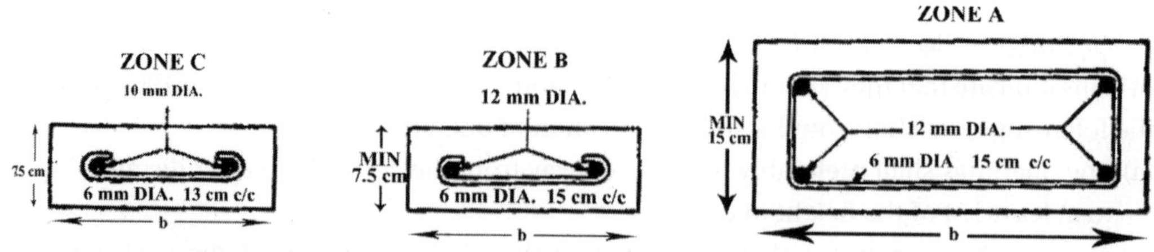

Fig.2: Cross-sections of RC lintel bands (Max span of wall 6 m)

The lintel band size depends upon the maximum span of the rooms and the design seismic coefficient. The width of the band may be kept not less than the thickness of the wall and minimum thickness, when of reinforced concrete, equal to the height of the masonry unit or multiple thereof. Alternatively, the reinforcing bars may be located in the bedding joints, increasing the thickness of the joints so as to have a minimum mortar cover of 6 mm around the bar. The sections of lintel band shown in Fig. 2 may be adopted for span of wall (distance between cross walls) of 6 m or smaller. For a longer span, the areas of longitudinal bars and links may correspondingly be increased.

The roof band need not be provided when floor or roof consists of reinforced concrete or reinforced brick slabs as they automatically bind the walls together and are capable of transmitting their inertial force to the shear walls. In other cases, sections similar to those of the lintel band may be adopted.

The *gable band* shall be provided on top of gable ends of walls and have the same section as the *lintel band.*

All bands shall have continuous connection at the corners and junctions of walls capable of developing positive and negative moments, Fig. 3 shows typical details where overlap length of 40 dia. excluding the hood is suggested.

Vertical Reinforcement in Walls: Use of vertical reinforcement in walls is desirable for increasing the strength of building against horizontal loads and for imparting ductility to them, thereby increasing their energy absorbing capacity. For reasons of economy and ease of construction, vertical reinforcement may

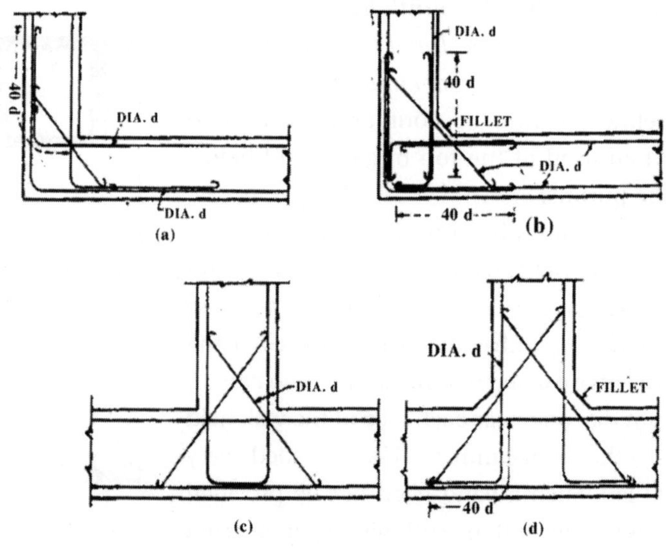

Fig.3: Band reinforcement detail at corner and junction of walls

Table 2:- Vertical Steel Reinforcement in Walls

No. of storeys	Storey	Zone B	Zone A
One	-	0.04 %	0.06 %
Two	Top	0.05 %	0.08 %
	Bottom	0.08 %	0.12 %
Three	Top	0.05 %	0.08 %
	Middle	0.08 %	0.12 %
	Bottom	0.10 %	0.15 %

be provided at the critical sections only, that is, corners and junctions of walls and jambs of openings.

No vertical steel need be provided in Zone C except where openings require it as discussed earlier.

The total amount of vertical steel may be adopted as indicated in Table 2 as percentages of gross area of load bearing walls. The total area so calculated may be proportioned to the various piers between openings and beyond them (solid portions seen in a sectional plan of the building passing through door and window openings) in the ratio of their cross-sectional areas. The steel area in any pier may be provided half and half on each end of it. Alternatively, the amount of vertical steel may be taken as per IS: 4326-1967.

Fig. 4 shows the method of placing of reinforcement in brick work and block masonry construction.

The vertical reinforcing bars should be embedded into the foundation concrete, pass through all intermediate bands, through floor slab reinforcement and be embedded into the roof slab or roof band at the top. Fig. 6 shows an

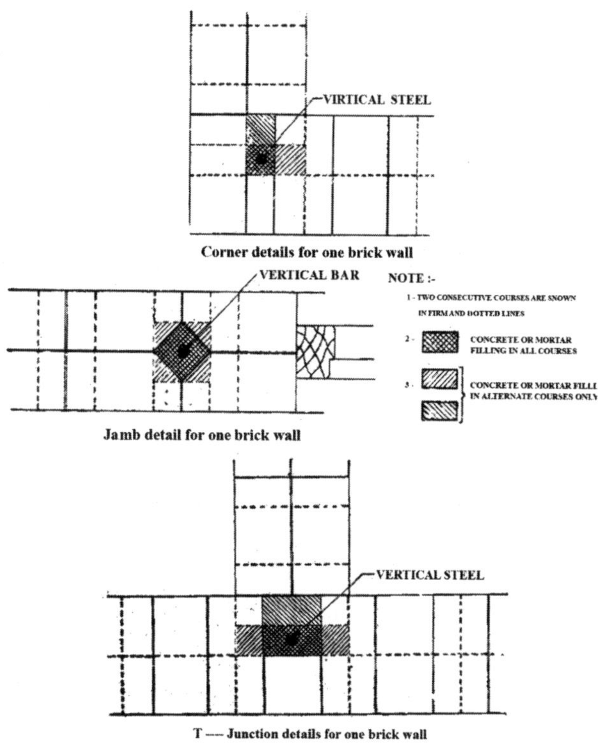

Fig.4: Providing vertical bars in brick work

* *The vertical bar diameters providing the required reinforcement were later spacified instead of percent area. See IS: 4326-1993.*

arrangement of horizontal bands and vertical steel.

Strengthening of Inferior Construction: At some places, due to limited economic resources or shortage of building materials, it may not be feasible to adopt the construction specified above and it may be necessary to continue construction in mud mortar. In such situations the following improvements are suggested.

The inferior construction consisting of inferior brick or dressed stone in mud mortar may be limited to two storeys in Zone C and to one storey in Zones B and A, the ground storey having 1½ brick thick and upper storey one brick thick walls. The lintel band reinforcement should be the same as specified earlier except that the reinforcing bars may be located in brick course laid in 1:3 cement sand mortar (see Fig. 5). Links may be omitted. The plinth band may be omitted. Roof band may be provided under similar conditions as indicated earlier. The vertical reinforcement may now consist of one bar 12 mm dia. at each corner and junction of walls in Zones B and A. Jamb steel may be omitted and openings restricted by the conditions stated earlier.

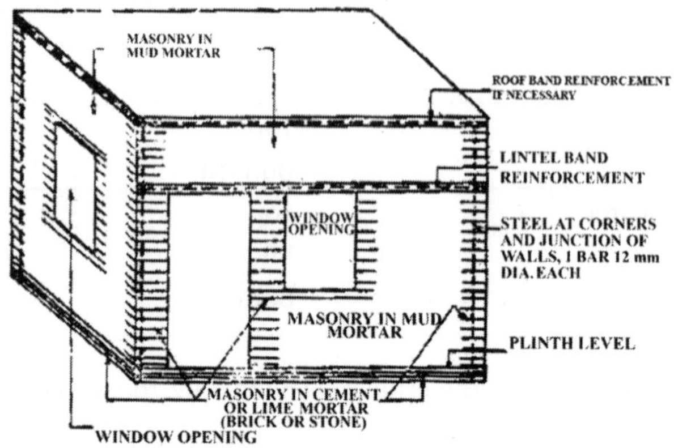

Fig.5: Strengthening brickwork in mud mortar

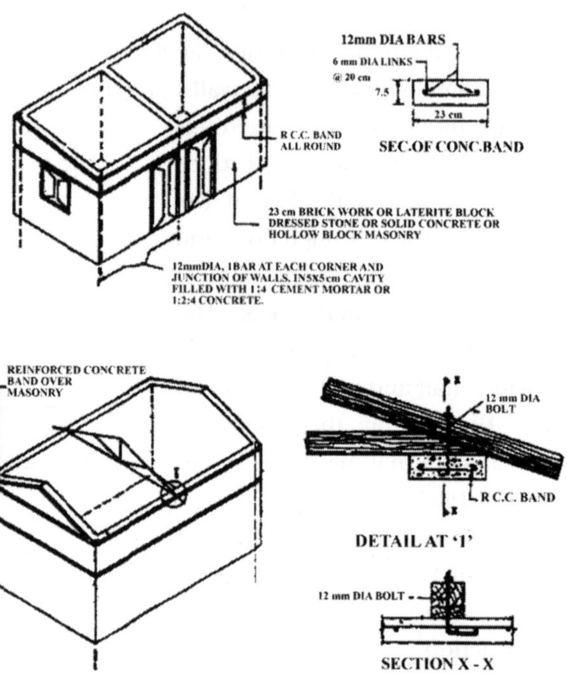

Fig.6: Typical details of masonry construction with band and vertical reinforcement

11.5 Timber Construction:

Timber has high strength per unit weight. It is, therefore, very suitable material for earthquake resistant buildings. IS: 883-1961 gives the specifications of structural timbers and may be referred to for detailed design.

The more important considerations in timber construction are:

(a) To minimise danger due to fire, the roof covering and cladding walls should be of non-combustible materials.

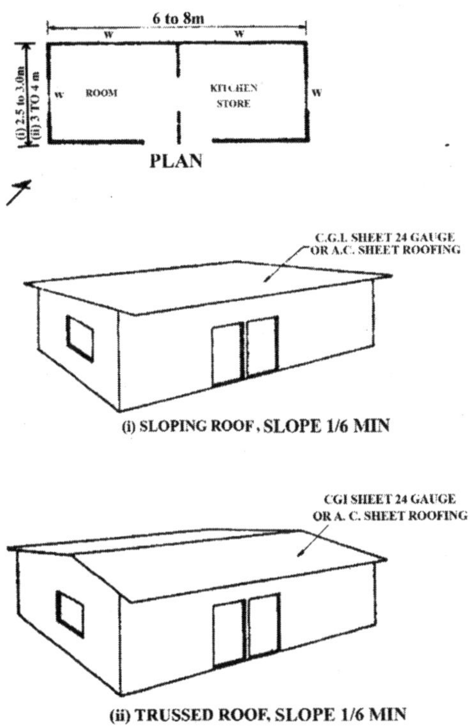

Fig.7(a): Single row of rooms

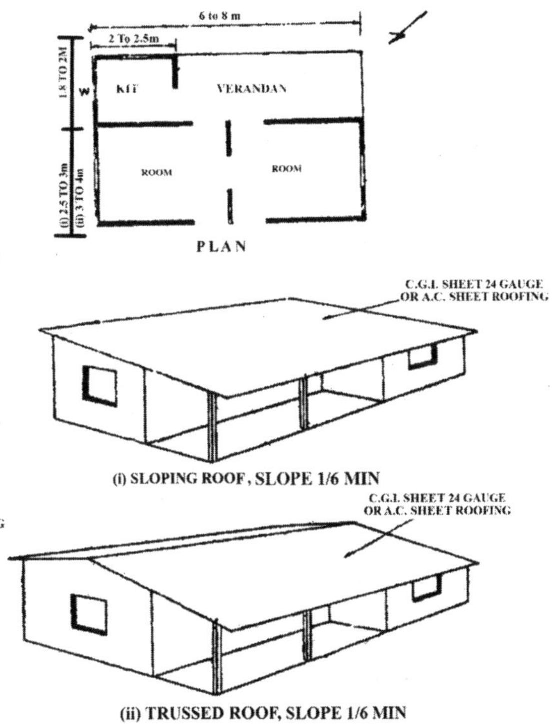

Fig.7(b): One row of main rooms with verandah or side room on side

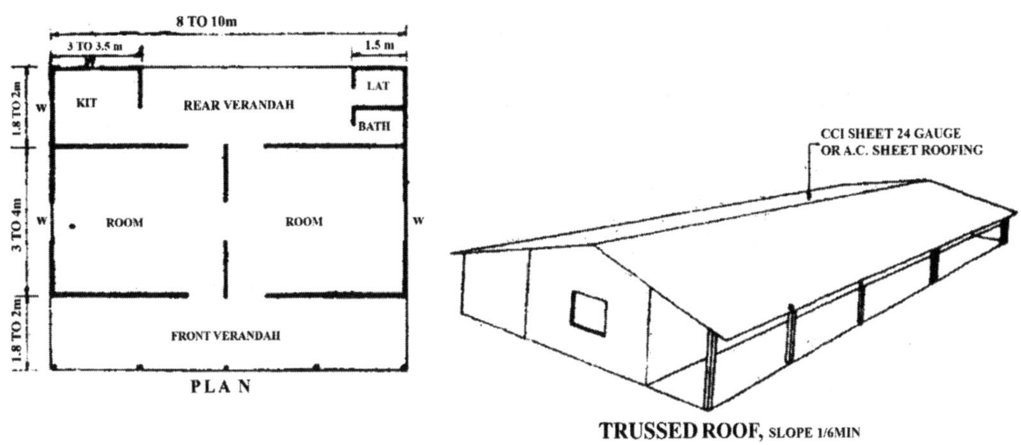

Fig.7(c): One row of main rooms with verandah or side rooms on both side

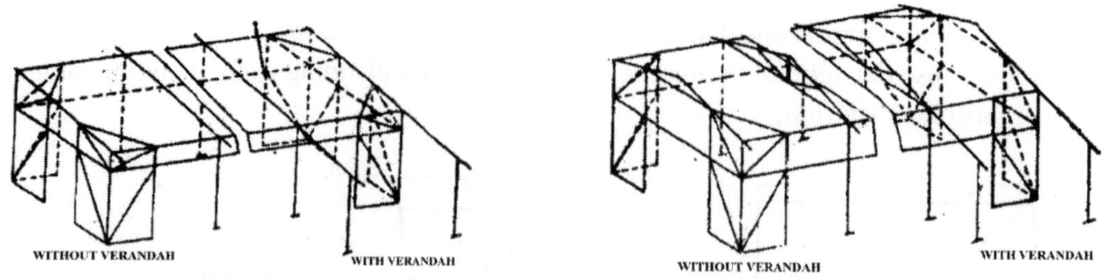

Fig.8 (b): Corner bracing arrangements

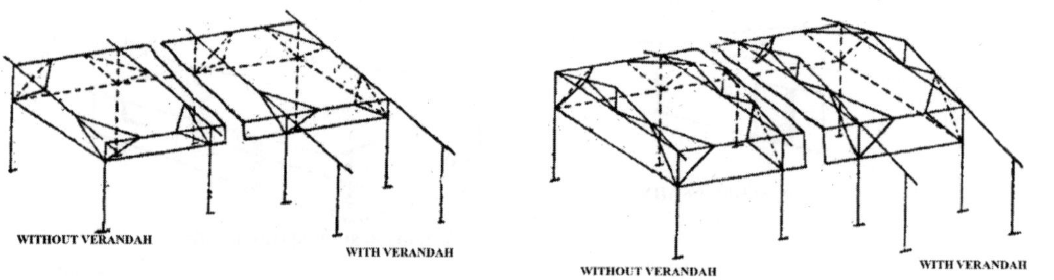

Fig.8 (b): Knee bracing arrangements

(b) The number of storeys may be limited to two, and the attic space may also be made use of.
(c) For imparting resistance to the lateral load, adequate diagonal bracing should be used in the plan and elevation.
(d) The connection between various members should be strong enough to take the load and kept tight by iron straps where possible.

Foundation: In order to avoid deterioration of timber, for permanent buildings the footings, below the plinth level, should be done in masonry or concrete and the columns of the superstructure should be securely connected with the footings either through sills or directly. For temporary huts, however, the timber poles, which act as posts for the superstructure, may be embedded in the ground for providing fixity. Painting the embedded portion with hot tar is very desirable.

Types of Construction: The commonly used types of construction in timber are the stud wall construction and brick nogged frame construction. Taking into account the considerations stated above for such constructions, the recommended details of construction are shown in Figs. 7 to 9.

(a) Three types of plans are considered:
   (i) Single row of rooms having one main span only (Fig. 7a)
   (ii) A row of main rooms with verandah or side rooms on one side (Fig. 7b).
   (iii) A row of main rooms with verandah or side rooms on both sides (Fig. 7c)

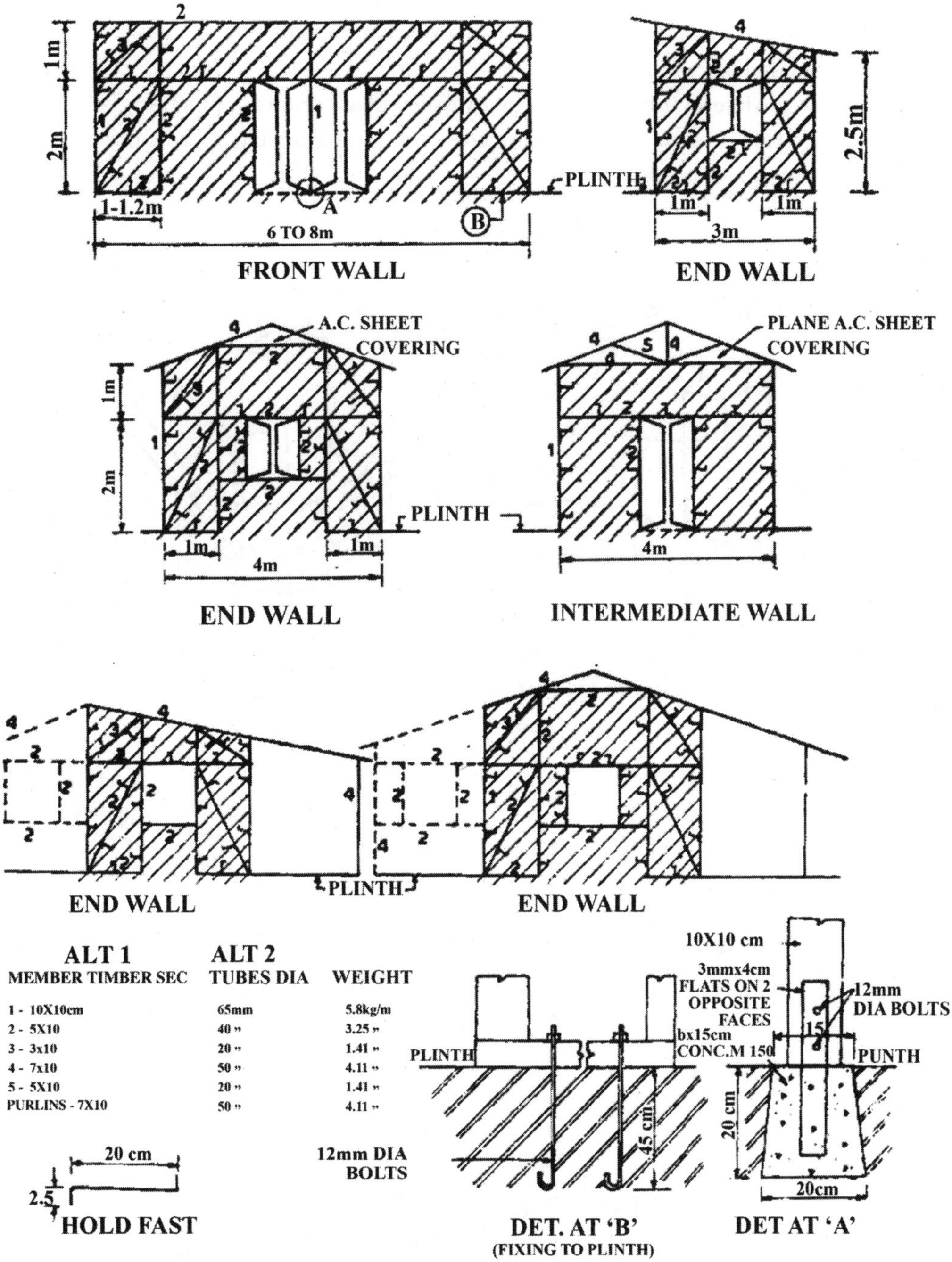

ALT 1 MEMBER TIMBER SEC	ALT 2 TUBES DIA	WEIGHT
1 - 10X10cm	65mm	5.8kg/m
2 - 5X10	40 "	3.25 "
3 - 3x10	20 "	1.41 "
4 - 7x10	50 "	4.11 "
5 - 5X10	20 "	1.41 "
PURLINS - 7X10	50 "	4.11 "

Fig.9 (a): Framing and joints detail

The bracing in each case may be arranged around the main rooms only but made sufficiently strong so as to take the inertia force of the whole house.

(b) Two types of bracing are indicated

(i) In one, the lateral strength has been provided near corners or junctions of walls. See Fig. 8a. This is more suitable where cross walls are available at short intervals, as is the case with residences.

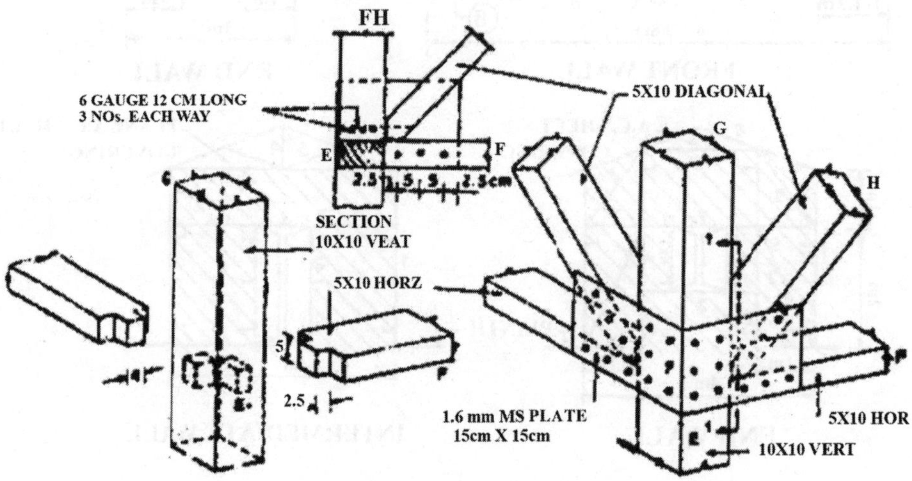

Joint Details at 'E'

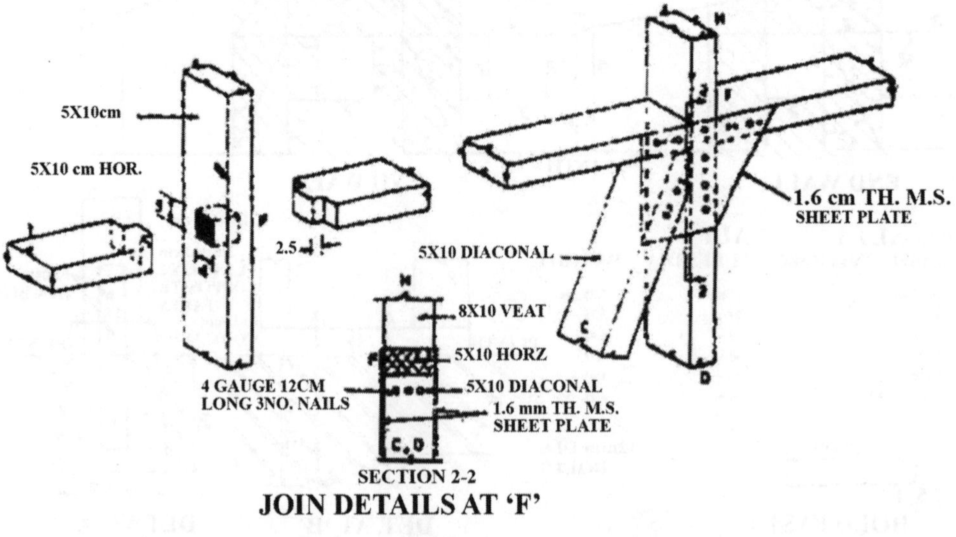

JOIN DETAILS AT 'F'

(ii) In the other, the strength is provided by knee bracing. See Fig. 8b. This is more

Note: Unless mentioned otherwise, all nails are of 6 gauge and 10 cm long

Fig. 9 (b) Framing and joints detail

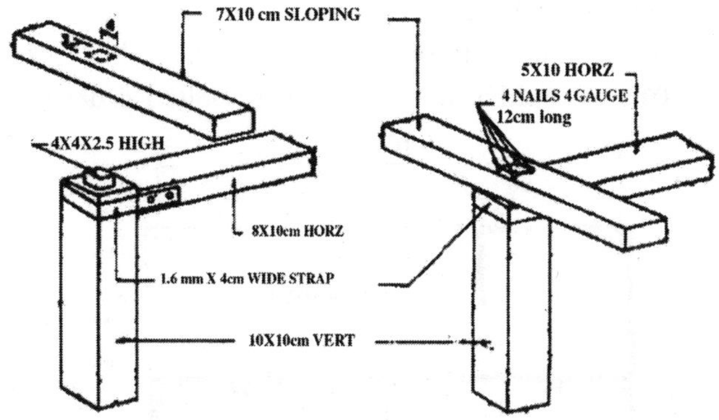

JOINT DETAILS AT 'G'

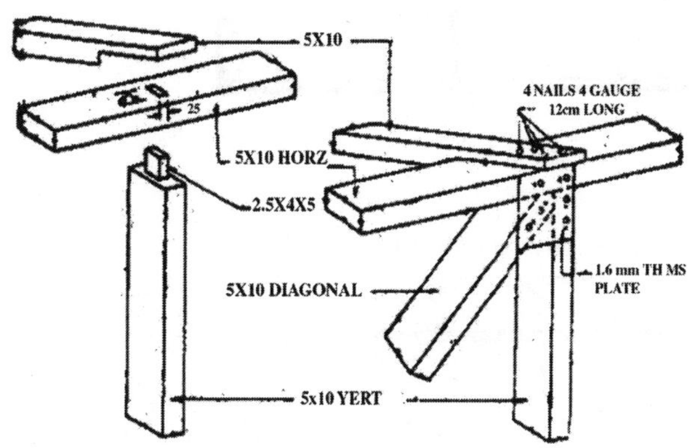

JOINT DETAILS AT 'H'

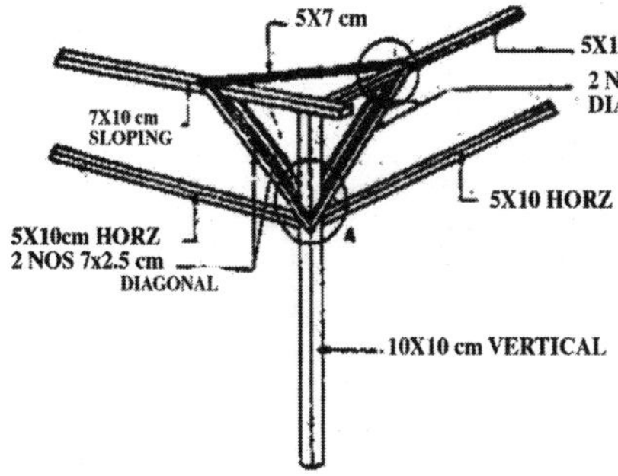

TYP DET OF KNEE BRACING

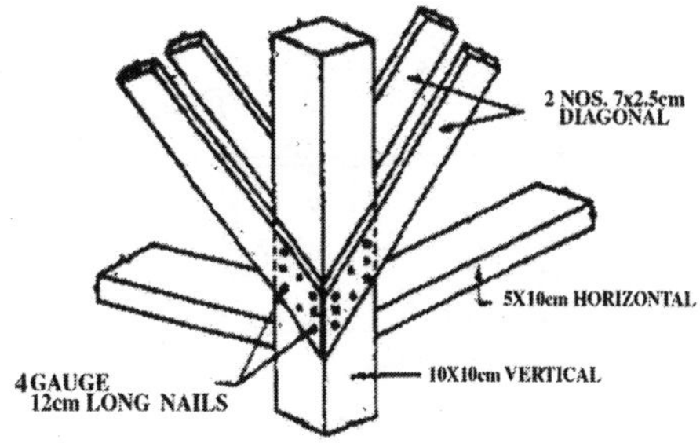

JOINT DET AT 'A'

Note: Unless mentioned otherwise, all nails are of 6 gauge, 10cm long

Fig. 9 (c) Framing and joints detail

suitable in the case of long halls, barracks type dormitories, godowns, etc.

(c) The roof is arranged in the form of timber rafters or timber trusses with A.C. or C.G.I. sheet covering. Trusses could be framed, nailed or bolted as desired. Tightness of joints is

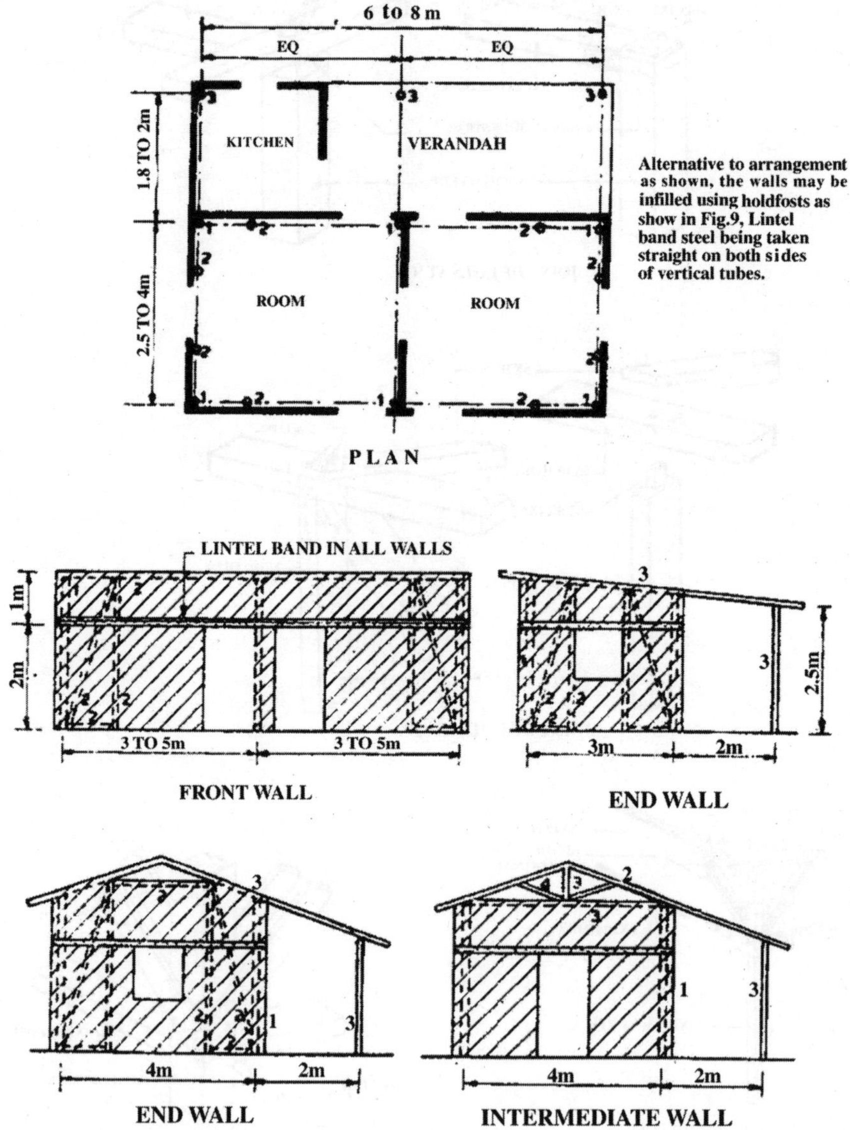

MEMBER	TUBE DIA	WEIGHT
1 -	65mm	5.8 kg/m
2 -	40 "	3.25 "
3 -	50 "	4.11 "
4 -	20 "	1.41 "

PURLINS _ 50mm DIA 4.11kg/m
LINTEL BAND_75mm TH. R.C.C.

Fig. 10(a) Framing in tubular construction

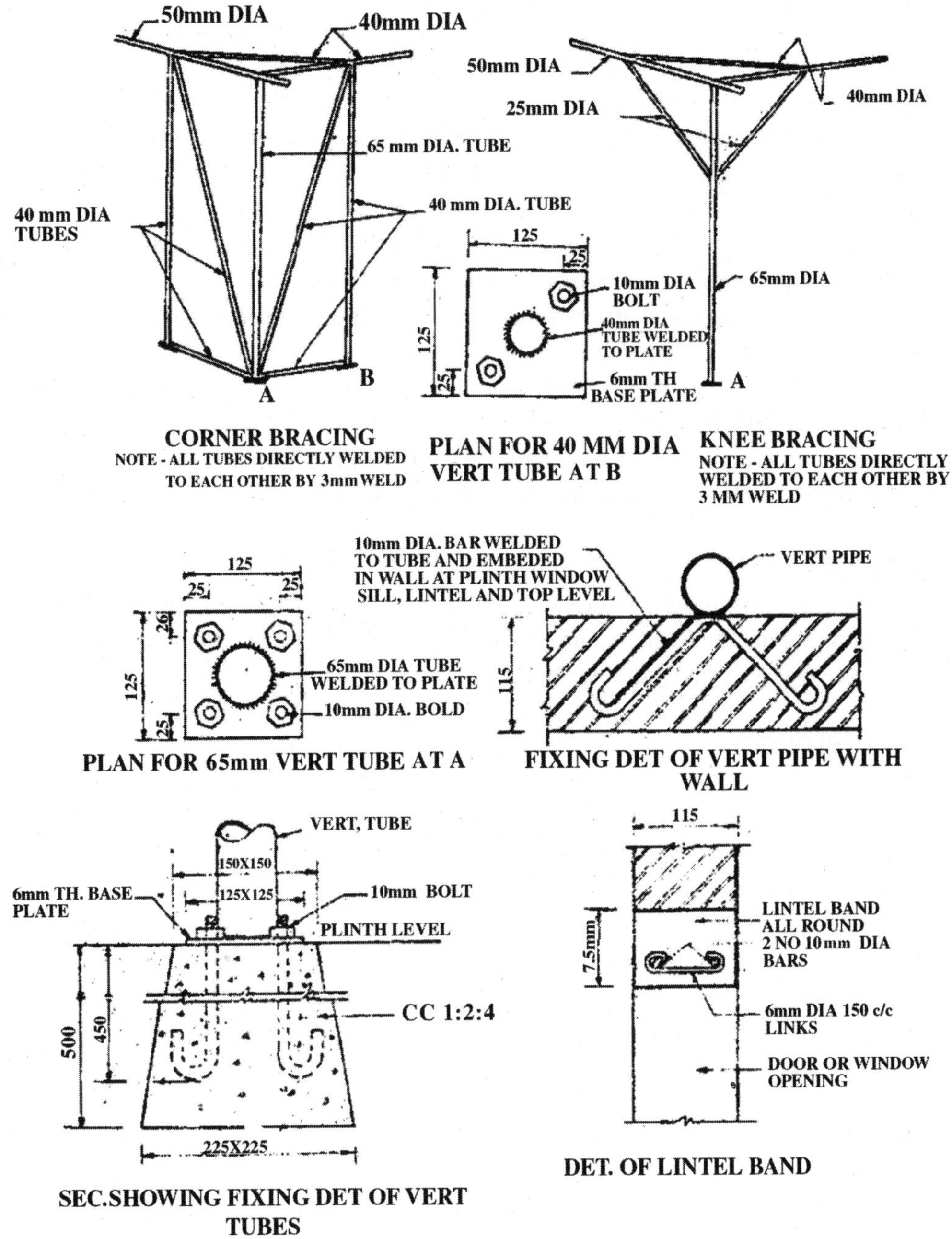

Fig. 10(b) Framing in tubular construction

to be emphasised which may be achieved with flat iron straps.

(d) The wall cladding may be in the form of A.C. or C.G.I. sheets, expanded metal with plaster or timber boards fixed to girts. No special details are necessary. Alternatively, the heavier elements like brick, squared stone, form a concrete block; hollow concrete block, etc. may be used for building the walls. Such panel walls must be constructed with care and according to principles given for masonry buildings. Whereas the sheeting will be suitable for temporary use, the walls will be necessary as a permanent measure and also for providing protection from heat and cold. Details are given in drawings for construction of such walls, which are assumed to be not more than 13 cm thick.

(e) Details for fixing the superstructure to footings columns and bracing members, joints and masonry infill are shown in Fig. (9)a.

11.6 Tubular Steel Construction

Steel pipes can be made use of on a mass scale with speed in place of timber members. Recommended details of this type of construction are shown in Fig. 10.

11.7 Conclusion

Summarising the recommendations for earthquake resistance, it may be stated that the construction should be light, strong against lateral loads, ductile to absorb energy and present its resistance against the earthquake as an integral unit. ■

11.8 References

1. Chandrasekaran, A.R., L.S. Srivastava and A.S. Arya, "Behavior of Structures in Koyna Earthquake of December 11, 1967" Indian Concrete Journal (under publication).
2. Krishna J. and A.S. Arya, "Building Construction in Seismic Zones of India" Second Symposium on Earthquake Engineering, University of Roorkee, November, 1962.
3. Krishna, J. and A.S. Arya, "Earthquake Resistant Design of Buildings" Journal of the Institution of Engineers (India) Vol. XLV, No. 7, March 1965.
4. Arya, A.S., "Design and Construction of Masonry Buildings in Seismic Areas" Bulletin of the Indian Society of Earthquake Technology, Vol. 4, No. 2, April 1967.
5. "Code of Practice for Earthquake Resistant Construction of Buildings" (IS: 4326-1967), ISI, New Delhi, 1967.
6. Agnihotri, V.K. Strength of Single-Storey Brick Shear Walls against Earthquake Forces", M.E. Thesis, University of Roorkee, 1962.
7. Code of Practice for Design of Structural Timber in Building's, (IS: 1966), I.S.I., New Delhi, 1966.

Chapter 12

Experimental Evaluation of Aseismic Strengthening Methods of Brick Buildings[1,2]

12.1 Introduction

The current reinforcing methods for brick or stone masonry buildings of a few storeys such as those given in IS: 4326-1975(7) are based on approximate analyses of shear and transverse walls (Agnihotri, 1962, Krishna &Arya, 1965, Arya, 1967), static lateral load tests on small size shear walls and one-room models (Chandra, 1963, Krishna, Chandra and Kanungo, 1966) and observed behaviour of such buildings in past earthquakes in India and abroad. However, the extent of effectiveness of these reinforcing methods has still not been investigated by dynamic testing nor fully established during severe earthquakes. The aim of this investigation is to test differently reinforced fairly large size specimens of brick enclosures to evaluate the efficiency of the reinforcing against severe base motions (Qamaruddin, 1978). A new concept of making the superstructure capable of sliding over the plinth, when the base acceleration exceeds the coefficient of friction, has also been investigated.

The dynamic testing of the specimens to ultimate failure has been carried out on a special shaking platform 6 m x 7 m in plan area built on an old railway wagon chassis (Keightley, 1977). The arrangement is shown in Fig. 1. Shock loading to this 'Shake Table' is provided by another wagon (loaded with boulder and sand) which rolls down a gentle incline (about 8% grade) giving an impact to the table through a set of springs. Under the first shock, the models are shaken one way and the table gets rolling and runs into collision with the second dead load wagon (Fig. 2). Thus, the structure constructed on the shake table receives several shocks through rebound from stationary wagons. The dead load wagon could be started to on the incline plane by a manually operationable winch and released instantly through a release mechanism.

12.2 Construction of Brick Test Structures

A total of eight half-scale models of single-storeyed brick structures (using half scale bricks) have been tested. The outside dimensions of the structures were 2.17 m x 1.75 m x 1.60 m

1 *Qmaruddin M., A.S.Arya, Chandra B., Deptt. Of Earthquake Engineering, University of Roorkee, Roorkee*

2 *VI Symposium on Earthquake Engineering, University of Roorkee, Oct. 3-7, 1976, vol I, pp 353-359.*

high above the plinth, with 7.6 cm thick reinforced concrete roof slab. Four different structures were symmetrically built on the table at one time. Steel angles 18 x 18 x 2 mm were welded to the table to act as a shear key for a bed of mortar and the lowest course of foundation bricks. The construction of two sets of four models with different strengthening arrangements is briefly described below.

First Set of Brick Structures

Four conventional brick structure models were constructed with reinforced lintel band and cast-in-situ roof slab as follows:

Model No. 1; Built using mud mortar,

Model No. 2: Built in mud mortar and strengthened with a 6 mm diameter vertical steel bar set in cement mortar at each corner and at the jambs of the openings. The model was further strengthened by a reinforced *lintel band.*

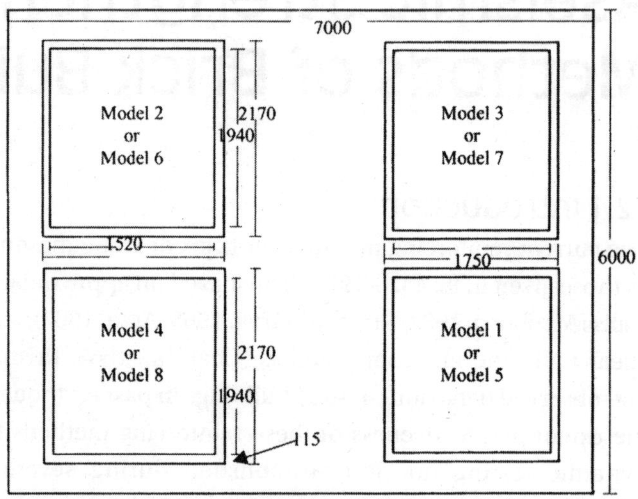

Fig.1:- Layout plan of brick building models constructed on railway wagon shake table facility

Model No.3: Built similar to model No. 1, but in 1:6 cement sand mortar.

Model No. 4: Built using 1:6 cement sand mortar and strengthened similarly as model No. 2.

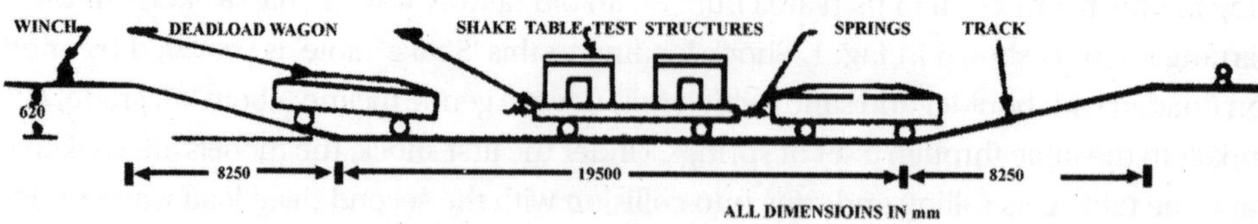

Fig.2:- Schematic view of railway wagon shake table test facility

Second Set of Brick Structures

Two models each of conventional and sliding type structures were made for the second series of tests as described below.

Model No. 5: Constructed in mud mortar with superstructure free to slide at the plinth level where a reinforced concrete band was provided. A reinforced concrete beam was cast above the plinth band by interposing a thin film of used up mobile oil so that a bond could not develop between them. A lintel band was also provided in addition to jamb steel of 6 mm dia. set in cement mortar in the door opening.

Model No. 6: Built similar to model No. 3 but in addition a reinforced concrete lintel band was also provided.

Model No. 7: This sliding type structure was made in 1:6 cement sand mortar. Other provisions in its construction were similar to model No. 5.

Model No. 8: The conventional type model was constructed using 1:6 cement sand mortar with plinth band in addition to other strengthening measures as provided in model No.4.

To keep uniform quality of construction, the same brick-layers were engaged throughout the construction programmes of the models.

12.3 Testing Procedure for Test Structures

The test structures were subjected to shock loads through the impact of the dead load wagon with the shake table, as described earlier. Initially, the amplitude of the shock was kept small by controlling the height of the rolling wagon and was then gradually increased in subsequent shocks by further pulling the dead load wagon up the incline before release. Accelerations were recorded at the table and the important locations of the structures.

After every shock, the damage of the structures was studied and recorded. The cracks developed in the walls were marked by black paint. The shock serial number was also marked along the cracks. Photographs of the damaged structures were also taken for further study.

12.4 Observations and Discussions

The progressive damage of the test structures as seen under increasing intensity of shocks is discussed below separately for each of them. For the first and second sets of four models, nineteen and eight shocks were applied respectively.

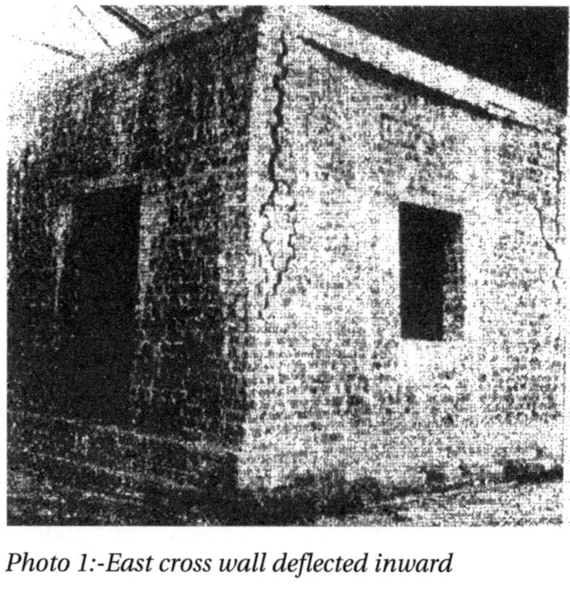

Photo 1:-East cross wall deflected inward

Model No. 1 (Mud mortar, no special reinforcing)

In shock No. 5, vertical cracks near the corners were significantly observed in the cross-walls while a few vertical and diagonal cracks were seen in the shear walls. It seems that the vertical cracks near the corners developed due to the inherent weakness at the corners due to the toothed joint and out of plane vibrations of the cross-walls which finally resulted in separating the adjoining walls during the next shock. The east cross-wall deflected inward by a maximum amount of about 75 mm at its top (photo. 1). During shock No. 8, the shear walls were very much damaged and were nearing collapse whereas the top spandrel with some portions of the piers of the cross-walls had fallen down such that the roof slab was freely supported on the top of the shear walls. The bonding effect of the roof slab was apparent, otherwise for other type of roof the walls would have collapsed much earlier.

Model No. 2 (Mud mortar, lintel band and vertical steel at corners and jambs)

The severe damage of this model began during shock No. 6. During subsequent shocks, the horizontal, vertical and diagonal cracks began to develop mostly in the portion of the wall built in mud mortar due to little tensile and shear strengths of the mud mortar whereas the localised zones of the model in which the reinforcements were provided in cement sand mortar remained unracked. It is surmised that the provision of the lintel band and reinforcements at the critical locations had transformed the beahviour of the structure completely compared to the unstrengthened model No. 1. After shock No. 9, the south-east and south-west corners were heavily damaged, resulting in big gaps (photo.2), between the separated portions as the corners moved outwards diagonally. This type of failure may be due to inadvertent artificial weakness as vertical reinforcement at the corners were first left out from the model Nos. 2 and 4 during their construction. The already set foundation was then broken at these locations for inserting the vertical steel. During the shock Nos. 12 to 14, gaps (of the order of about 200 mm) between the separated portions at the four corners were very much widened, so further testing was stopped.

Photo 2:-Gaps at bottom of South-East and South-West Corners

Model No. 3 (Cement mortar, no strengthening)

After shock No. 6, it was observed that a horizontal crack had developed at the junction of the walls with the foundation and also a very slight sliding of the superstructure was noticed at the plinth level though the model was of conventional construction. This may have been due to poor workmanship as this structure was not fully bonded with the foundation at the plinth level.

Under subsequent shocks, the walls exhibited further horizontal cracks and less vertical and diagonal cracks. A major part of the spandrel and left pier of the west cross-wall had fallen under shock No. 10 and the walls were divided into large separate chunks (Photo. 3) and so further observation of the model was abandoned.

Photo 3:-Major part of west cross-wall fallen

Model No. 4 (Cement mortar, lintel band, vertical steel at corners and jambs)

Under shocks after shock No. 6, mainly horizontal cracks developed while a few diagonal cracks were also observed. Horizontal cracks appeared at the top and bottom of the corners which may be due to the presence of vertical steel at these locations which came into action to rescue the brickwork against tension. After shocks 11 and 12, mostly horizontal cracks of small width developed, well distributed in all the walls. The major damage of the structure was witnessed after shock No. 16. The north-east and north-west corners were pushed outward during shock No. 17 by about 80 and 40 mm respectively whereas these values increased to about 150 mm during the last shock (Photo. 4). Upto the last shock, though the top spandrels of the walls were also damaged, they were not at the verge of collapse, as was exhibited by the portion below the lintel band.

Photo 4:-Badly damaged eorth-east and north-west corners

Model No. 5 (Mud mortar, sliding type, lintel and plinth bands and bond beam)

This sliding type structure in mud mortar exhibited slight damage during the very first shock along with sliding of the superstructure over the plinth band towards west, measured as 10 mm at the south-east corner. During the next shock, which was accidentally severe, the walls developed very wide cracks, resulting in severe damage to the model (crack width ranged from 5 to 20 mm). The model was very heavily damaged under the fourth shock as wide open gaps were observed in the walls. A major portion of the north wall collapsed during the fifth shock and so observations were stopped at this stage.

Model No. 6 (Cement mortar, lintel band)

Severe damage of the walls was observed during the second shock which was rather severe. In the south wall, a vertical crack opening of about 50 mm was seen upstream at the south - west corner during the fourth shock while the other walls also exhibited wide area cracks but to a lesser extent. Under shock No. 5, the bottom portion of the piers of the north wall fell down and the wall broke into many separate blocks. The condition of the other walls was similar and, therefore, further observation was suspended.

Model No. 7 (cement mortar, sliding type, lintel and plinth bands and bond beams)

No damage was observed in the first shock during which the superstructure slid towards west by 4.5 mm. Slight damage was observed in the walls during the second severe shock. The superstructure slid towards west by 10. The model had performed excellently due to its sliding behaviour under a table acceleration of 2.15 g, with only few minor repairable cracks. Since only 0.19g of acceleration was recorded at the roof top, it may be stated that on account of sliding, the superstructure had attracted less lateral force. During the subsequent shocks, the damage to the model had increased with the increase in shock number. Its further observation was abandoned after shock No. 8.

Model No. 8 (Cement mortar, lintel band, vertical steel at corners and jambs)

Slight damage was observed in the walls after shock No. 2 and up to the fifth shock, the fine cracks did appear but to a moderate extent. The bottom regions of the four corners were significantly damaged during the sixth shock. All the walls were damaged due to the development of horizontal and vertical cracks. The presence of vertical steel at the critical sections and lintel and plinth bands, perhaps, may be the cause for such crack pattern. Under the next shock, the model was severely damaged as the old cracks opened upto a width ranging from 5 to 10 mm. During shock No. 8, the bottom portion of the north-east corner shifted outwards by about 20 mm while the north-west corner displaced by about 30 mm leaving a wide gap between the separated portions and, therefore, further testing was stopped.

12.5 Damage Level of the Test Structure

Based on the above qualitative analysis, the relative competence of the models to resist shock has been presented through histograms (Fig. 3) showing the extent of damage during the shocks.

It is seen from Fig. 3(a) that the unstrengthened structure in mud mortar collapsed only at shock 7 while the corresponding strengthened structure exhibited its total damage during the fourteenth shock and the unstrengthened model in cement sand mortar collapsed under shock 10. Thus, the strengthened model in mud mortar appears to perform better than the unstrengthened structure in cement mortar under dynamic loads. On the other hand, the strengthened model in cement sand mortar behaved best as its collapse condition occur after shock 19.

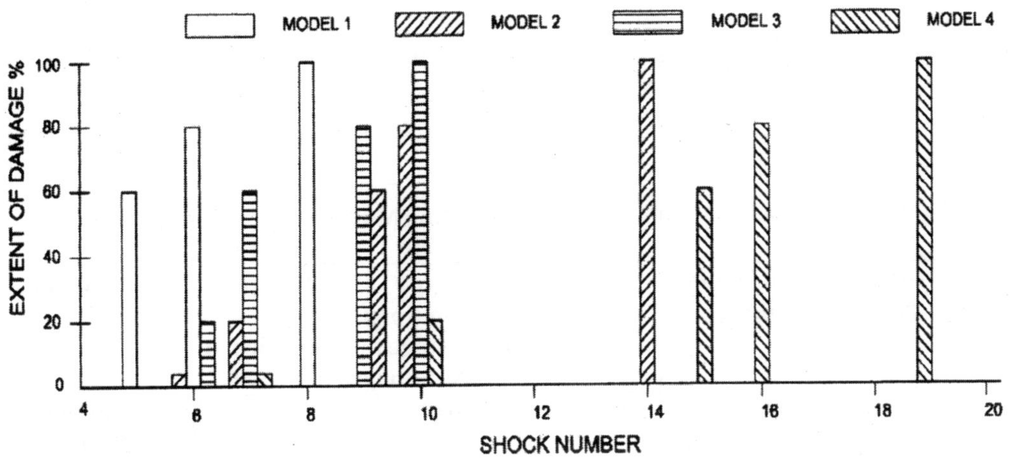

Fig.3(a):-Histogram for extent of damage of first set of structures (damage estimated from the extent of cracking)

The relative performance of the second set of test structure is described in Fig. 3(b). The sliding type structure in mud mortar competed well with that model in cement mortar with the lintel bond but without vertical steel. It is seen from the figure that the sliding type model in cement mortar and the strengthened structure in cement mortar did not collapse up to the last shock though they were very badly damaged. Both structures performed more or less similarly under subsequent shocks.

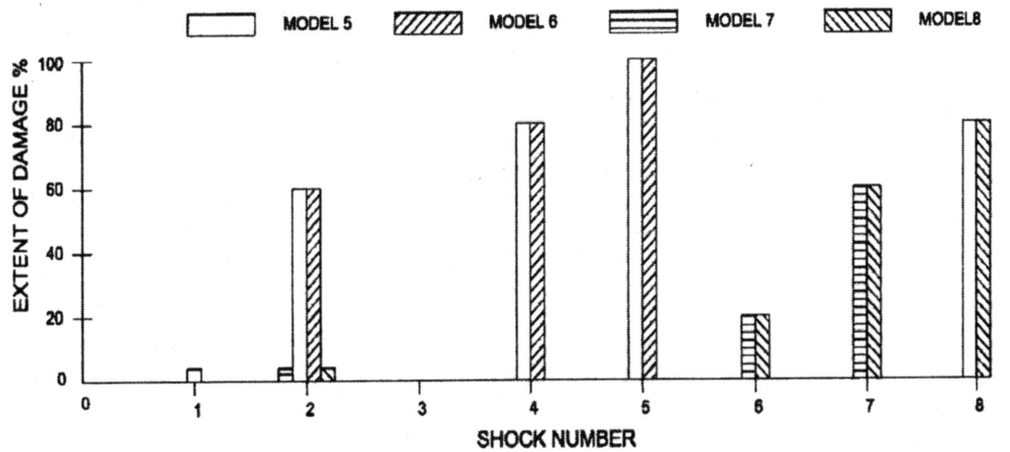

Fig.3(b):-Histogram for extent of damage of second set of structures

12.6 Cost-Benefit Ratio

Cost ratios of the different test structures with their response have been indicated in Table 1 to have a broad quantitative evaluation about the benefit that may be derived from the various types of construction tried during this test programme of the brick buildings*. A test structure in 1:6 cement mortar is chosen as a basis for such comparison. The following significant observations can be made from this table:

Table 1 - Results Of Half Scale Brick Building Model Tested On Railway Wagon Shake Table

Model No.	Description of Model		Cost Ratio	Table Acceleration (in g) at the stage of severe damage	Acceleration (in g) recorded at roof level	Remarks
	Mortar	Reinforced				
1	Mud	No Steel	0.947	0.627	N.O.	Conventional Type Model
2	Mud	LVJ	0.992	1.458	N.O.	Conventional Type Model
3	1:6 Cement Sand	No Steel	1.000	1.000	0.34	Showed sliding behaviour due to faulty construction (conventional type model)
4	1:6 Cement Sand	LVJ	1.045	2.229	0.91	Conventional Type Model
5	Mud	LPB	1.037	S	N.O.	(Sliding type model) Non-uniform sliding observed under walls, max. sliding 18 mm
6	1:6 Cement Sand	L	1.025	S	N.O.	Conventional Type Model
7	1:6 Cement Sand	LPB	1.073	2.843	0.89	Sliding type model, max. sliding 35 mm
8	1:6 Cement Sand	LPVJ	1.078	2.843	2.13	Conventional Type Model

Note: L- Lintel band with 3-6 mmϕ, P = Plinth band with 3-6 mm ϕ, V = Vertical steel at corners 1-6 mm (0.016 % of walls), J = Jamb steel 1-6 mm ϕ (~0.032 % of walls), Bond beam just above plinth band with 3-6 mm ϕ, N.O. = Not observed. S = Severe damage would have occurred at the table acceleration less than 2.17g as shock No. 2 was accidentally severe.

* *These are based on material and labour costs in laboratory and could be much different in construction sites.*

i) Unstrengthened brick structure in mud mortar is not dependable under shock loads.
ii) Performance of strengthened brick structure in mud mortar is better than unstrengthened one in cement mortar with almost the same cost of construction.
iii) Sliding type structure in cement sand mortar is economical vis-à-vis its performance compared to other structures.

12.7 Conclusion

The following conclusions could be drawn from the present study:

(a) A building in mud mortar with strengthening provisions and that with sliding arrangement performed for better than unreinforced one.
(b) A strengthed building even with mud mortar and that with sliding arrangement behaved in a better way than the building built a cement sand mortar but without reinforcing.
(c) The sliding type building in cement mortar compared well with the fully strengthed building in cement mortar.
(d) The tests have strengthened the belief that the brick structure built in cement mortar with reinforcing steel provided at critical sections can withstand major shocks without collapse. ■

12.8 References

1. Agnihotri, V.K. (1962), "Strength of Single-Storey Brick Shear walls Against Earthquake Forces", M.E. Thesis University of Roorkee, Roorkee.
2. Arya, A.S. (1967), "Design and Construction of Masonry Buildings in Seismic Areas", Bull. Indian Society of Earthquake Technology, Roorkee, India, Vol. 4, No. 2, June, pp. 25-37.
3. Chandra, Brijesh (1963), "Strength of Single-Storey Unreinforced Brick Buildings Against Earthquake Forces," M.E.Thesis, Civil Engineering Department, University of Roorkee, Roorkee.
4. Keightley, W.O., (1977), "Report of Indo-U.S. Subcommission on Education and Culture", School of Research and Training in Earthquake Engineering, University of Roorkee, Roorkee.
5. Krishna, J. and Arya, A.S. (1965), "Earthquake Resistant Design of Buildings", Civil Engineering Journal, Institution of Engineers (India), March, pp. 610-627.
6. Krishna, J., Chandra, B. and Kanungo, S.K. (1966), "Behaviour of Load Bearing Brick Shear Walls During Earthquakes", Proceedings of the Third Symposium on Earthquake Engg., Vol. II, University of Roorkee, India, pp. 19-25.
7. IS: 4326-1975, Indian Standard Code of Practice for Earthquake Resistant Construction of Buildings, Indian Standard Institution, New Delhi.
8. Qamaruddin, M. (1978), "Development of Brick Building System for Improved Earthquake Performance," Doctoral Thesis in progress, Unversity of Roorkee, Roorkee.

Chapter 13

Dynamic Testing of Brick Building Models[1,2]

13.1 Introduction

Experience from past earthquakes has repeatedly shown (1-3) that masonry buildings have suffered the maximum catastrophic damage and have also accounted for the maximum loss of life, as compared with other building systems, except for adobe buildings. This suffering is mainly on account of the *heavy weight of masonry* coupled with *poor tensile and shear strengths*. In the recognition that masonry construction will, for a long time to come continue to be the most common mode of building construction in all countries, except perhaps in the relatively more developed ones, it is necessary to evolve methods which incorporate suitable features in masonry buildings so that the latter become adequately earthquake resistant, if not earthquake proof. A great deal of research has been conducted in this area during the last 25 years, and simple means of providing reinforcement for critical sections of masonry buildings have been recommended (4-14). These have not, however, been tested experimentally for actual or simulated ground motion inputs. The experimental investigations described here have been carried out to study the dynamic behaviour of brick building models constructed with different aseismic arrangements.

Some research efforts have lately been directed towards finding isolation techniques whereby the effects of an earthquake on buildings could be minimised. No similar effort had been made until recently in respect of the seismic isolation of masonry buildings, which are usually smaller in size and height but are very widespread. Observations made during the Dhubri Earthquake in Assam in 1930, and the Bihar-Nepal Earthquake in 1934, had indicated that those buildings in which the possibility of movement existed at their points of support suffered less damage than those buildings in which no such freedom of movement existed (15-16). From these observations, a new concept for the construction of brick buildings has

---

1 *M.Qamaruddin,Reader, Department of Civil Engineering, Aligarh Muslim University, Aligarh; B.Chandra and A.S.Arya, Professor, Department of Earthquake Engineering, University of Roorkee, Roorkee.*

2 *Proc. Institution of Civil Engineers, U.K., Part-2, 1984, Paper 8788, 77, September, pp. 353-365*

been proposed in which a clear smoothened surface is created just above the damp-proof course at plinth level, without any mortar, and the superstructure simply rests at this level and is free to slide except for frictional resistance. This type of construction is referred to in this paper as the *'sliding type'.*

The concept of the sliding joint centres around two ideas; firstly, that the force transmitted to the superstructure will be limited to that required to slide at the joints, and secondly, that energy will be dissipated during frictional sliding and will not be passed on to the superstructure. This concept has been examined by theoretical dynamic analysis, using the recorded data of two severe earthquakes, as well as by using half-scale models tested on a shake table up to ultimate strength (17). In this paper, the experimental studies and the results thus obtained are presented.

13.2 Pilot Tests on Sliding Type Building Models

13.2.1 Details of models and testing arrangement

As a first step, an attempt has been made to study the feasibility of the sliding joint concept through one-fourth scale models, as well as through computation of the effect of earthquakes on the sliding type single storey buildings; (17) and, as a second step, large-scale models have been tested.

Two single storey, one-roomed models were constructed in 1:6 cement sand mortar, using bricks of dimensions 114 mm x 57 mm x 38 mm (Fig. 1). A steel channel frame with vertical flanges and a flat web was used to serve as a base for the superstructure. The whole model could be lifted up, for the purpose of inserting layers of different frictional materials (graphite powder and dry and wet sand), with the help of four bolts inserted in the threaded holes. Acceleration pick-ups were mounted at the base of the model, at the top of the roof slab and the shake table top. One displacement pick-up, to measure the

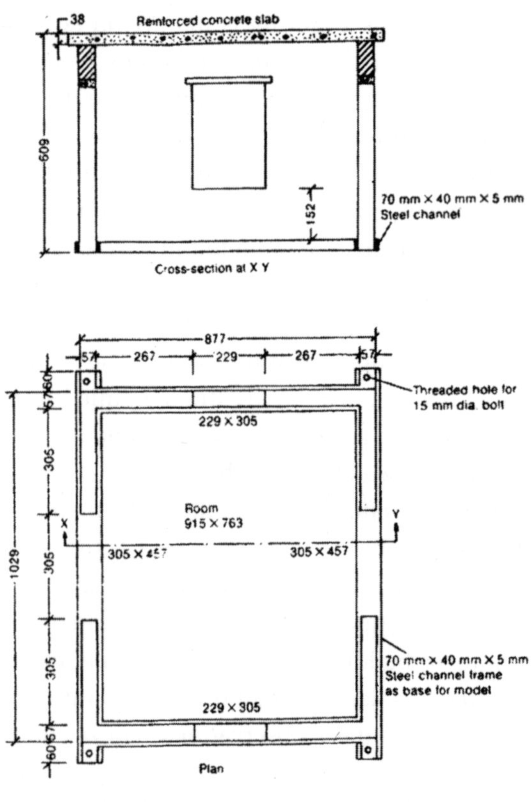

Fig.1: Sliding type, ¼ scale model in cement mortar. (All dimensions in millimeters)

relative displacement between the model base and the shake table, was connected with the model at base level.

The model was tested first with the base free to slide in response to the steady-state horizontal table motion between frequencies 8 Hz to 26 Hz. The accelerations and the amount of sliding were recorded. The testing of the model was continued by applying different dynamic forces as well as by using different sliding materials. After these tests had been completed, the same model was fixed to the table top. Steady-state tests were then performed on the model with its base fixed, using the same method as for the first tests.

13.2.2 Observations

Acceleration records were not taken in the case of model 1, as preliminary tests were performed for qualitative study. The results of dynamic tests performed on model 2 are listed in Table 1.

Table 1. Results of Pilot Test Model (one-fourth scale)

Shock No.	Type of model base	Materials used to permit sliding	Coefficient of friction	Acceleration Record at		Ratio of roof to base acceleration	Amount of sliding at table base, mm
				Table base g	Roof top g		
1	Free	Graphite powder	0.25	0.32	0.20	0.625	2.0
2	Free	Dry sand	0.34	0.86	0.60	0.70	0.5
3	Free	Wet sand	0.41	0.86	0.77	0.90	0.5
4	Fixed	-	-	0.38	0.89	2.34	-

This table shows that in the case of the model with the free base, roof acceleration in each instance is less than the corresponding table acceleration. Also, as the coefficient of friction increases from 0.25 to 0.41, the ratio of roof to base acceleration increases from 0.625 to 0.90, and in the case of the fixed base, it becomes 2.34. These observations indicate that the isolation resulting from sliding had an appropriate influence and, also, that the input energy at the base was dissipated to some extent in the sliding of the model. In the case of the model with the fixed base, roof acceleration is greater than table acceleration, which clearly indicates amplification of the motion in response to stiffness of the moael.

For a given dynamic force, no damage has been observed to those models with a free base while the opposite is true in the case of the model with a fixed base.

Thus, these preliminary tests on sliding type models strengthen the idea that by introducing a sliding joint at the plinth level, the effective seismic force can be reduced as compared with fixed base structures.

13.3 Tests on Half-Size Building Models

13.3.1 Test models

Eight brick building models were tested, six of these being conventional, i.e. with usual mortar joints, and two with a sliding joint at the plinth level. These models were tested in two sets of four, each set simultaneously. Fig. 2 shows dimensions, openings in walls and the position of horizontal and vertical steel bars.

The mortar and reinforcement were varied in the different models as presented in Table 2. To achieve a proper sliding joint, a reinforced concrete band was constructed at the plinth level, to replace the damp-proof course as part of the foundation masonry, and it was given a smooth finish at the top. Used mobil oil was then painted over it and a reinforced concrete band, hereafter termed *'bond beam'*, was cast all round except through the door opening. The superstructure masonry was constructed on the *bond beam* (Fig. 3). All models had a reinforced concrete slab roof bonded to the masonry by mortar, and the vertical corner steel was anchored into the roof slab.

13.3.2 Shaking platform

Fig. 4 is a schematic view of the testing facility. The shaking platform was fabricated on an old railway wagon chassis placed in the middle of a level portion of track. The platform, 7 m x 6 m in plan, carries helical springs on both sides to moderate the shock from a loaded wagon on either side which roll down an incline from a predetermined height. In the actual testing programme, four models were constructed on the platform.One weighted wagon was stationed separate from the platform but close to it, and the other weighted wagon was pulled up the incline by means of a

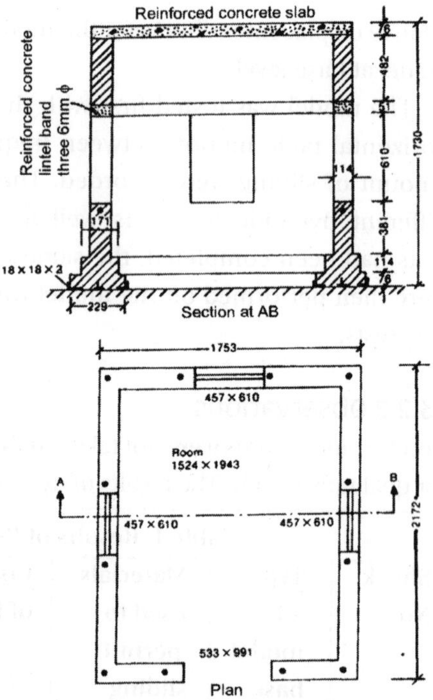

Fig.2: Brick building, ½ scale model in cement/mud mortar with lintel band and steel at corners and jambs

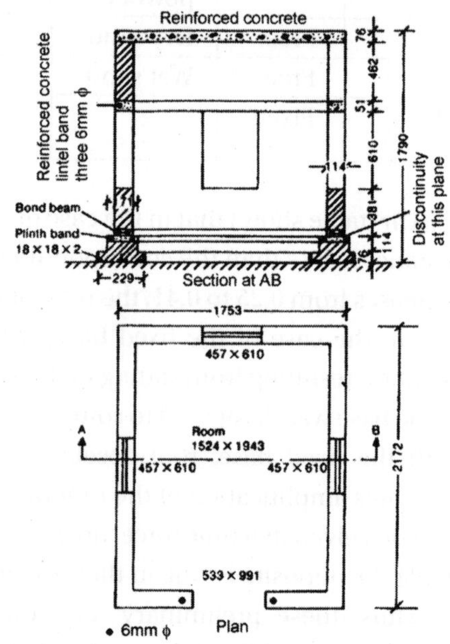

Fig.3: Sliding type brick building, ½ scale model in cement/mud mortar.

Table 2:- Details of half-scale models			
Model No.	**Type**	**Details of mortar and reinforcement**	**Remarks**
1	Conventional	Mud mortar. No. reinforcing	(a) Good burnt bricks made to half scale, i.e. approx. 112 mm x 56 mm x 38 mm, were used in the construction of models. Concrete in band and roof slab was M 150 (b) Lintel and plinth bands consisted of three bars 6 mm dia. Placed in 38 mm thick concrete over all walls. (c) Bond beam contained reinforced as lintel and plinth bands. (d) Vertical steel at corners anchored in foundation masonry and roof slab. (e) Vertical steel at jambs anchored into lintel band and bond beam in sliding type, otherwise in plinth band.
2.	Conventional	Mud mortar. Lintel band. 6 mm dia. vertical bars at corners of walls and at jambs of openings set in cement.	
3.	Conventional	Cement sand mortar 1:6. No reinforcing.	
4.	Conventiona	Cement sand mortar. Lintel band and vertical steel as in model 2	
5.	Sliding	Mud mortar. Lintel band, 6 mm dia. vertical bars at jambs of openings. Plinth band and bond beam.	
6.	Conventional	Cement sand mortar 1:6. No reinforcing.	
7.	Sliding	Cement sand mortar 1:6. Reinforcing similar to model 5.	
8.	Conventional	Cement sand mortar 1:6. Lintel and plinth bands. Vertical steel bars at corners of walls and at jambs of openings.	

winch and released through a clutch arrangement. The wagon imparted a *half-sine pulse* to the platform which after rolling received another *half-sine pulse in the reverse direction* from the stationary wagon. Horizontal accelerations were measured at the base and top levels of the models. The dynamic action was increased by increasing the height of fall of the loaded wagon. The strikes were applied from two opposite directions alternatively. In this way, the models built on the platform received several shocks through initial impacts and subsequent rebounds.

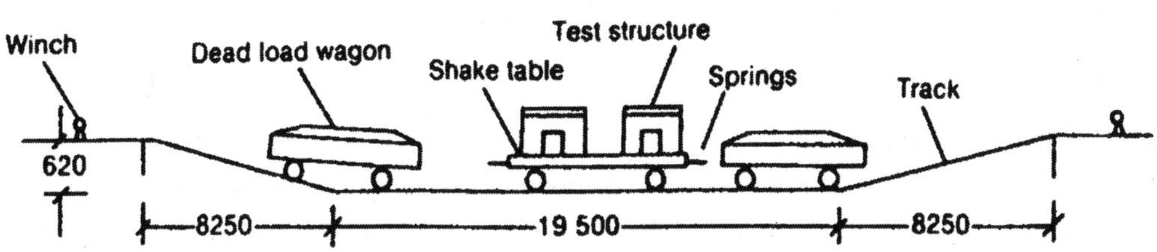

Fig.4: Shake table testing arrangement.

13.3.3 Damage study

The model structures were tested to an ultimate state through gradually increased shocks. The damage to the models, i.e. cracking deformation and displacements, was studied and recorded. The cracks which developed in the walls were given an identifying mark. The shock serial number was also marked along side the cracks so as to differentiate them from subsequent cracks and also to study their progressive widening in further shocks. The displacements of the sliding type models over the plinth band were measured from a reference line marked on the four corners of the plinth band after every shock. Photographs of the damaged models were also taken for further reference in the study.

13.4 Relative Ability of the Models to Resist Shock Loads

Two parameters are identified for a quantitative measure of the damage resisting capability of any of the model structures.

(a) The input energy is chosen as the parameter for defining the dynamic action on the models. Its merit is that the effect of all the shocks, rebounds, etc. could be taken into account by scaler addition. As the base motion imparted to the table consisted of distinct half-wave shocks, the motion almost dying completely before the other shock or rebound was given, the input energy, U, in each shock could be computed as

$$U_t = 0.5m_t V_t^2 \quad (1)$$

where m_t is the total mass of the shaking platform including models, and V_t is the velocity of the platform attained at the end of the shock. Thus, the energy input per unit mass U is

$$U = 0.5V_t^2 \quad (2)$$

(b) For defining the damage level of the model, the parameter chosen is the ratio of the area of cracks in a wall to the total area of the wall in elevation. The area of a crack is equal to its length multiplied by the wall thickness. In this definition, the crack width which would have indicated the relative movement of masonry blocks and the energy absorption over and above the energy absorbed in cracking, has been neglected because the precise determination of the crack widths as well as of the movement of masonry blocks posed an insurmountable problem. To obtain an idea of the relative damage, the adopted definition was considered to be adequate.

13.5 Observations

The two parameters as defined above have been computed for each shock and for all the models. The cumulative extent of the damage is plotted against the cumulative input energy per unit mass for both sets of test structures in Fig. 5, from which the following significant observations are made.

(i) The extent of damage to all the models increases with the increase in input energy, whether this is applied in one big shock or in a number of smaller shocks, provided that each shock exceeds the damage threshold.

(ii) From the models of the first set given in Fig. 5(a), it can be seen that for a given amount of input energy, the percentage of damage is the least in the case of the strengthened structure built in cement sand mortar (model 4); then comes the strengthened structure in mud mortar (model 2); which is followed by the unstrengthened structure in cement mortar (model 3). The unstrengthened model in mud mortar (model 1) is the weakest structure among the first set of structures.

(iii) From the second set of models given in Fig. 5(b), it is observed that in the initial stages, as the input energy per unit mass increases, the performance of the sliding type and strengthened structures in cement mortar (models 7 and 8) is similar, whereas the sliding type model built in mud mortar (model 5) shows less damage compared to the conventional structure in cement (model 6). But in the final stages of shocks, models 5 and 6 show a similar extent of damage, while the sliding type model 7 in cement mortar shows less damage compared to the strengthened model 8.

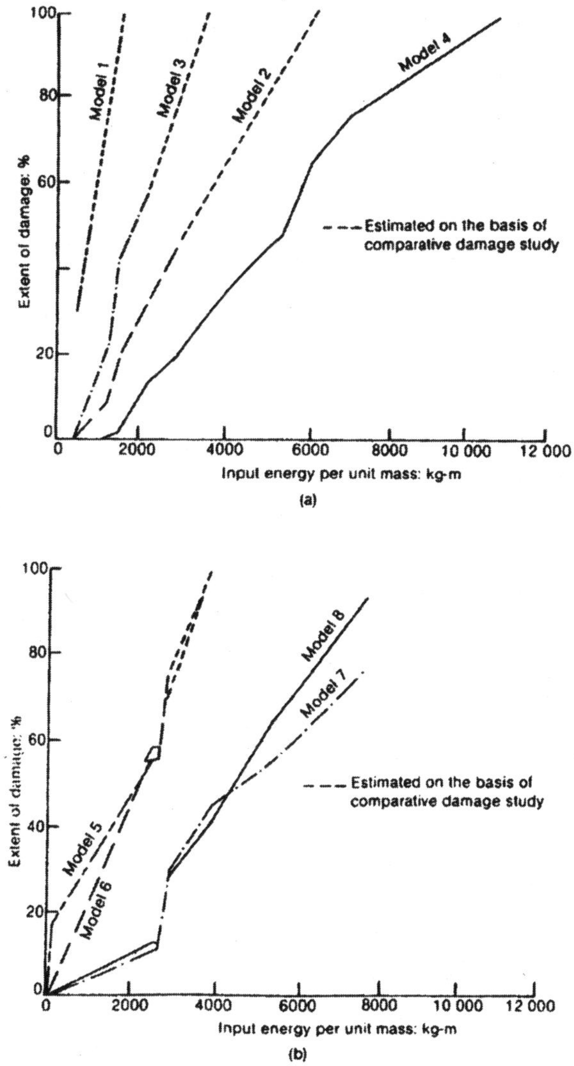

Fig.5: Extent of damage against input energy in ½ size brick building models for (a) first set of models: (b) second set of models

The accelerations recorded under various shocks at the base and roof of models 7 and 8 showed clearly that under all shocks, the roof acceleration of the sliding type model is generally much smaller than that of the conventional model. It was observed that during the main shocks and rebounds, the residual displacement in the sliding type model alternated with the alternating shocks and there was no cumulative displacement in any direction. During an actual earthquake, which generally consisted of alternating impulses, similar displacement behaviour is expected.

13.6 Prediction of Earthquake Shock for Prototype Buildings

An attempt has been made here to predict the peak ground acceleration of an earthquake which would cause the same threshold damage to the prototype structure as was caused to the model structures by the shake table motion. For this purpose, it is assumed that the brick work of the model has a tensile strength 10% higher than that of the prototype (17). Model 8 has been selected for this study because its response acceleration values are available in an undamped condition under recorded table motion. For the elastic condition, the spectral acceleration values of the prototype earthquake motion are also available. Prediction equations for the seismic coefficient applicable to the prototype are derived as follows.

The pier method of analysis (18) is employed here for computing tensile stresses in the building elements of the prototype and model structures. The net tensile stress in a pier is determined by superimposing the axial stresses resulting from dead and live loads, and the bending and overturning stresses (19) resulting from an earthquake load. A typical pier of the model and the corresponding pier of the prototype are considered here for the analysis. The analysis is assumed to be applicable to other piers. The different stresses are computed for model 8 and by superimposing them, the net tensile stress (in kg/cm^2) p_{tm}, is

$$p_{tm} = (3.14\ C_m - 0.25) \qquad 3)$$

where C_m is the seismic coefficient for the model. Similarly, the net tensile stress p_{tp} in the prototype is computed as

$$p_{tp} = \lambda\ (3.14\ C_p - 0.25) \qquad (4)$$

where C_p is the seismic coefficient for the prototype and λ is the scale ratio, 2.0 in this case.

Using equations (3) and (4), the following prediction equation is obtained for the seismic coefficient of the prototype

$$C_p = 0.46C_m + 0.043 \qquad (5)$$

The spectral acceleration S'_a of the prototype for the shake table may be computed by

$$S'_a = \frac{C_p}{C_m} S_{am} \qquad (6)$$

where S_{am} is the spectral acceleration of the model for the table motion.

Knowing the time period and the appropriate damping value, the spectral acceleration of the prototype S_{ap} can be found for the prototype earthquake. The ratio of S'_a to S_{ap} will indicate the scaling of the peak ground acceleration of the prototype accelerogram to suit the threshold damage of the prototype. Thus, the peak ground acceleration for the prototype structure could be related to the maximum base acceleration of the model for the threshold damage as follows:

$$\alpha_p = \frac{C_p}{C_m}\left(\frac{S_{am}}{S_{ap}}\right)\alpha \qquad (7)$$

in which α_p = scaled peak ground acceleration of the prototype earthquake accelerogram corresponding to the threshold damage of prototype; and α = actual peak ground acceleration of the prototype earthquake motion.

As an illustration, the Koyna earthquake is used here for predicting the peak ground accelerogram for the threshold damage of the prototype of model 8. For the Koyna shock, the peak ground acceleration, α = 0.63 g. Using the prediction equation (5), the value of Cp for the prototype is evaluated as 0.312. With the help of equation (7), the value of α_p is computed for the prototype. It works out to be 0.16 g.

13.7 Summary of Significant Results

The percentage damage in a strengthened structure in cement mortar (model 4) was much less compared to that in a similarly strengthened model in mud mortar (model 2) for the same amount of input energy.

The strengthened structure in mud mortar (model 2) was not only stronger than the corresponding unstrengthened model, but was also capable of withstanding four times as much input energy as the unstrengthened structure.

The strengthened model in mud mortar (model 2) performed better than the unstrengthened structure in cement mortar (model 3) although their respective construction costs work out almost the same.

The sliding base (model 7) and fully strengthened (model 8) structures had not reached their total damage level even after the last shock had been imparted during the tests.

As the number of shocks and the damage level increased, the amplification of acceleration in model 4 generally decreased, indicating decreasing stiffness and elongation of period. Similar trends were also observed in models 5 to 8. The roof acceleration of the sliding type model was noticeably less compared to the base motion and the roof acceleration of other models.

The dynamic behaviour of those structures strengthened with steel or provided with a sliding facility at the plinth, was considerably improved both in terms of resistance to base acceleration and capacity to dissipate input energy. This improved behaviour was achieved at little increase in the cost of their construction.

The attempt made to predict the damage threshold acceleration of the prototype building from the observed damage threshold of models, indicated that a Koyna type accelerogram, with the maximum peak scaled down to 0.16 g, would be needed just to cause cracking in brick buildings of good quality construction in 1:6 cement sand mortar.

13.8 Conclusions

Unstrengthened brick buildings of conventional construction show a varied strength and energy capacity depending on the mortar bond strength. Also, the damage threshold acceleration is seen to be quite low as predicted for the prototype buildings from the model tests. Hence, without strengthening measures they could be used only in areas of low to medium seismicity, and in low height construction.

The provision of a lintel band and steel at the corners and junctions of walls as well as at the jambs of openings is generally adequate in imparting the desired strength and energy absorbing capacity, even for the probable maximum earthquakes.

In buildings with the sliding arrangement provided at plinth level, with not too high a coefficient of friction, there is a significant reduction in the effective seismic force acting on the superstructure, to the extent that the cracking and damage are much reduced and the behaviour is better than that of the buildings with strengthening measures. The maximum dynamic and permanent displacements of the superstructure relative to the base, even during the maximum probable earthquakes, are likely to be small. ■

13.9 References

1. Arya, A.S. et al, "Influence of Natural Disasters (earthquakes) on Educational Facilties, Part I-Collection and analysis of available information-annotated bibliography". Report of School of Research and Training in Earthquake Engineering, University of Roorkee, Roorkee, 1977.
2. Monge J.E., "Seismic Behaviour and Design of Small Buildings in Chile", Proc. 4th World Conf. on Earthquake Engineering, Chile, 1969.
3. Streinbrugge K.V. and Flores R, "The Chilean Earthquake of May 1960, a Structural Engineering View Point", Bull. Seism. Soc., Am., 1963, 53, No.2.
4. Moinfar A.A, "Earthquake Resistant Design of Brick Masonry Buildings", European Symposium on Earthquake Engineering, Imperial College, London, 1972.
5. Plummer H.C. and Blue J.A., "Reinforced Brick Masonry – Lateral Force Design". Structural Clay Products Institute, Washington, 1953.
6. Yorkdale A.H., "Masonry Building Systems – State of the Art Report No. 9", Technical Committee No. 3. Int. Conf. on Planning and Design of Tall Buildings, Lehigh University, 1970.
7. Krishna J. and Arya A.S., "Earthquake Resistant Design of Buildings", J. Instn. Engrs India, 1965, 7, March, 610-627.

8. Arya A.S., "Design and Construction of Masonry Buildings in Seismic Areas", Bull. Indian Soc. Earthquake Technol., 1967, 4, No. 2, June, 25-37.

9. Arya A.S., "Construction of Small Buildings in Seismic Area", Bull. Indian Soc., Earthquake Technol., 1968, 5, Nos. 3 and 4.

10 Krishna J. and Chandra B., "Strengthening of Brick Buildings Against Earthquake Forces", Proc. 3rd World Conf. on Earthquake Engineering, New Zealand, 1965.

11. Krishna J. and Chandra B., "Strengthening of Brick Buildings in Seismic Zones. Proc." 4th World Conf. on Earthquake Engineering, Chile, 1969.

12. Chandra B. and Kumar K., "Earthquake Resistant Construction of Brick Buildings." Jai Krishna 60th Birth Anniversary Commemoration Volume, Sarita Prakashan, Meerut, 1974.

13. International Association of Earthquake Engineering, "Earthquake Resistant Regulations - a world list." International Association of Earthquake, New York, 1973.

14. Indian Standards Institution., "Code of Practice for Earthquake Resistant Design and Construction of Buildings.," New Delhi, 1977, IS: 4326-1976.

15. Joshi R.N., "Striking Behaviour of Structures in Assam Earthquakes," Proc. 2nd World Conf. on Earthquake Engineering, Tokya and Kyoto, 1960.

16. Gee E.R., "Dhubri Earthquake of 1930," Mem. Goel. Surp. India, 1934, No. LXV.

17. Qamaruddin M. "Development of Brick Building Systems for Improved Earthquake Performance," Ph.D. Thesis, University of Roorkee, 1978.

18. "Analysis of Small Reinforced Concrete Buildings for Earthquake Forces," Portland Cement Association, Chicago, 1955.

19. Blume J.A. et al., "Design of Multistory Reinforced Concrete Buildings for Earthquake Motions," Portland Cement Associations, Illinois, 1961.

Chapter 14

A New Concept for Resistance of Masonry Buildings in Severe Earthquake Shocks[1,2]

14.1 Introduction

Experience of past earthquakes shows(1) that masonry buildings suffer the maximum catastrophic damage and account for the maximum loss of life compared to other building systems, except the adobe buildings. This is mainly due to the heavy weight of masonry coupled with poor tensile and shear strength. In spite of the observed earthquake behaviour, masonry continues to be the most popular building material in almost all part of the world because of the easy availability of material, tradition of construction, economy as well as its excellent insulation properties. It is, therefore, necessary that methods should be evolved for incorporating suitable features in masonry buildings to render them adequately earthquake resistance, if not earthquake proof. Also, these features should be inexpensive. Many research efforts have been made in this direction in the last 20 years and simple means of providing reinforcement at critical sections of masonry buildings have been recommended (2-7). These recommendations include provision of (i) horizontal ring beams (or variously called bands or bond beams) to be installed at plinth, lintel and roof levels of the buildings, and (ii) vertical steel bars at the corners and junctions of wall, and around door and window openings. In some countries dowel bars are provided at the corners instead of, or in addition to, the continuous horizontal bands (8). Detailed reinforcing provisions have been formulated in many codes of the world, the most notable among them being the Indian Code (9). These provisions are aimed at achieving collapse proof buildings during the probable earthquakes in the various seismic zones and do not guarantee protection against severe damage which could occur in these buildings during large earthquakes.

Recently, some research efforts have been directed towards finding isolation techniques so that the earthquake effects on tall buildings could be minimised. No such effort has been made for the seismic isolation of masonry buildings which are usually small in size and height but large in number. Recognising this, the authors have tried a new technique for

---

1 *A.S.Arya, B. Chandra, Professors, Univers. of Roorkee, Roorkee and M.Qamaruddin, Reader. Aligarh Muslim University*

2 *Journal of Institution of Engineers (I), Vol.61, Pt. Cl.6, May 1981, pp 302-308.*

reducing the earthquake force on masonry buildings by providing a well defined sliding joint between the superstructure of the building and the foundation at the plinth level. Observations during the Dhubri earthquake in 1930 in Assam and the Bihar-Nepal earthquake in 1934 had indicated that some buildings in which possibility of movements existed at their points of support suffered less damage than those in which such freedom of movement was not available (10,11). The concept of the sliding joint centres around the ideas that (i) the force transmitted to the superstructure will be limited to those required to slide the joints, i.e., a sort of isolation will be incorporated in the building; and (ii) energy will be dissipated during frictional sliding and will not be passed on to the superstructure. The concept was examined by theoretical dynamic analysis using two actually recorded severe ground motions as well as by half size specimens tested on shake table up to the ultimate strength (12). This paper presents the analytical and experimental studies and their results.

14.2 Mathematical Idealisation of Sliding Type Building

The concept of the sliding joint at plinth level is shown in Fig. 1(a), through a single-storeyed brick building. It is assumed that a layer of suitable material with a known coefficient of friction is laid between the contact surfaces of the bond beam of the superstructure and plinth in the substructure. For computing the earthquake response of this type of buildings, the building is idealised as a two degrees of freedom discrete mass model, as shown in Fig. 1(b). The spring action in the system is assumed to be provided by the shear walls, that is, those resisting shear forces parallel to the direction of the earthquake shock. Internal damping is represented by a dashpot in parallel with the spring. The mass of the roof slab and of one-half the height of walls is lumped at the roof level and one-half the mass of walls is lumped at the level of the bond beam. The two-mass spring system is then permitted to slide at the plinth level.

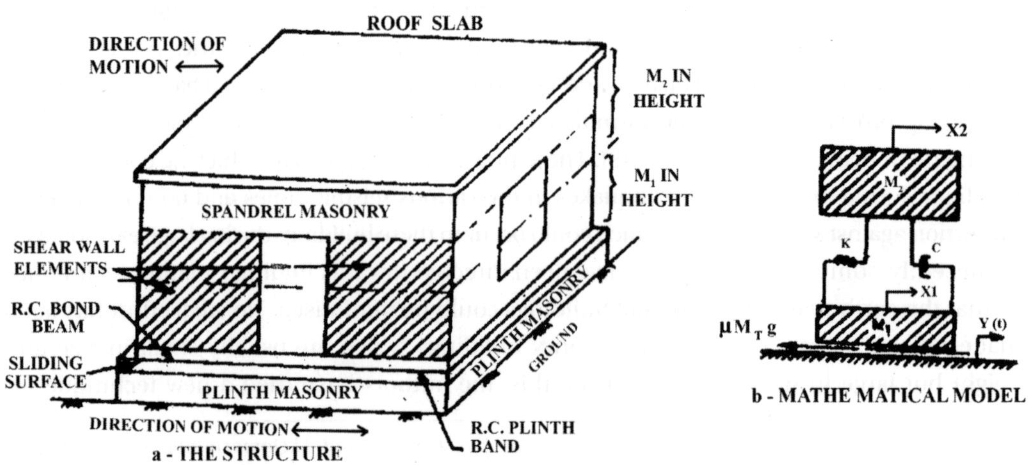

Fig. 1:- The sliding concept

14.2.1 Assumptions

The following assumptions are made for analysing the system of Fig. 1(b):

i. The coefficient of friction between the sliding surfaces remains constant throughout the motion of the system, right from the start to the stop, that is, a rigid plastic idealisation is assumed.

ii. The building material is assumed to be elastic and remaining within the limit of proportionality. Thus, the idealised spring is linear-elastic. Its stiffness is worked out by considering bending as well as shear deformations in the wall elements.

iii. The building is subjected to only one horizontal component of ground shaking at a time.

iv. Unlimited sliding displacement can occur at the contact surface without overturning or tilting. In a real structure, the displacement so calculated will be a parameter for design that is, suitable arrangement will have to be made to accommodate the same.

14.2.3 Equations Of Motion

There are three different phases in the complete motion history of the sliding system due to the frictional resistance at its base. The equations of motion are derived as follows.

Phase I

Initially, so long as the acceleration of the moving system does not overcome the frictional resistance, mass Mb moves with the base since these is no sliding, and the system behaves as a single degree freedom system. Therefore, the equation of motion is

$$M_t \ddot{x}_t + C_s (\dot{Z}_t - \dot{Z}_b) + K_s (Z_t - Z_b) = 0 \quad (1)$$

or,

$$\ddot{Z}_t + 2p\xi(\dot{Z}_t - \dot{Z}_b) + p^2 (Z_t - Z_b) = -\ddot{y}(t) \quad (2)$$

Phase II

The sliding of the bottom mass begins when the frictional resistance at the plinth level is overcome by the force which causes sliding. The force to cause sliding (S_F) is given by

$$S_F = C_s (\dot{Z}_t - \dot{Z}_b) + K_s (Z_t - Z_b) - M_b \ddot{x}_b \quad (3a)$$

Sliding of bottom mass occurs if

$$|S_F| > \mu M_T g \quad (3b)$$

The system now acts as two degrees of freedom system for which the equations of motion can be written in a simplified form as

$$\ddot{Z}_b - 2p\xi\theta(\dot{Z}_t - \dot{Z}_b) - p^2\theta(Z_t - Z_b) + F = -\ddot{y}(t) \tag{4}$$

and

$$\ddot{Z}_t + 2p\xi(\dot{Z}_t - \dot{Z}_b) + p^2(Z_t - Z_b) = -\ddot{y}(t) \tag{5}$$

where

$$F = \mu g(1+\theta)\,\mathrm{sgn}(Z_b) \tag{6}$$

In equation (6),

$\mathrm{sgn}(\dot{Z}_b) = +1$ if $\dot{Z}_b$ is positive,

$\mathrm{sgn}(\dot{Z}_b) = -1$ if $\dot{Z}_b$ is negative.

Phase III

At any instant of time during the motion of the system, if $|S_F| < \mu M_T g$, then the sliding of the bottom mass is stopped whereas the top mass continues to vibrate. Therefore, again the system becomes a single degree of freedom system and, hence, its motion would be expressed by equation (2).

Throughout the time history of the ground shaking, the bottom mass of the system either stops or continues to slide according to the conditions enumerated earlier. The equations of motion for the different phases were solved numerically for obtaining the complete seismic response. The Runge-Kutta fourth order method was used for numerical integration.

14.3 Parametric Study of Sliding Type Building

For estimating realistic forces and displacements of sliding type buildings, the response has been computed for two actually recorded severe accelerograms – the El Centro shock of May 18, 1940 (N-S component) and the Koyna shock of December 11, 1967 (longitudinal component). A range of parametric values representing the physical properties of the single storey building has been used to arrive at generalised results. The results of this study are then compared with those of associated conventional non-sliding buildings subjected to the same ground shaking.

In equations (2), (4) and (5), there are three dimensionless parameters p (or inversely time period T), ξ and θ which determine the dynamic response of the system to a given ground motion. In actual buildings, these values vary depending on the conditions of planning,

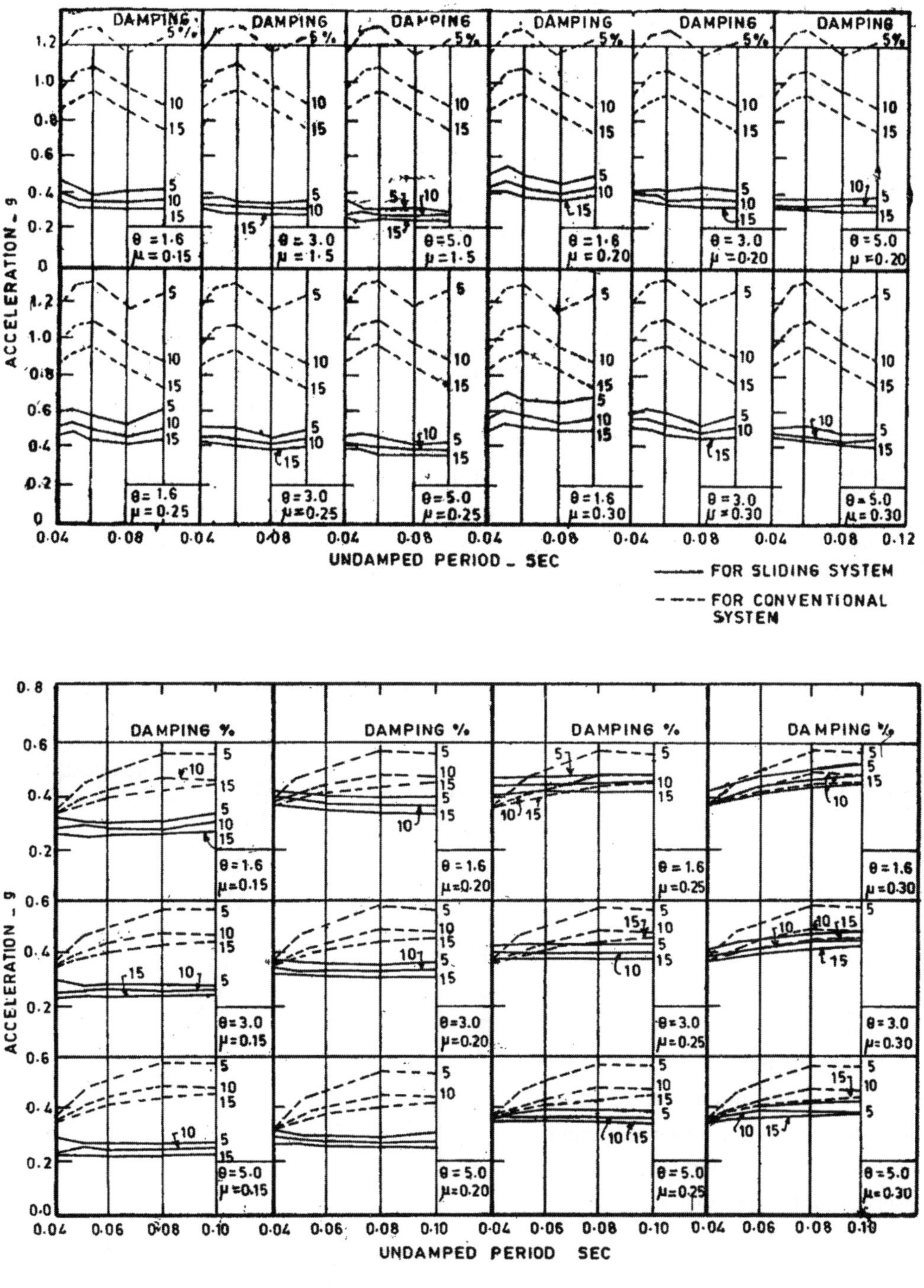

Fig.2: Frictional response acceleration spectra: Koyna Earthquake of December 11, 1967 (top) and El Centro earthquake of 1940 (bottom)

materials and quality of construction. It is estimated that the following range of values should cover a wide variety of single storeyed masonry structures.

Time period, T = 0.04, 0.05, 0.06, 0.08, 0.10 sec.

Damping value, ξ = 0.05, 0.10, 0.15

Mass ratio, θ = 1.6, 1.8, 2.0, 3.0, 4.0, 5.0

The coefficient of friction μ is an important parameter which affects the response of the sliding system considerably. In this study, the values of μ used were 0.15, 0.20, 0.25, 0.30 and 0.40.

It is assumed that a coefficient of friction less than 0.15 in sliding will be difficult to obtain and for a value greater than 0.40, no sliding motion may occur in most real earthquakes and the system may just act like a fixed system.

The response of the system was computed for the various combinations of the above parameters. Some of the results are presented in Fig. 2, in terms of the absolute acceleration of the top mass M_t and undamped natural period T for various parameter values. These may be termed *'frictional acceleration response spectra'*. The acceleration responses for similar conventional type single degree of freedom systems are also plotted for direct comparison. The maximum and residual relative displacements of the bottom mass are shown in Fig. 3.

14.3.1 Acceleration Response

The frictional acceleration response spectra (Fig. 2) show the influence of various parameters on the maximum response of the systems when subjected to the Koyna and El Centro shocks. The dashed lines in these figures show the acceleration spectra for conventional buildings having the same period and damping but fixed at the base. Unlike the conventional systems, the frictional spectra are generally flat and the value does not change much as the period of the system as well as other parameters are varied. Only slight variation is observed for higher range of coefficients of friction, that is $\mu > 0.3$

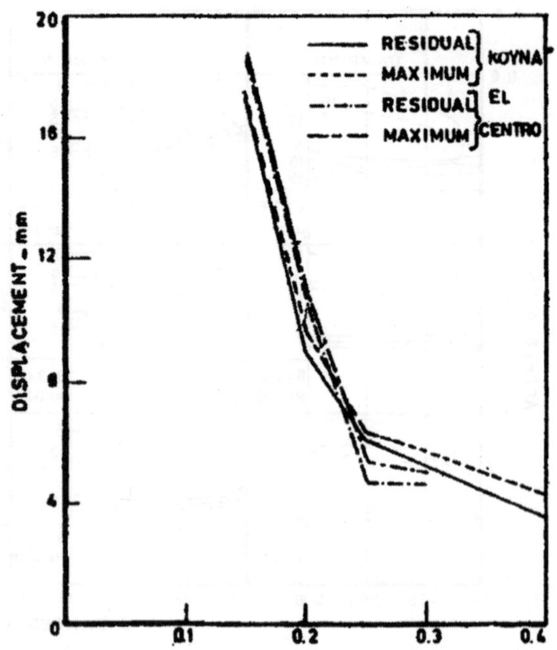

Fig.3: Variation of maximum and residual displacement with friction coefficient.

In a friction mounted rigid system, if ground acceleration coefficient at any instant of time exceeds the coefficient of friction, the rigid body begins to slide, the limiting force remaining equal to mass times this threshold acceleration (13). Thus, the response of the system is independent of its time period for different values of coefficient of friction. Since only short period structures have been studied, it is logical that the acceleration response of such structures should not depend much on their period and only slight variation can be expected.

Influence of Viscous Damping

An increase in viscous damping decreases the spectral acceleration of the sliding systems. This is the usual result as for conventional spectra and indicates the increasing energy dissipation in internal friction of the system as the damping coefficient increases.

Effect of Coefficient of Friction

In all parameter combinations, the spectral acceleration decreases as the friction coefficient decreases. This is logical since the resistance against the sliding of the system decreases as the coefficient of friction between the sliding surfaces decreases, and a build up of larger inertia force in the superstructure gets restricted. Thus, the spectral acceleration of such a system decreases. Theoretically, if the coefficient of friction is reduced to zero, no inertia forces will be transmitted to the building.

Effect of Mass Ratio

As the ratio of total mass to the base mass increases, the spectral acceleration decreases in all cases of parameter combinations for both the earthquakes. This means that for a given time period and damping, an increase in the top mass relative to the base mass will lead to reduction in spectral acceleration.

14.3.2 Comparison of Response Of Conventional And Sliding Systems

From Fig. 2 it is observed that the spectral acceleration of the sliding system is much less than that of the corresponding conventional system subjected to the Koyna shock for all parameter combinations. The same trend is generally seen in the case of the El Centro shock too, the only exceptions are the systems having period less than 0.05 sec and coefficient of friction nearer to the peak ground acceleration of the El Centro shock. A plausible explanation of this case is: so long as the bottom mass does not move, the inertia force is contributed by the top mass only but as the bottom mass moves, the inertia force due to bottom mass is also added to the system and the flexibility of the system increases which tends to reduce the dynamic amplification. Generally, the flexibility effect predominates and the response is reduced due to sliding, but when the coefficient of friction approaches the fixity condition with respect to the peak ground acceleration, the inertia force of the bottom mass seems to become more influential and the response increases.

The sliding arrangement is of greater advantage in the case of the Koyna shock where the predominant frequency in the accelerogram was higher than that of El Centro as well as the peak ground acceleration was much higher (0.63 g against 0.32 g).

14.3.3 Displacement Response

Fig. 3 shows the computed maximum and residual relative displacements of the base-mass for the two earthquake shocks considered as a function of coefficient of friction. From these results, the following general observations are made.

i. The residual or permanent displacement is only slightly smaller than the maximum displacement reached during, earthquake motion. Therefore, maximum dynamic displacement may be used for practical design without undue conservatism.

ii. The maximum value of the displacement goes on decreasing as the coefficient of friction is increased from 0.15 to 0.4. The value of maximum displacement in both the earthquakes is 18.8 mm which occurs for $\mu = 0.15$.

14.4 Tests on Half-Full Size Building

In order to verify the theoretical results, laboratory testing of building specimens with sliding joints at the plinth level was carried out on a special shaking platform under shock loading. Half size of single room building was adopted, as shown in Fig. 4 and half size bricks (112 x 56 x 38 mm) were used in the building construction. The shaking platform was 7 m x 6 m in plan and could accommodate four building specimens simultaneously.

14.4.1 Test Specimens

Four specimens (numbered 5-8) were tested simultaneously on a shock type shaking table. Two of these were conventional, that is, with usual mortar joints, and two with sliding joint at the plinth level. Fig. 4 shows the masonry dimensions, openings in walls and the position of horizontal and vertical steel bars. The mortar and reinforcement was varied in different specimens as given in

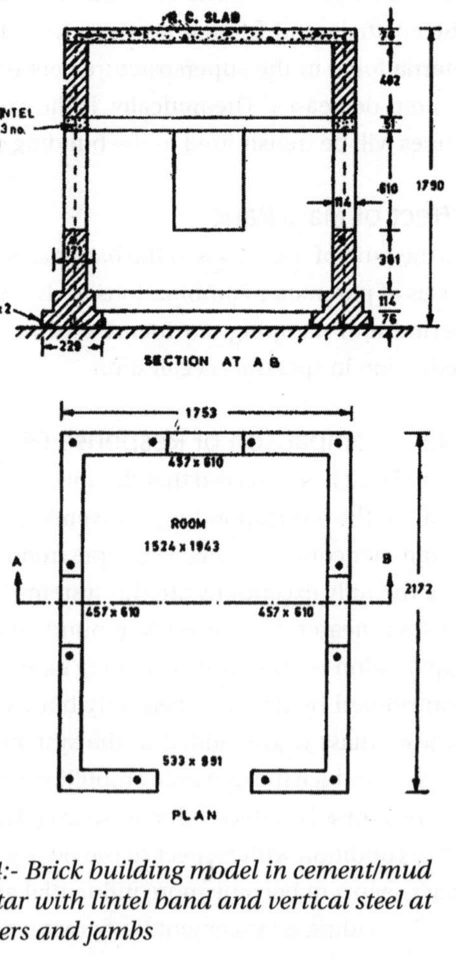

Fig.4:- Brick building model in cement/mud mortar with lintel band and vertical steel at corners and jambs

<table>
<tr><th colspan="4">Table 1:- Details of Specimens</th></tr>
<tr><th>Specimen</th><th>Type</th><th>Details of Mortar and Reinforcement</th><th>Remarks</th></tr>
<tr><td>5</td><td>Sliding</td><td>Mud mortar. Lintel band, 6 mm dia vertical bars at jambs of openings. Plinth band and bond beam</td><td rowspan="4">(i) Lintel and plinth bands and bond beam consisted of 3 bars mm placed in 38 mm thick M 1 5 concrete over all walls
(ii) Vertical steel at corners anchored in roof slab and foundation masonry in the conventional specimen but in the bond beam in the sliding specimen
(iii) Vertical steel at jambs anchored into lintel band and bond beam in sliding type, otherwise in plinth band</td></tr>
<tr><td>6</td><td>Conventional</td><td>Cement sand mortar 1:6. Lintel band. No other reinforcing</td></tr>
<tr><td>7</td><td>Sliding</td><td>Cement sand mortar 1:6. Reinforcing similar to specimen 5</td></tr>
<tr><td>8</td><td>Conventional</td><td>Cement sand mortar 1:6. Lintel and plinth bands. Vertical steel bars at corners of walls and jambs of openings</td></tr>
</table>

Table 1. To achieve a proper sliding joint, a reinforced concrete band was constructed at the plinth level to replace the damp-proof course as part of the foundation masonry. It was finished smooth at top. Then used mobile oil was painted over it and a reinforced concrete band, termed hereafter as *'bond beam'* for bonding the vertical steel in the superstructure masonry, was cast all around, except through the door opening. The superstructure masonry was constructed on the bond beam and all vertical bars anchored to the bond beam. All specimens had, a reinforced concrete slab roof bonded to the masonry through mortar and vertical corner steel was anchored into the roof slab.

14.4.2 Shake Table

Fig. 5 shows a sketch of the testing facility. The shaking platform was built on an old railway wagon chassis placed on the level portion of the track in the middle.

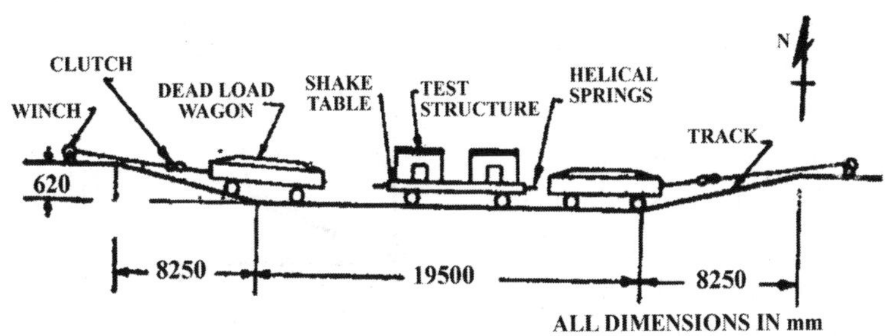

Fig.5: Schematic view of railway wagon shake table test facility

The platform carries helical springs on both sides to take the shock from a loaded wagon on either side which rolls down an incline from a predetermined height. The wagon imparts a half-sine pulse to the platform which after rolling gets another half-sine pulse in the reverse direction from the stationary wagon. The dynamic action is increased by increasing the height of fall of the loaded wagon.

14.4.3 Relative Competence of Models to Withstand Dynamic Loads

To have a quantitative measure of the damage resisting capability of any of the model structures, two parameters need to be identified: (i) to define the extent of damage and (ii) to define the input effort to cause damage.

The 'input energy' is chosen as the parameter for defining the dynamic action on the specimens since it has the merit that effects of all the alternating shocks can be taken into account by scalar addition.

For defining the damage level of the model, the parameter chosen is the ratio of the area of cracks in a wall to the total vertical area of the wall. The area of a crack is equal to its length multiplied by the wall thickness.

The two parameters as defined above were computed for each shock and for all the models. The cumulative extent of damage against cumulative input energy per unit mass for the test structures is shown in Fig. 6 from which the following significant observations are made.

(i) The extent of damage of all the models increases with the increase in input energy whether given in one big shock or a number of smaller shocks provided that each shock exceeded the damage threshold.

(ii) In the initial stages, as the input energy per unit mass increased, as the sliding type and the stronger structures in cement mortar (specimens 7 and 8) performed similarly whereas the sliding specimen built in mud mortar (specimen 5) showed less damage compared to the conventional structure in cement mortar (specimen 6). But in the

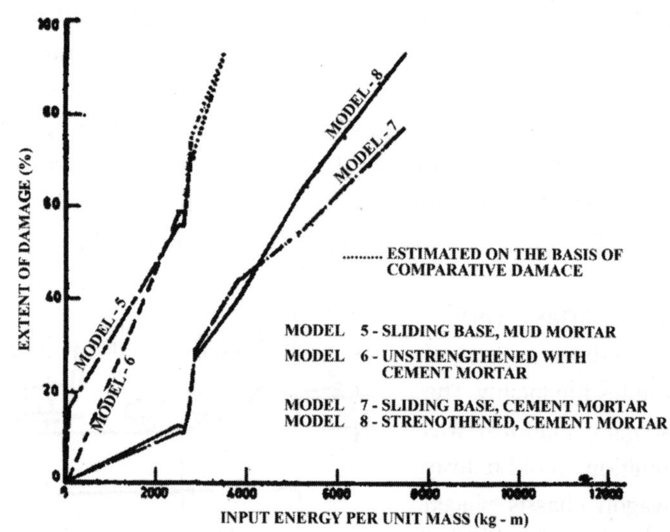

Fig.6: Progressive damage and relative competence of set of brick test structures subjected to shock load

Shock	Table 2:- Acceleration Response Of Specimens		
		Acceleration, g	
	At Table Base	At Roof of Sliding Specimen No.7	At Roof of Conventional Specimen no.8
1	0.51	0.55	0.59
2	2.15	0.59	1.37
3	0.53	0.12	0.55
4	0.66	0.12	0.53
5	0.58	0.12	0.45
6	1.08	0.93	2.00
7	2.36	0.89	2.13
8	2.45	1.28	1.79

final stages of shocks, these two showed a similar extent of damage, while the sliding specimen 7 in cement mortar showed less damage compared to the conventional specimen 8.

The accelerations recorded under various shocks at the base and roof of specimens 7 and 8 are presented in Table 2. Under all shocks, small and large, the roof acceleration of sliding specimen is generally much smaller than that of the conventional specimen. This observation corroborates qualitatively very well with the analytical results and proves the effectiveness of the sliding arrangement in reducing the seismic response of the superstructure.

During the main shocks and rebounds it was noted that the residual displacement in the sliding specimen alternated with the alternating shocks and there was no cumulative displacement in any one direction. In an actual earthquake which in general consists of alternating pulses similar displacement behavior is expected.

14.5 Conclusions

1. The sliding arrangement provided at the plinth level, with not too high a coefficient of friction, will be very effective in reducing the earthquake acceleration acting on the building superstructure.
2. The maximum dynamic and permanent displacements of the superstructure relative to the base even during the probable maximum earthquakes are likely to be small enough which could be accommodated easily over the plinth projections.

Thus, the new concept has the potential of not only avoiding complete collapse of masonry buildings but also minimising the cracking damage due to reduced earthquake forces and could be the cheapest possible solution to the safety of masonry buildings in severe earthquake zones.

14.6 NOTATION

C_s = coefficient of the viscous damper

g = acceleration due to gravity

K_s = spring constant

M_b = bottom mass

M_t = top mass lumped at the roof level

M_T = $M_b + M_t$

p = natural circular frequency of the system = $\sqrt{\frac{K_s}{M_t}}$

$\ddot{x}_b$ = absolute acceleration of bottom mass

$\ddot{x}_t, \ddot{Z}_t$ = absolute and relative accelerations of the top mass Mt, respectively

Z_b, Z_t = lateral relative displacements of masses M_b and M_t, respectively

$\dot{Z}_b, \dot{Z}_t$ = relative velocities of masses M_b and M_t, respectively

$\ddot{Z}_b$ = relative acceleration of M_b

θ = mass ratio = $\frac{M_t}{M_b}$

m = coefficient of friction

ξ = fraction of critical damping = $\frac{C_s}{2pM_t}$

14.7 References

1. A.S. Arya, B. Chandra and P Rani, "Influence of Natural Disasters (Earthquakes) on Educational Facilities, Part I- Collection and Analysis of Available Information-Annotated Bibliography," Report of School of Research and Training in Earthquake Engineering, University of Roorkee, Roorkee, 1977.
2. J. Krishna and A.S. Arya, "Earthquake Resistant Design of Buildings," Journal of the Institution of Engineers (India) vol.7, 1965.
3. A.S. Arya, "Design and Construction of Masonry Buildings in Seismic Areas," Bulletin of the Indian Society of Earthquake Technology, vol 4, no.2, 1967.
4. A.S. Arya, "Construction of Small Buildings in Seismic Areas," Bulletin of the Indian Society of Earthquake Technology, vol. 5, nos. 3 and 4, 1968.
5. J. Krishna and B. Chandra, "Strengthening of Brick Buildings Against Earthquake Forces," Proceedings of the Third World Conference on Earthquake Engineering, New Zealand, 1965.
6. J. Krishna and B. Chandra, "Strengthening of Brick Buildings in Seismic Zones", Proceedings of the Fourth World Conference on Earthquake Engineering, Chile, 1969.
7. B. Chandra and K. Kumar, "Earthquake Resistant Construction of Brick Buildings", Jai Krishna 60th Birth Anniversary Commemoration Volume, Sarita Prakashan, Meerut, 1974.

8. "Earthquake Resistant Regulations – A World List." International Association of Earthquake Engineering, 1973.
9. IS: 4326-1976, "Code of Practice for Earthquake Resistant Design and Construction of Buildings", Indian Standards Institution, New Delhi, 1977.
10. R.N. Joshi, "Striking Behaviour of Structures in Assam Earthquakes", Proceedings of the Second World Conference on Earthquake Engineering, Tokyo and Kyoto, 1960.
11. E.R. Gee, "Dhubri Earthquake of 1930," Memoirs of the Geological Survey of India, no LXV, 1934.
12. M. Qamaruddin, "Development of Brick Building Systems for Improved Earthquake Performance", Ph.D. Thesis, University of Roorkee, 1978.
13. J. Krishna, A.S. Arya and K. Kumar. "Determination of Isoacceleration Lines by Sliding and Overturning of Objects", Proceedings of the Fifth World Conference on Earthquake Engineering, Rome, 1973.

Chapter 15

Simplified Guidelines for all New Buildings in the Seismic Zones of India[1,2]

15.1 Introduction

Usually new construction of buildings under IAY, Prime Minister Rojgar Yojana and buildings under various other national and state schemes get started in the month of May. The Ministry of Home Affairs is keen that **all new buildings** should be made earthquake resistant in the first instant so that we do not add to the stock of existing unsafe buildings. Since most of the buildings are constructed using brick work or, solid hollow concrete blocks with flat roofs, very simple illustrated guidance is provided in this paper for incorporating the earthquake resistant features suitable for seismic Zone V. Of course, these simplified guidelines will be equally applicable to various similar buildings to be constructed by the private sector for residences, business, schools, dispensaries, clinics, etc.

15.2 Essential Elements for Earthquake Safety

15.2.1 Horizontal and vertical elements

The essential elements required to make a building earthquake safe are as shown in Fig. 1. Some additional requirements are detailed in the following paragraphs. However, all elements shown in Fig.1 may not be required in some cases as mentioned in the following paras.

15.3 Good Cement Mortar

Minimum wall thickness will be kept 200 mm for CC blocks and 230 mm for bricks. The cement mortar should be used in the ratio of 1 part of cement with 4 parts of sand (1 sack of cement mixed with 4 equal sacks of sand) in Zone V and 1 part of cement with 6 parts of sands in Zones IV and III. Cement may be 33 or 43 grade with or without puzzolona. 53 grade need not be used in construction of walls.

---

1 *Anand S. Arya and Jananjan Panda, GoI (MHA) - UNDP DRM Programme.*

2 *Published as Information Brochure for all District Collectors in Seismic Zone III, IV & V in 2005*

Note:- The details given here are extract from IS:4326-1993 Code of Practice as applicable to buildings of brick/CC block.

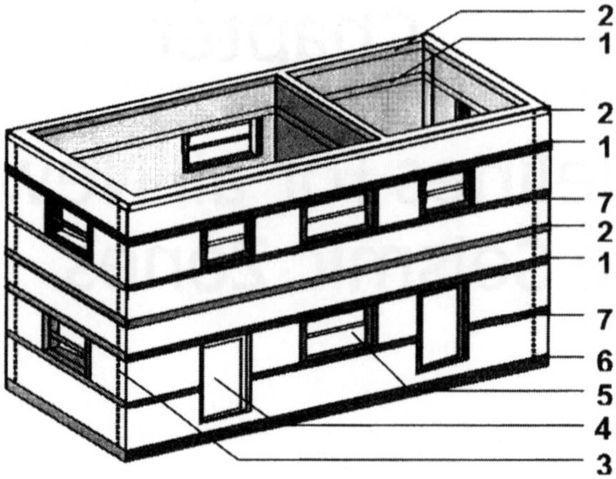

1. Lintel Band
2. Roof/ Floor Band
3. Vertical reinforcing bar at corner
4. Door
5. Window
6. Plinth Band
7. Window Sill Bands

Fig.1: Essential internal elements in walls of buildings for earthquake Safety

15.4 Horizontal Seismic Bands

A seismic band consists of reinforced concrete flat runner through **all external and internal masonry walls** at the following levels in the building:-

a. at the plinth level of the building in all soil conditions in foundation (may be deleted on rocky ground).

b. at the levels of lintels of doors and windows in all storeys.

c. at the ceiling level of roofs consisting of wooden joists or, prefabricated reinforced concrete beams or, planks. *(Such band will not be necessary if the roof consists of reinforced concrete or reinforced brick slabs cast on the walls covering a minimum of 2/3 of the thickness of the wall).*

d. At window sill level in all storeys in important buildings in Zone IV and all buildings in Zone V.

Table-1: Recommended Size and Longitudinal Steel in Seismic Bands

Internal length of wall	Zone III Residential		Zone III Important and Zone IV Residential		Zone IV Important and Zone V Residential		Zone V Important	
	No.	Dia.	No.	Dia.	No	Dia.	No.	Dia.
5 m or, less	2	8 mm	2	8 mm	2	10 mm	2	12 mm
6 m	2	8 mm	2	10 mm	2	12 mm	4	10 mm
7 m	2	10 mm	2	12 mm	4	10 mm	4	12 mm
8 m	2	12 mm	4	10 mm	4	12 mm	4	14 mm

Note:- Important buildings will include schools, hospitals, clinics, dispensaries, aanganwadis, meeting halls, congregation halls, VIP residences, etc. See IS: 1893 (Part 1) 2002.

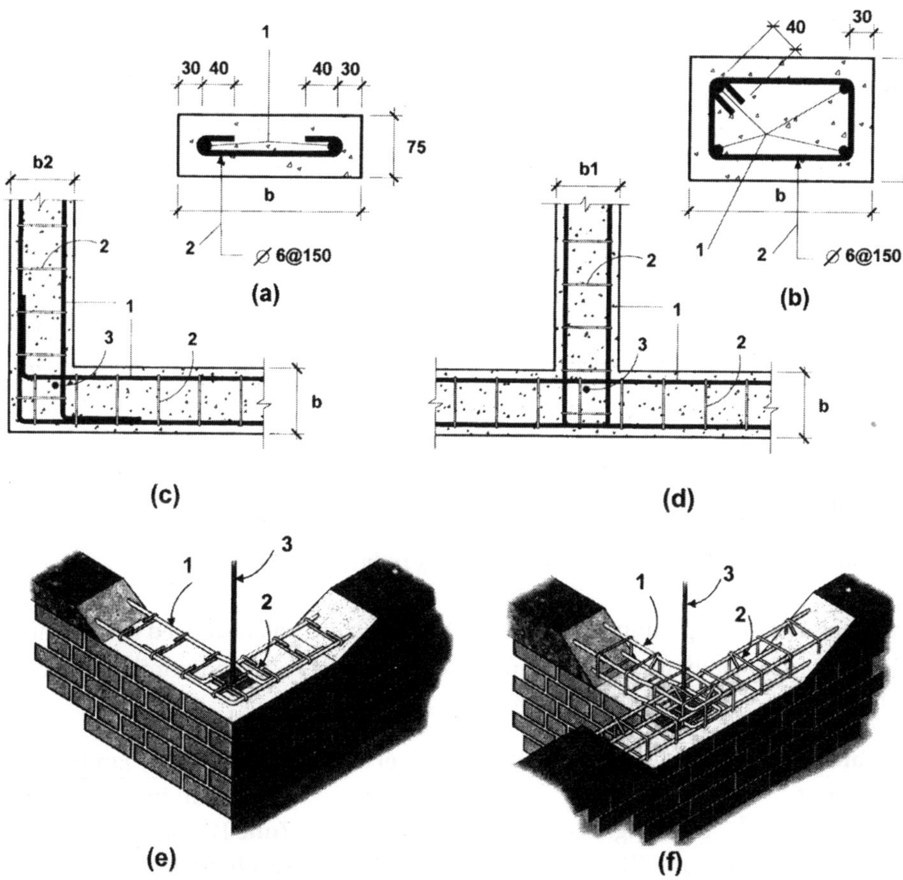

(a) Section of the Band with 2 longitudinal steel bars
(b) Section of the Band with 4 longitudinal steel bars
(c) Structural Plan at L- type wall junction
(d) Structural Plan at T- type wall junction
(e) 3 Dimensional view of the L - type wall junction
(f) 3 Dimensional view of the T - type wall junction

1. Longitudinal reinforcements
2. Lateral Ties
3. Vertical reinforcement at corners

b, b1, b2 Wall thickness

Fig.2: Reinforcement and bending details of seismic bands

The dimensions of the band and the reinforcement inside depend upon the length of the walls between the perpendicular cross walls. The table above ***(Table-1)*** shows the dimensions to be adopted for the seismic bands and the internal reinforcement details to be provided. The reinforcement and bending details of seismic bands are given in ***Fig. 2.*** Reinforcing bars will be of Fe 415 type [TOR or, High Yield Strength Deformed, i.e. HYSD bars]

15.5 Vertical Reinforcement in the Brick Walls

For earthquake safety in the seismic zones, reinforcing bars have to be embedded in specified cases in brick or block masonry at the corners of all the rooms and the sides of the

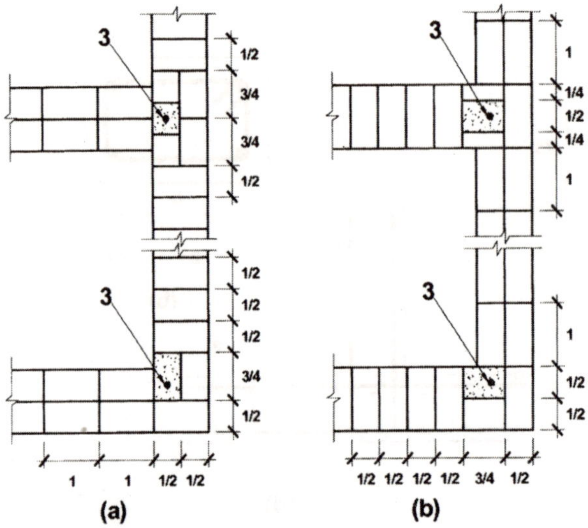

a & b : Alternate courses in one brick wall
1 : One brick length
½ : Half brick length
¼ : Quarter of a brick length
¾ : Three quarters of a brick length
3 : Vertical reinforcement bars with Concrete/ mortar filling in pocket of M20 grade (1:1½:3 nominal mix)

Fig.3: Typical details of providing vertical Steel Bars in brick masonry

door openings. Window openings larger than 60 cm in width will also need such reinforcing bars ***(Figure -3)***. The diameter of the bar depends upon the zone and number of storeys in the building. The recommendations are given in ***Table-2.***

Table-2: Recommended Size of Vertical Steel at Corners and Door Jambs

No. of storeys	Floor or Roof	Residential buildings Zone III	Important Buildings * in Zone III and residential in Zone IV	Important in Zone IV and residential in Zone V	Important in Zone V
		Dia of Single HYSD (TOR) Bar at corners of room (mm)			
One	-	-	10	12	14
Two	Top	-	10	12	14
	Bottom	-	12	16	18
Three	Top	10	10	12	14
	Middle	10	12	16	18
	Bottom	12	12	16	18
Four	Top	10	10	Four storey buildings with bearing walls not permitted for these cases.	
	Third	10	12		
	Second	12	16		
	ottom	12	20		

* Important buildings as stated under Table 1

Providing the vertical bars in the brick work and concrete blocks requires special techniques which could be easily learnt by the supervising engineers and the masons will need to be trained.

Table-3: Recommended Joint Details with the Vertical Reinforcement at Corner for Masonry Walls Using Different Kind of Materials

Type of Joint	Corner reinforcement in case of Brick Masonry	Corner reinforcement in case of Solid Concrete Block Masonry	Corner reinforcement in case of Hollow Concrete Block Masonry (see the hole and slit made)
L- Joint	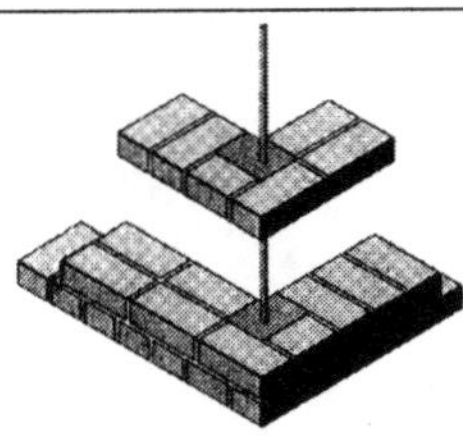	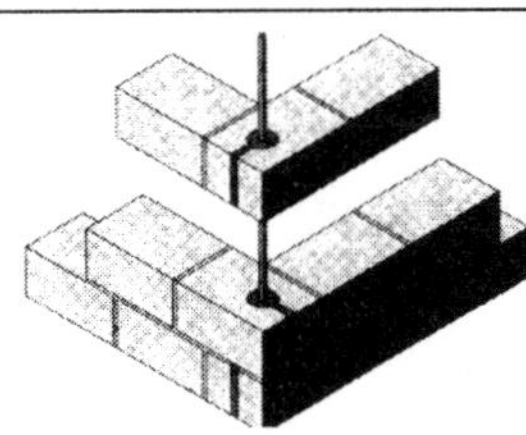	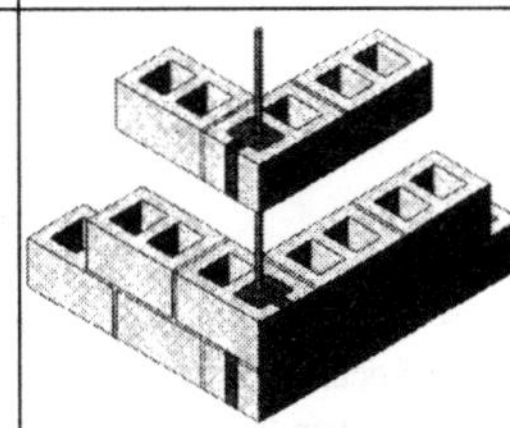
T- Joint	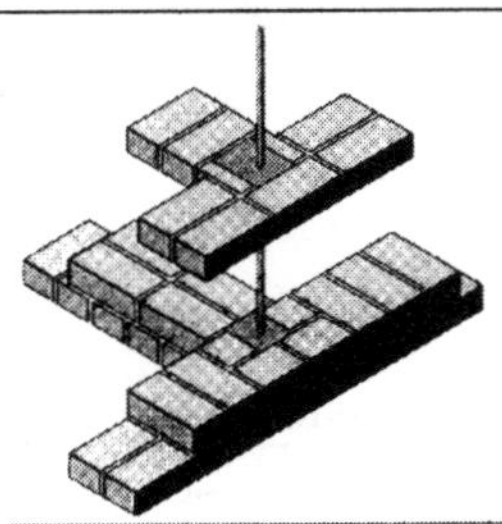	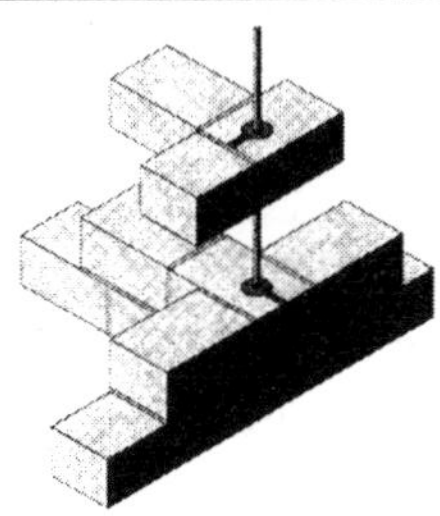	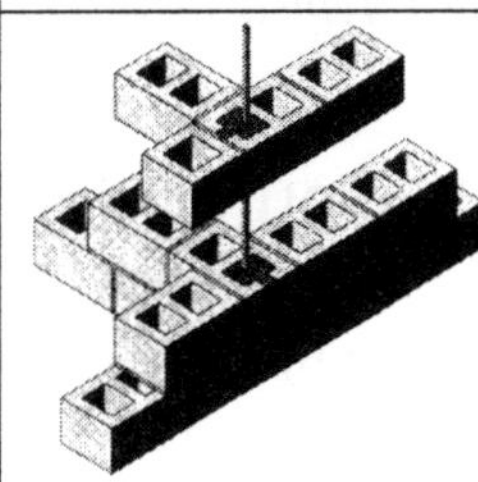

These vertical bars have to be started from the foundation concrete, pass through all seismic bands where they will be tied to the band reinforcements using binding wire and embedded into the ceiling band/roof slab as the case may be, using a 300 mm long 90° bend. Sometimes, the vertical bars will not be made in one full length. In that case, the extension of the vertical reinforcement bars required, an overlap of minimum of 50 times the bar diameter should be provided. The two overlapped reinforcement bars should be tied together by using the binding wires.

15.6 Vertical Reinforcement at Jambs of Openings

All doors and the window openings wider than 600 mm will have vertical reinforcement in the jambs as shown in ***Fig.4.*** Such reinforcement may not be provided around small ventilators and windows less than 600mm in size.

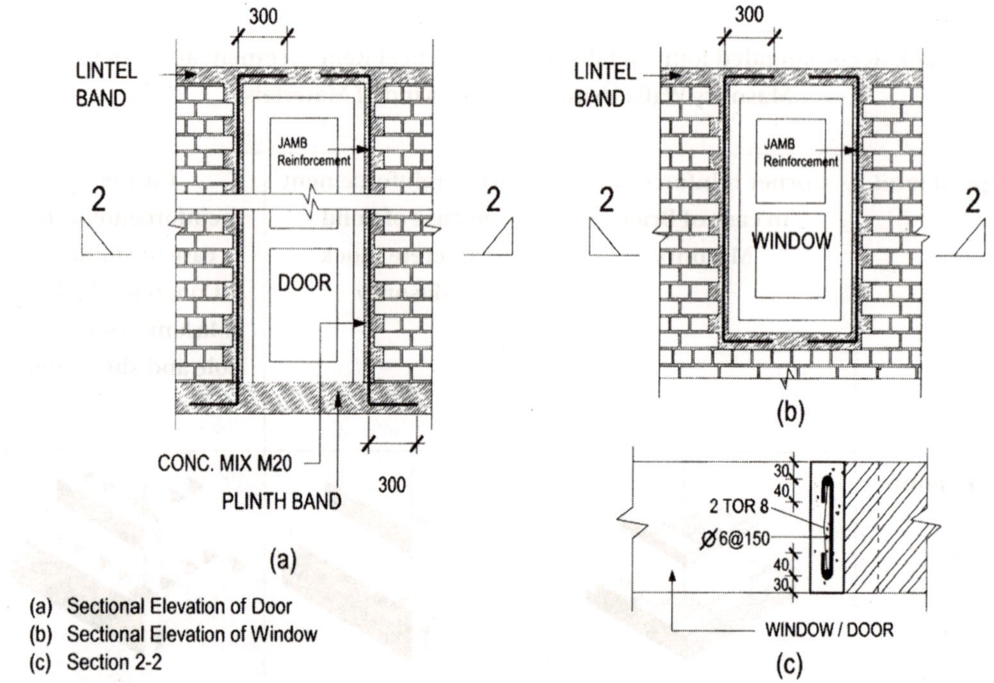

Fig.4: Typical details of providing vertical steel bars around doors/windows

15.7 Foundation

Foundation width 'B' should be decided by the load coming on the foundation and the bearing capacity. Masonry width may be reduced by 1/2 times T in every step of 150 mm height.

15.7.1 Foundation in liquefiable soil

In sandy soils with a high water table within 8 m depth below ground level, which may get liquefied during an earthquake of MSK intensity VIII to IX, pile foundation needs to be used in consultation with the structural/ geotechnical engineer.

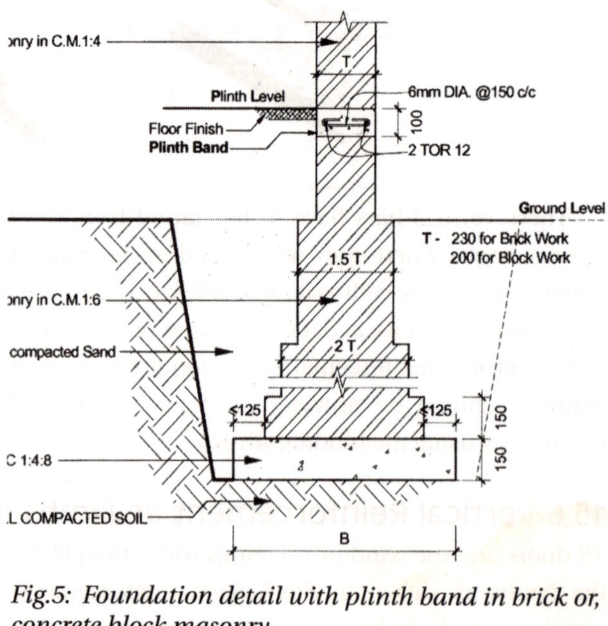

Fig.5: Foundation detail with plinth band in brick or, concrete block masonry

15.8 Costs of Improving Seismic Resistance of Buildings

The earthquake resisting features specified above to be incorporated while constructing any new building depend on the seismic intensity zone in which the building is located, the base soil and the functional use of the building, whether considered important or ordinary. The extra cost of these resisting features will vary accordingly. Now reasonably accurate information is available on the percent extra cost in the case of masonry buildings built in cement mortar in various seismic zones of India and can be taken as follows:

Residential buildings in Zone III	2 -2.5 %
Important building in Zone III and residential in Zone IV	3 - 4%
Important building in Zone IV and any building in Zone V	5 - 6%

The percent cost will be higher for the weaker informal buildings using mud mortar since the cost of the resisting features will be about the same, but the cost of the building being lower, the percentage will be higher. For guidance, the increase in cost may be assumed as additional 1/2%.

If the building is not constructed as per the earthquake resistant codes, and has to be retrofitted later, the extra cost will be higher and may be assumed about 2.5 to 3 times the above percentages.

15.9 Some Real Earthquake Observations

An interesting study of Mean Damage Ratios (MDR) as related to the Seismic Resistance (SR) present in buildings damaged during recent earthquakes in Italy and Algeria has been made by Tiedemann[2] as shown in Tables 3, 4 & 5. The damage ratio is worked out as a percentage of the cost required for repairing to that of rebuilding, and the seismic resistance of the buildings is estimated in terms of acceleration due to gravity, g. Since both the terms will have possible variations, the values cannot be taken as exact but the relative values will definitely indicate the quantitative trends. From Tables 3 to 4, it is seen that even in MM VIII-IX areas, the mean damage ratio decreases from 76.9% to 2.33% as the seismic resistance increases from 1.5%g to as low a design value a 6-7% g and collapses are eliminated. The importance of these results is to be realised where we find that the design seismic coefficient of buildings in Categories B, C, D & E, in IS:4326-1993 are 5 to 6 %, more than 6% to less than 8%, 8% to less than 12% and more than 12% of g respectively. If the code provisions are used properly in design of buildings, these will be adequate to bring down the MDR values to very low values. MDR is seen to increase in Table 5 as the number of storeys increases from 1 to 4, pointing out to the need of greater strengthening care in construction of higher buildings or, restricting the building height to two storeys.

In the Uttarkashi earthquake of October 20, 1991, the simple provision of one bar of 16 mm dia vertically at the four corners of two-storeyed Maneri Hydroelectric Project colony block masonry buildings, saved all these two-storeyed buildings from collapse, although the first

Table 3 : Mean Damage Ratio vs. Seismic Resistance in Italian Earthquake of November 23, 1980, (M=6.5, MMI=VIII)

SR, % of g	4-6	3	2	Total 2-6
No. of Buildings*	16	22	18	56
MDR, %	1.33	18.14	53.22	31.46

* Buildings of old traditional construction

Table 4:- Mean Damage Ratio vs. Seismic Resistance in Al- Asnam, Algeria, Earthquake of October 10, 1980, (M=7.5, MMI=VIII-IX)

SR, % of g	6-7	4-5	2.5-3.5	2	1-1.5	Total 1-6
No. of Buildings*	22	52	103	68	63	308
MDR, %	2.33	10.15	24.72	41.09	76.9	26.9
No. of Building collapses	0	0	8	1	31	40

*Buildings of modern construction after 1954 earthquake of M = 6.8

Table 5:- Mean Damage Ratio vs. Number of Storeys, Al-Asnam Earthquake of October 10, 1980

No. of Storeys	1	2	3	4	Total
No. of Buildings*	25	98	19	3	145
MDR, %	18.2	28.1	33.95	57.5	27.69

* *Unspecified modern buildings, tentative SR = 3% of g*

storey suffered damage. These 450 buildings were as close as 5 km from the epicentre. Also, they could be repaired at an average cost of about 15% of the reconstruction cost. Not even one life was lost in these buildings. Besides these buildings, the two and three storeyed masonry buildings built by CPWD for the India-Tibetan Border Policy at Mahitanda following the provisions of the Indian codes IS:1893-1975 and IS:4326-1976 were only slightly cracked and did not even require any costly repairs.

15.10 Conclusion

Practical experience of the behaviour of buildings in the recent earthquakes in India, namely, Uttarkashi 1991, Latur 1993, Jabalpur 1997, Chamoli 1999 and Kachchh 2001 shows clearly that those buildings in which the simple measures of seismic bands and vertical steel at corners was provided, did not collapse. Such critically reinforced buildings have their seismic resistance varying between 5% to 12% of g and as shown by the experience in the earthquakes in Italy and Algeria, such buildings have not collapsed and their damage ratio to the replacement cost also got reduced to less than 5%. ■

15.11 References

1 Anand S. Arya and Jananjan Panda, "Earthquake Safe Construction, Safe Construction of Masonry Building in Zone III, Zone IV and Zone V, published by National Disaster Management Division, Ministry of Home Affairs, GoI, New Delhi 2005.

2 Tiedemann H., "Building Quality and Earthquake Damage: Some Statistics of the South Italian and Algerian Earthquake of November 23 and October 10, 1980," Proc., Symp. on Earthquake Disaster Mitigation, University of Roorkee, March, 1981, pp.9-34

Chapter 16

Seismic Analysis of Experimental Buildings[1,2]

16.1 Introduction

The National Buildings Organisation, Government of India, has put up four and five storeyed buildings with single brick load bearing constructions and using different experimental components of building in seismic Zones III and IV of India in the last few years. The main objectives of the new techniques employed in the buildings were economy and speedy construction. Table 1 and Figs. 1 to 5 give the details of the buildings with respect to their location, organisation of units and the year of construction. The experimental techniques used in various buildings are shown in Table 2. A study was undertaken to examine the suitability of such buildings in the seismic environment with a view to improve upon the building design/performance in future constructions. The present paper describes the results of seismic analysis of these buildings and includes recommendation for improvement in such constructions.

Table 1:- The Buildings Studied

S. No.	Place	Location	Seismic Zones	No. of Storeys	No. of Units per Floor	Year of completion
1	Roorkee	Gandhi Vatika	IV	4	4	1975
2	Lucknow	Dilkusha	III	4	4	1982
3	Parwanoo	-	IV	4	2	Under constr.
4	Delhi	Dhaula Kuan	IV	4	4	1974
5	Calcutta	Maniktala	III	5	10	1974

16.2 Seismic Response of Buildings

The seismic response of the structure is a function of expected ground motion and the properties of structures incorporating the soil properties at the site. The structural properties

1 *A.S.Arya, Professor, Brijesh Chandra, Professor, Satendra P. Gupta, Professor and Pankaj, Lecture, Deptt. Of Earthquake Engineering, University of Roorkee, Roorkee*

2 *International Conference on Low Cost Housing for Developing Countries, Central Building Research Institute, Roorkee, Nov.1984, pp 623 - 636*

include distribution of stiffness and mass and damping characteristics which are inherent properties of the structural system. In addition, the elastic and elasto-plastic properties control the dynamic behaviour to a great extent. As a matter of fact, a structure by virtue of its inelastic behaviour is able to shake down earthquake forces considerably although it may undergo inelastic deformations. In structures or structural systems where elasto-plastic or inelastic behaviour is possible, one can take advantage of such reduction in seismic forces. The IS Code of Practice for Earthquake Resistant Design IS: 1893-1975 envisages that structures would

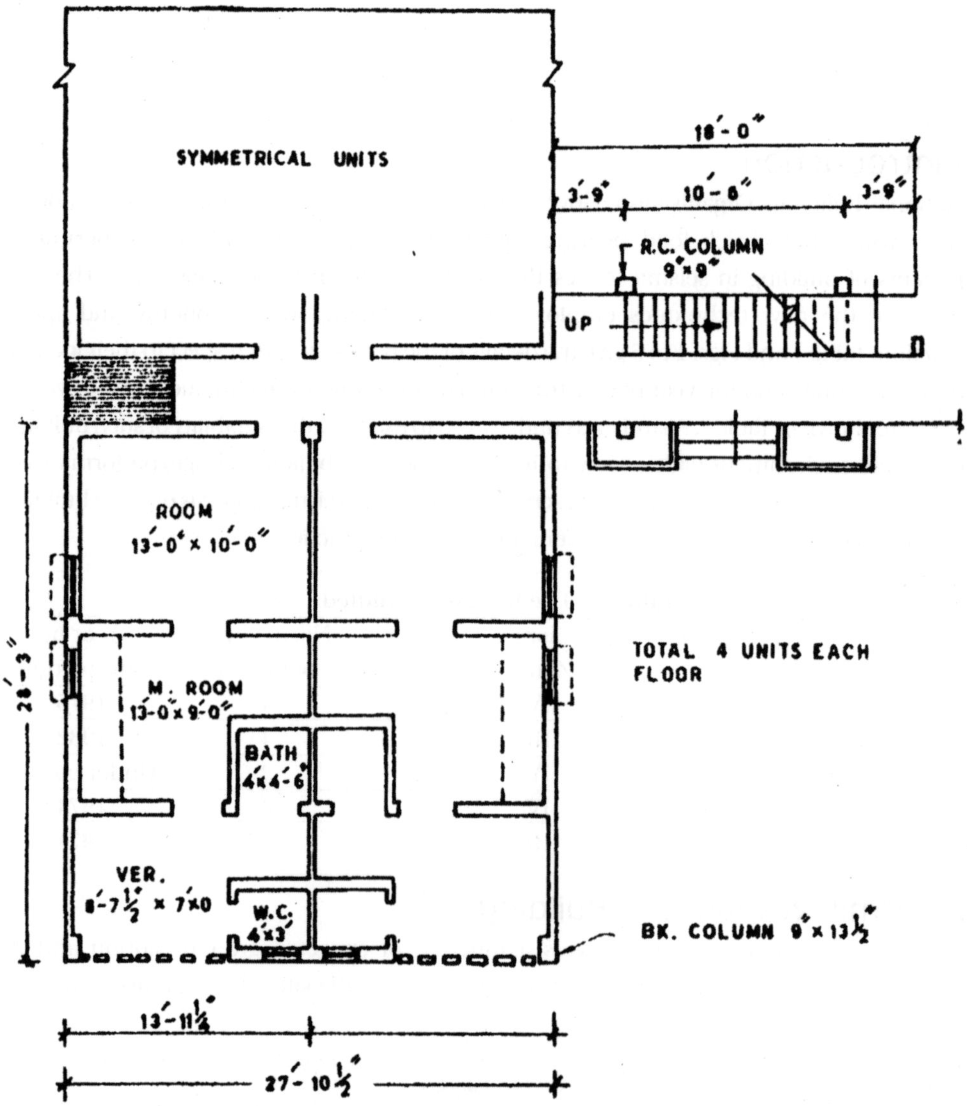

Fig.1: Plan of an experimental building at Roorkee

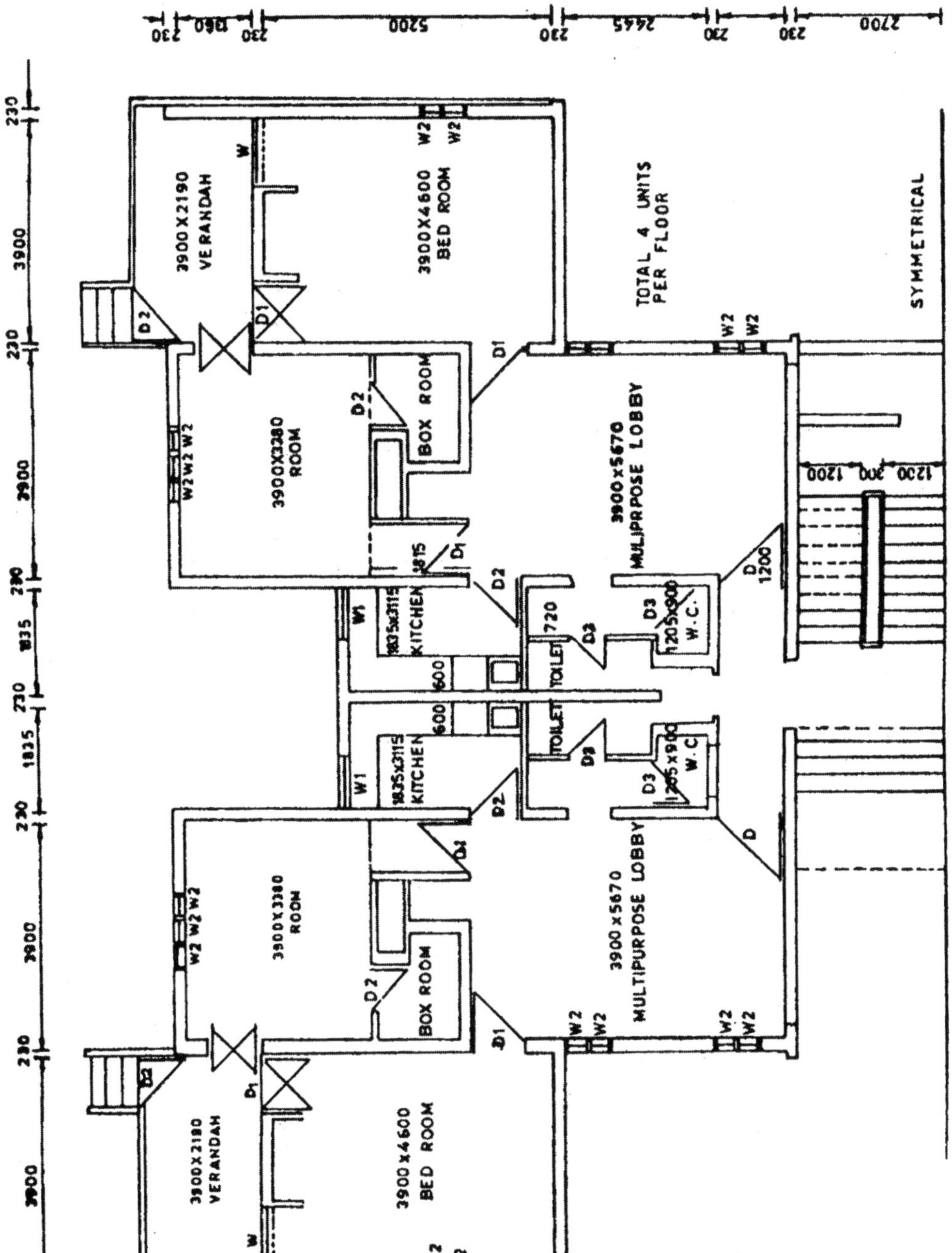

Fig.2: Plan of an experimental building at Lucknow

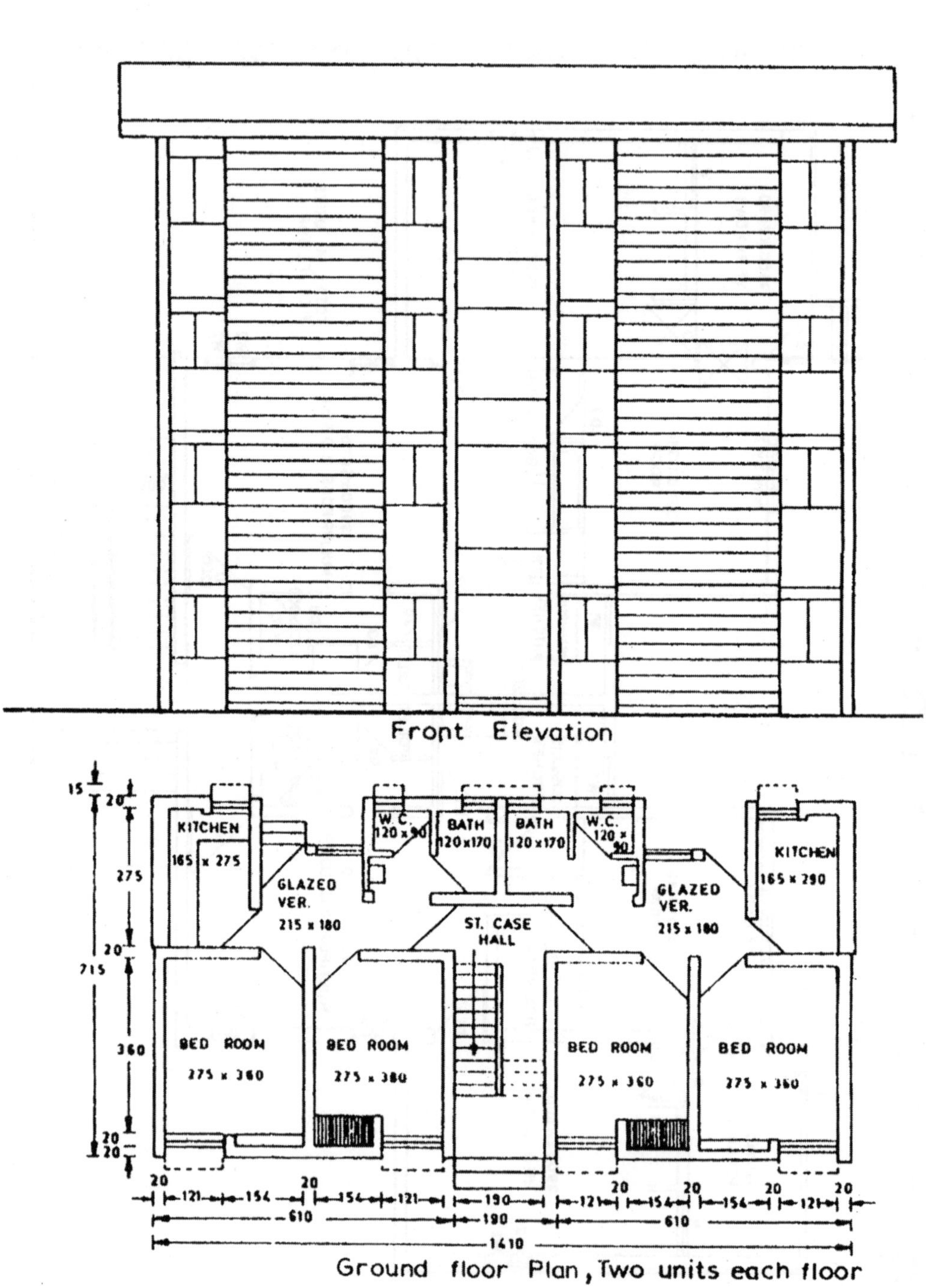

Fig.3: Ground floor plan and front elevation of LIG house, Parwanoo

normally exhibit such behaviour and, therefore, the seismic coefficients given in this code are much smaller. However, in brittle structures like brick buildings such behaviour would not be present and, therefore, such buildings would attract larger seismic forces, making it imperative to check their safety against a seismic coefficient much higher than the codal value. The increased value may be taken about 4 times the codal value to represent the order of seismic force attracted by a brick building in the maximum probable design earthquake.

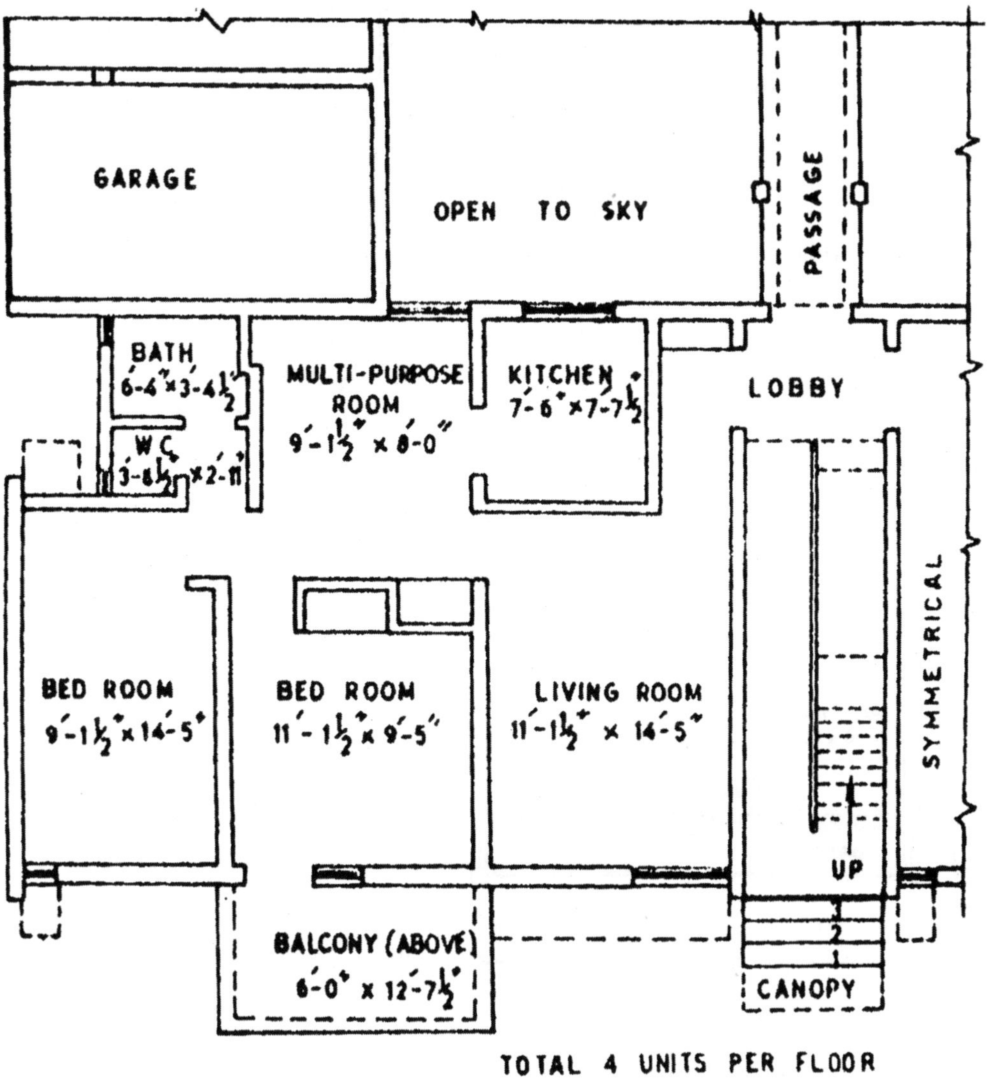

Fig.4: Ground floor plan of an experimental building at Delhi (Dhaula Kuan)

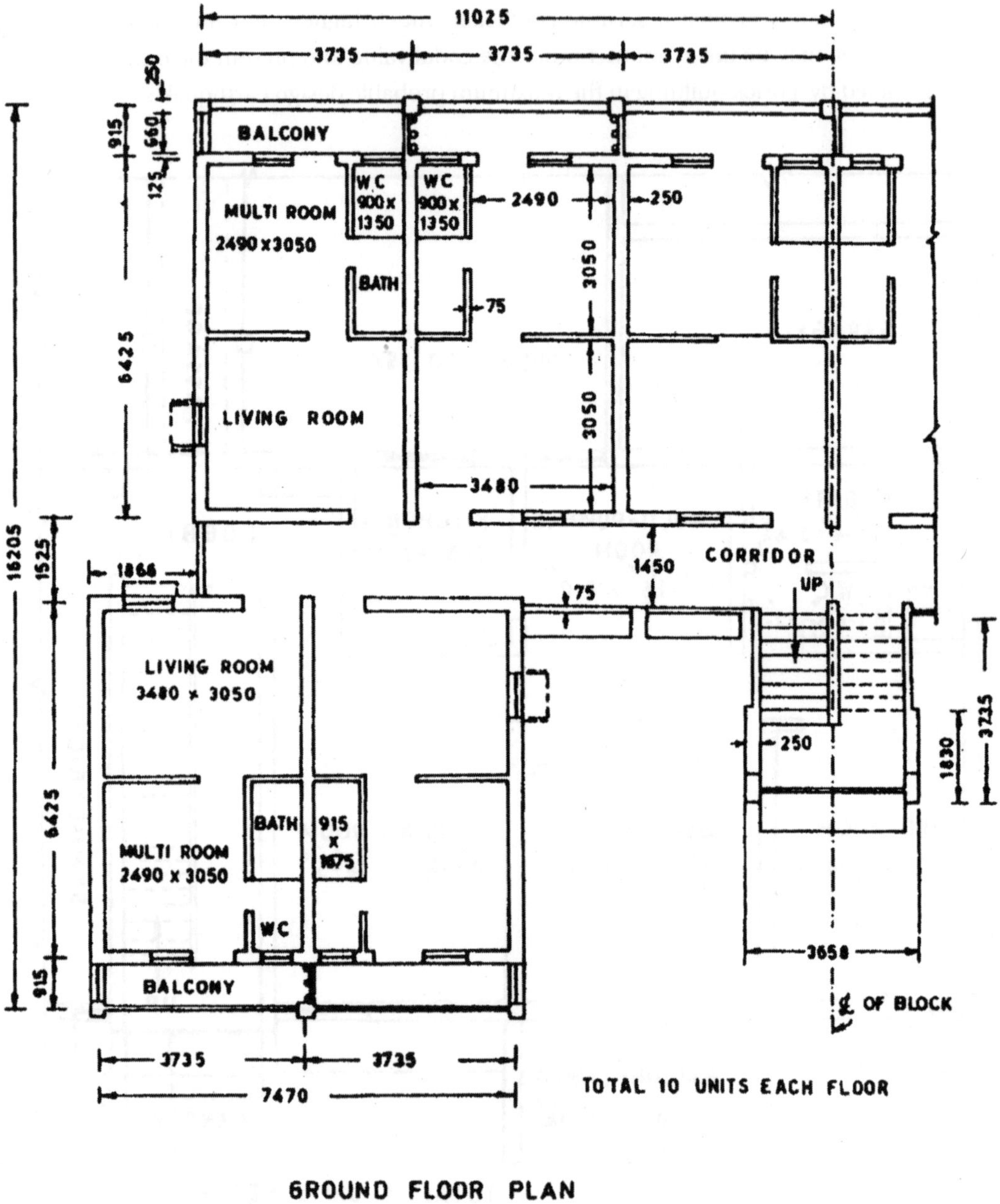

Fig.5: Layout of an experimental building at Maniktala, Kolkatta

16.3 Seismic Analysis

As explained earlier, the seismic analysis of the buildings has been carried out using two values of seismic coefficient, one corresponding to the codal value and the other *four times* this value. Therefore, 0.05 and 0.20 for buildings in seismic Zone IV and 0.04 and 0.16 for buildings in seismic Zone III have been sued. The smaller coefficients 0.05 and 0.04 are those specified in IS: 1893-1975 for elastic design and may be considered to take care of minor frequent earthquakes. The higher coefficients taken as *four times the codal values* are adopted here to check the ultimate strength of the buildings using yield stress in the reinforcing steel.

The buildings are analysed using the pier method of lateral load analysis in which the various piers or the wall itself share the lateral shear in proportion to their shear rigidities. The bending moments in the piers and consequent bending stresses are computed at the critical points of the piers, that is, top and bottom sections. Stresses on account of overturning effect of lateral forces as well as those on account of dead load are worked out separately, using the solid wall and pier areas and assuming linear distribution of the stresses as in simple bending of beams. The combined stresses are worked out at critical points (corners of piers) and the total stress is then compared with maximum permissible stress (compression as well as tension) to examining competence of the section.

16.4 Permissible Stresses in Brick Work

(a) *Compressive Stress:* The permissible stresses in compression in brick work were computed according to the specifications laid down in the National Building Code of India (1970). These normal working stresses were increased by a factor of 1.33 for the smaller seismic coefficient based on IS: 1893-1975 and a factor 2.0 for the higher seismic coefficient, representing probable maximum value of the earthquake. These allowable stresses were compared with computed compressive stresses in each pier.

(b) *Tensile Stress:* The National Building Code is silent about the allowable tensile stress in brickwork under seismic conditions. It is, however, realised that brick work in any mortar does have some tensile strength which would be mobilised before cracking during earthquakes. Various tests for strength of brick work (couplets, columns, walls) have been carried out (1,4). Table 3 gives the brick masonry properties worked out in these references for six different mortars.

For checking the tensile stresses in the piers, two values of the permissible stresses may be adopted; for the smaller seismic coefficient, the allowable tensile stress was taken as one-third of the ultimate strength given in Table 3. For the larger seismic coefficient, the actual tensile stresses developed in the piers may be compared with the ultimate tensile strength in the bending, as given in this table.

16.5 Results of the Analysis

(a) The compressive and tensile stresses corresponding to code seismic coefficients characterising frequent earthquakes, are found to be within allowable limits for Roorkee,

Table 2:- Experimental Techniques Used in Various Buildings

S.No.	Item	Roorkee	Lucknow	Parwanoo	Delhi	Calcutta
1.	Under-reamed pile foundation	*				
2.	Single Brick Load-Bearing Wall	*	*	*	*	*
3.	Precast RCC Frames for Doors/Windows	*			*	
4.	Precast RC Planks and Joists for Floors	*	*			
5.	Precast Channel Units for Floors/Roof			*	*	*
6.	Precast L-Pan Units for Roof			*		
7.	Thin Precast Lintels	*				
8.	Single Stack System of Plumbing	*			*	
9.	Improved Method of Brick Laying/ Plastering	*				
10.	Part Replacement of Cement by Fly Ash				*	
11.	Composite Cement Lime Mortar				*	*
12.	Bitumen Paper Terracing				*	
13.	Blending of Sand for RCC Work				*	
14.	PVC Pipes				*	

* Use of the technique

Table 3:-Brick Masonry Properties

Sl.No.	Mortar used in Brickwork	Compressive strength (crushing) kg/cm^2	Tensile strength (Bending)kg/cm^2	Shearing Strength (Max. stress) kg/cm^2
1.	1:3 Cement-sand	94.4	7.07	10.70
2.	1.6 Cement-sand	61.3	2.50	3.87
3.	1:12 Cement - sand	53.9	0.36	1.05
4.	1:2 Lime Surkhi	58.7	0.95	2.28
5.	1:2 Lime Cinder	53.0	0.60	1.45
6.	Clay Mud (Dry State)	47.6	0.32	1.13

Lucknow, Parwanoo and Dhaula Kuan buildings (Tables 4,5,6 and 7). Thus, one brick thick load bearing walls show no adverse behaviour for codal seismic coefficients. However, Calcutta building shows distress in walls as the permissible tensile stresses are exceeded in the XX-direction in all five storeys. (Table 8) although cracking may not occur since stresses are within the ultimate values. The analysis also shows that the state of compressive stress in masonry in the XX-direction of this building is within permissible limits in all five storeys. The YY-direction is seen to be safe, as the stresses do not exceed the permissible limits in tension as well as compression. The significantly different behaviour in XX and YY directions can be attributed to the fact that most door and

Table 4:- Results of Seismic Analysis of Building (Zone IV) at Roorkee (12 mm dia Vertical Bars at Every Corner and Junction Provided at Every Storey)

Direction	Storey	Seismic Coefficient-.05			Seismic Coefficient-.20		
		State of compressive stresses in masonry	State of tensile stresses in masonry	Remarks	State of compressive stresses in masonry	State tensile stresses in masonry	Remarks
X-X	1	within permissible	No tensile stress	Safe	within permissible	exceeds ultimate strength at many points.	cracking will occur, (1)
	2	-do-	-do-	-do-	-do-	exceeds ultimate strength at some points.	-do-
	3	-do-	-do-	-do-	-do-	-do-	cracking will occur,(2)
	4	-do-	-do-	-do-	-do-	within ultimate tensile stress	No cracks
Y-Y	1	-do-	-do-	-do-	within permissible	-do-	No cracks.
	2	-do-	-do-	-do-	-do-	-do-	-do-
	3	-do-	-do-	-do-	-do-	-do-	-do-
	4	-do-	-do-	-do-	-do-	-do-	-do-

Few up to 20%, Some : 20% to 50% , Many : 50% and above

(1) But 12mm dia bars will take up tension though some of them many yield

(2) -do-

window openings have been provided in the load bearing walls in the XX-direction, thereby making this direction weaker.

It may appear that there was no need to provide lintel band/reinforcing steel at corners and junctions of walls in Roorkee, Lucknow, Parwanoo and Dhaula Kuan buildings. However, as the earthquake force increases, the situation with respect to tensile stresses changes very fast, because whereas the compressive stress due to dead load remains constant, both σ_o and σ_b increase linearly with the seismic coefficient.

Therefore, tensile stresses do occur at larger seismic coefficients. This fact was realised while recommending the reinforcing measure in IS: 4326 which are indeed meant to safeguard the building against collapse during probable maximum earthquakes. Therefore the higher chosen seismic coefficients only are true measure of the efficiency of the reinforcing details.

(b) For the maximum probable earthquake condition characterised by seismic coefficients taken as four times those specified in IS: 1893-1975, the state of tensile stresses in masonry exceeds ultimate strength of brickwork at many points in almost all buildings in XX as well as YY direction indicating the need for vertical reinforcements.

It is seen that Roorkee building will be subject to cracking, especially in the lower three storeys (Table 4). In this building 12 mm dia. bars have been provided at corners and junctions in every storey and they will be able to take the tension, though some of them in the bottom two storeys may go into yielding.

In case of Lucknow building (Table 5) in which no vertical reinforcement has been provided, cracks are indicated to occur when it is subjected to maximum probable earthquake. This building lying in zone III and therefore, not requiring vertical reinforcement bars as per IS: 4326-1976, was found in need of such reinforcing due to the use of weaker mortar than specified. Exterior walls in the top two storeys where 1:3 lime-surkhi mortar has been used will be more susceptible to cracks. Use of 1:6 cement sand or 1:2:9 cement-lime-sand would have met the requirements. Collapse, is however, not considered probable.

The effect of the higher seismic coefficient on Parwanoo and Dhaula Khuan buildings will be cracking in different storeys as indicated by Tables 6 and 7 respectively. However, the presence of 16 mm dia. bars in the bottom two storeys and 12 mm dia. bars in the upper two storeys will help in taking up most of the tension, thus preventing collapse.

As seen from Table 8 Calcutta building is highly susceptible to damage especially in the XX-direction of analysis. In this building, the openings in the long wall violate the specifications given in IS: 4326 and leave the piers too narrow. On the other hand the unpierced (without openings) load bearing walls span in YY-direction making this direction strong and capable of taking seismic forces corresponding to smaller as well as higher seismic coefficient. Moreover, no vertical reinforcement has been provided in this building. Thus the situation of this building in XX-direction is rather unsafe.

The compressive stresses also increase with higher seismic coefficients but the brickwork does not show crushing since the ultimate strength of brick-work will be about six times the

Table 5:- Results of Seismic Analysis of Building (Zone III) at Lucknow (No vertical reinforcement provided in building)

Direction	Storey	Seismic Coefficient-.05			Seismic Coefficient-.20		
		State of compressive stresses in masonry	State of tensile stresses in masonry	Remarks	State of compressive stresses in masonry	State of tensile stresses in masonry	Remarks
X-X	1	within permissible	No tensile stress	Safe	within permissible	within ultimate strength.	No cracks.
	2	-do-	-do-	-do-	-do-	-do-	No cracks.
	3	-do-	-do-	-do-	-do-	exceeds ultimate strength at many points	Cracking will occur at many points.
	4	-do-	-do-	-do-	-do-	within ultimate strength	No cracks
Y-Y	1	within permissible	No tensile stress	-do-	within permissible	exceeds ultimate strength at few points.	cracking will occur at few points.
	2	-do-	-do-	-do-	-do-	-do-	-do-
	3	-do-	-do-	-do-	-do-	exceeds ultimate strength at many points.	cracking will occur at many points.
	4	-do-	within permissible	-do-	-do-	exceeds ultimate strength at some points.	cracking will occur at some points.

Few upto 20%, Some : 20% to 50 , Many : 50% and above

allowable working stresses under normal loads. In the cases considered, the compressive stresses are seen to remain within twice the normal working stresses for brick work in all buildings except the Calcutta building (Table 8) in which under the action of the probable maximum earthquake, the bottom two storeys would be overstressed in compression also.

(c) Thus, it is seen that the sections at the corners of piers in Zones IV and V must be strengthened with steel reinforcement to take care of the excessive tensile stresses. This would prevent collapse of the structure even though cracking may show up at many

Table 6:- Results of Seismic Analysis of Building (Zone IV) at Parwanoo (Vertical Reinforcement at Corners and Junctions of Walls Provided) (16 mm in storey 1 and 2; 12 mm in storey 3 and 4)

Dire - ction	Sto - rey	Seismic Coefficient-.05			Seismic Coefficient-.20		
		State of compressive stresses in masonry	State of tensile stresses in masonry	Remarks	State of compressive stresses in masonry	State of tensile stresses in masonry	Remarks
X-X	1	within permissible	within permissible	safe	within permissible	Exceeds ultimate strength at many points	Cracking may occur, but (1)
	2	-do-	-do-	-do-	-do-	-do-	-do-
	3	-do-	-do-	-do-	-do-	-do-	Besides cracking, (2)
	4	-do-	-do-	-do-	-do-	Exceeds ultimate strength at some points	12 mm bars will be within their yield strength.
Y-Y	1	within permissible	No tensile stress	safe	within permissible	exceeds ultimate strength at many points.	Besides cracking (3)
	2	-do-	-do-	-do-	-do-	-do-	-do-
	3	-do-	-do-	-do-	-do-	exceeds ultimate strength at few points.	Besides cracking, 12 mm bars may yield.
	4	-do-	-do-	-do-	-do-	within ultimate strength	Besides cracking, 12 mm (4)

Few upto 20%, Some : 20% to 50% , Many : 50% and above

Note: (1) 16 mm dia bars will take up tension though some of them may yield

(2) Some of the 12mm dia bars may yield

(3) 16 mm bars will take up the tension though some of them may yield.

(4) Bars will be within their yield strenght.

Table 7:- Results of Seismic Analysis of Dhaula Kaun Building (Zone IV) at Delhi (Vertical Reinforcement at Corners and Junctions of Walls Provided 16 mm dia. in Storey 1 and 2; 12 mm dia in Storey 3 and 4)

Direction	Storey	Seismic Coefficient-0.05			Seismic Coefficient-.20		
		State of compressive stresses in masonry	State of tensile stresses in masonry	Remarks	State of compressive stresses in masonry	State of tensile stresses in masonry	Remarks
X-X	1	within permissible	No tensile stress	safe	within permissible	Exceeds ultimate strength at few points	Cracking will occur, (1)
	2	-do-	-do-	-do-	-do-	Exceeds ultimate strength at some points	Cracking will occur, (2)
	3	-do-	within permissible	-do-	-do-	exceeds ultimate strength at few points.	Cracking will occur, (3)
	4	-do-	-do-	-do-	-do-	Within ultimate tensile strength	No cracks
Y-Y	1	within permissible	No tensile stress	Safe	within permissible	Within ultimate tensile strength	No cracks.
	2	-do-	-do-	-do-	-do-	exceeds ultimate strength at few points.	Cracking will occur, (4)
	3	-do-	-do-	-do-	-do-	Within ultimate tensile strength	No cracks.
	4	-do-	-do-	-do-	-do-	-do-	-do-

Few upto 20%, Some : 20% to 50%, Many : 50% and above

Note: (1) But 16 mm dia bars will be within their yield strength.

(2) But 16mm dia bars will take up tension, though some of them may yield.

(3) But 12mm dia bars will take up tension though some of them may yield

(4) But 16mm dia bars will be within their yield stregth.

Table 8:- Results of Seismic Analysis of Calcutta Building (Zone III)
(No Vertical Reinforcement has been Provided in this Building.)

Direction	Storey	Seismic Coefficient-0.04			Seismic Coefficient-0.16		
		State of compressive stresses in masonry	State of tensile stresses in masonry	Remarks	State of compr-essive stresses in masonry	State of tensile stresses in masonry	Remarks
X-X	1	within permissible	exceeds permissible at few	cracking may not occur as stresses are within ultimate stress	exceeds permissible at some points	Exceeds ultimate strength at many points	Cracking will occur, masonry overstr-essed in comp-ression
	2	-do-	-do-	-do-	exceeds permissible points	-do-	-do-
	3	-do-	exceeds permissible at some points	-do-	within permissible	do-	cracking will occur, compr-ession within safe limits.
	4	-do-	-do-	-do-	-do-	-do-	-do-
	5	-do-	-do-	-do-	-do-	-do-	-do-
Y-Y	1	within permissible	No tensile stress	safe	within permissible	no tensile stress	Cracking will not occur
	2	-do-	-do-	-do-	-do-	within perm-issible	-do-
	3	-do-	-do-	-do-	-do-	-do-	-do-
	4	-do-	-do-	-do-	-do-	-do-	-do-
	5	-do-	-do-	-do-	-do-	-do-	-do-

Few upto 20%, Some : 20% to 50%, Many : 50% and above

places. Another advantage not obvious from the calculations is the energy absorption through the yielding of reinforcement which brings down the effective value of the dynamic force if it tends to exceed the force corresponding to the yield level of steel.

(d) Another point of importance is the size and placing of window and door opening in the walls. In this respect IS: 4326-1976 specifications should be carefully followed.

(e) *Seismic Performance of Floors/Roofs:* All the three types of the systems for flooring/roofing viz. precast channels, joist-cum-planks and L-pan roofs have an inherent deficiency against earthquake ground motion affects. That is, since they are not usually integrated with each other in an effective manner, those units are not capable of providing the horizontal diaphragm action which integrates the behaviour of all the four walls of a room against lateral loads. In fact, whenever there is shaking, they will tend to vibrate separately and the joints between them will have a tendency to open out. For integrating the action of the precast units with each other and also incorporating in it the shearing resistance required for diaphragm action, certain details have been proposed. These are shown in Fig. 6.

Similarly, integral action will be required for the adequate performance of the L-pan roofing system. Details for the same have also been worked out, as shown in Fig. 7.

It may be mentioned that if topping concrete of a minimum thickness of 35 mm with 0.3% of the concrete area as steel reinforcement in each direction in the form of a mesh (preferably welded wire fabric) is used on the flat floors or roofs, it will ensure diaphragm action of floors besides leak proofing of the floor/roof and also be more effective as sound and heat insulation. It may be emphasised that in new constructions, this topping concrete should be fully embedded in the walls. If it is desired to improve the existing floor, the topping concrete mesh should be connected with the walls.

16.6 Conclusion

On the basis of seismic analysis of the buildings shown herein, the following conclusions may be drawn:

16.6.1 Use of one-brick load bearing walls:

No adverse behaviour of one brick thick load bearing walls is observed in the experimental buildings at Roorkee, Delhi and Parwanoo. For the normal frequent earthquake condition the compressive stresses in all walls and piers remain within the enhanced permissible stresses for this condition and the reinforcement provided in these buildings in found to be adequate to take care of the tensile stresses generated in the large earthquake condition.

The Lucknow building lying in Zone III is found in need of reinforcing due to the use of lime mortar. Use of 1:6 cement-sand or 1:2:9 cement-lime-sand would have met the requirements of tensile stresses in this building. Provision of vertical steel, though in a somewhat smaller cross-sectional area, at corners and junctions of walls in four storeyed buildings even in Zone III would be desirable.

The five storeyed building at Calcutta shows weakness in its longitudinal direction for the codal value as well as the probable maximum seismic coefficient used in the analysis. This is mainly because the provision of a series of openings constituting about 2/3 of the wall length instead of less than 50% as specified in IS: 4326. The reinforcement, which should have been provided in the piers around openings when opening's size or its position violated IS: 4326 specification, was not provided. The result is that the piers develop large tensile as well as

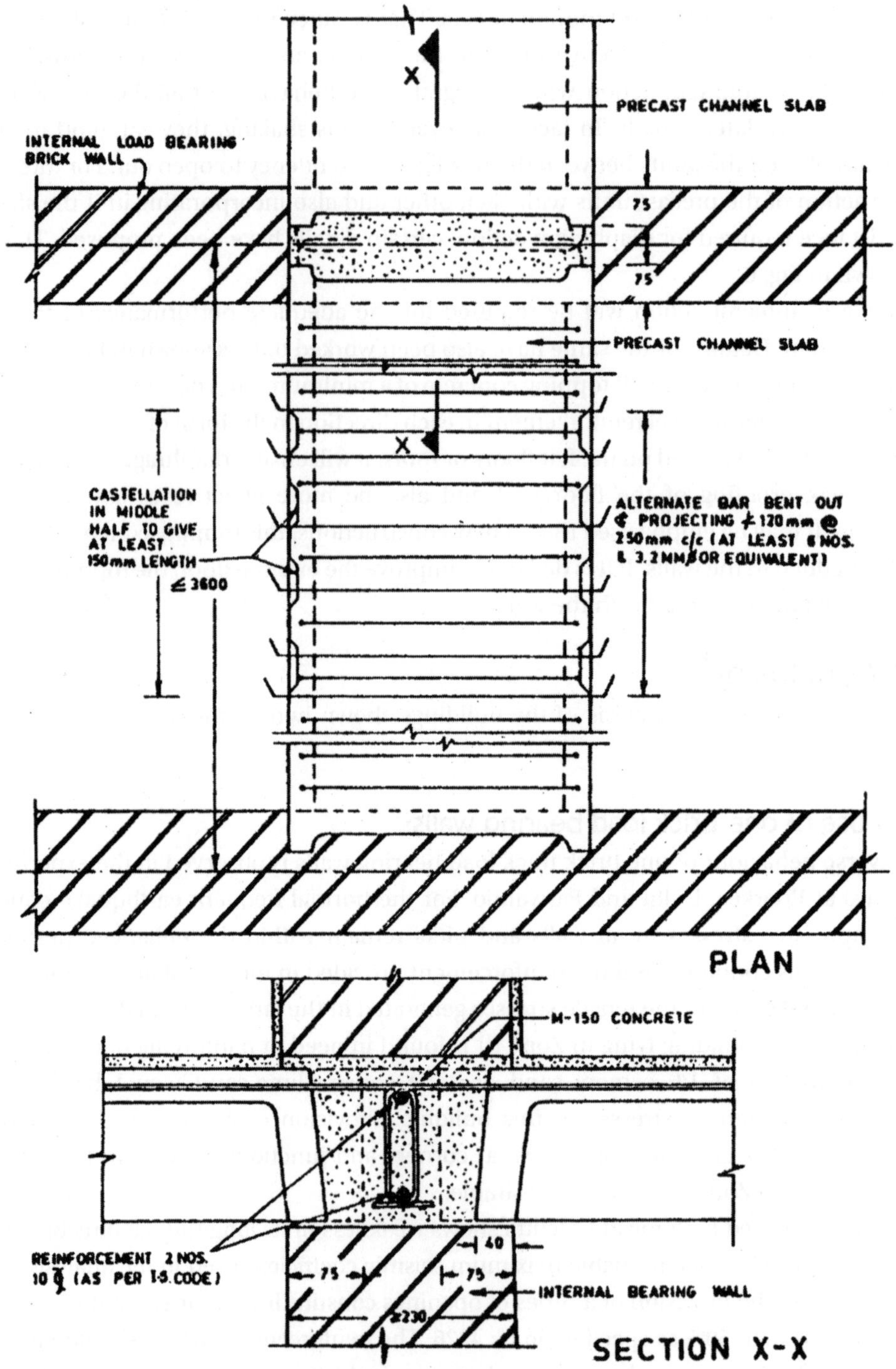

Fig.6: Details for integral action of precast units

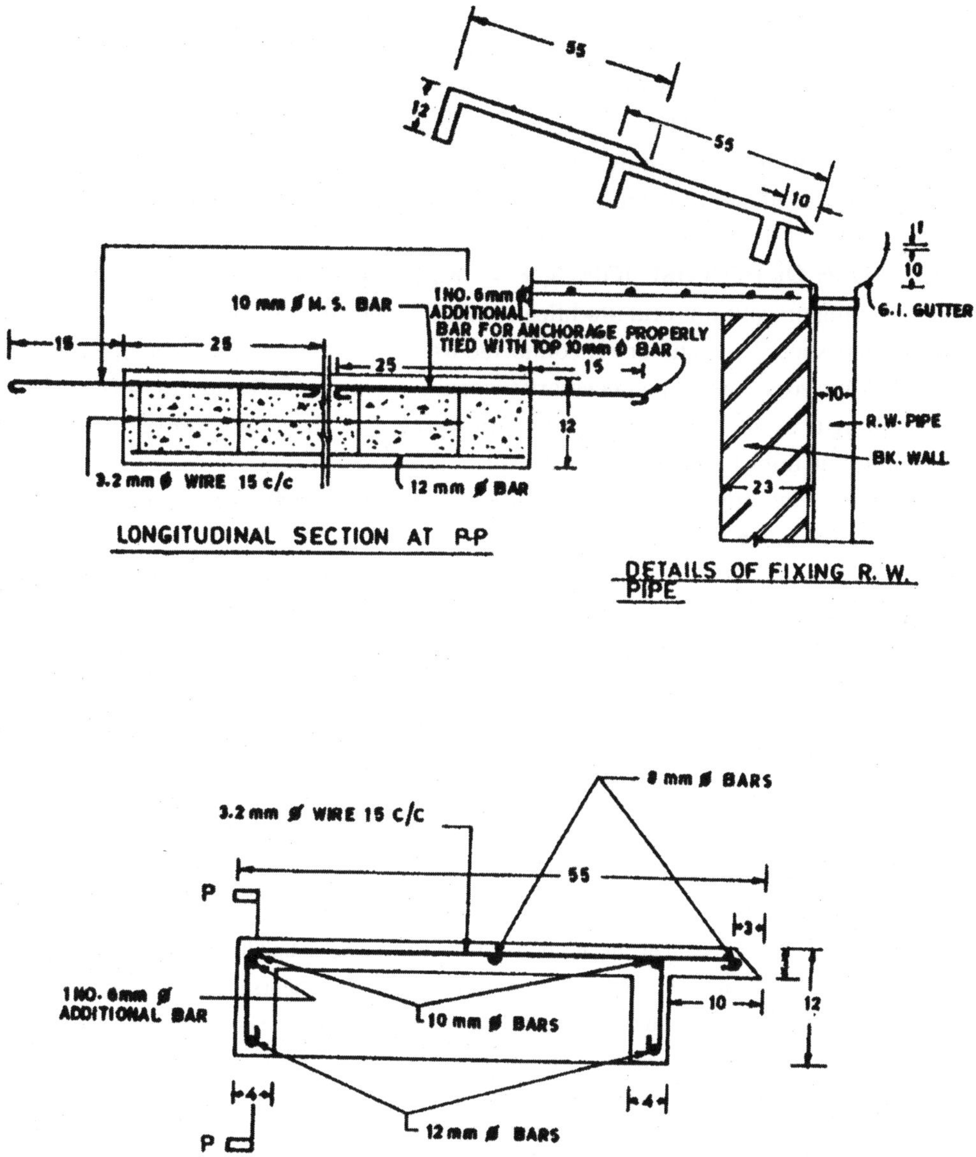

Fig.7:- Detail of L-plan roofing

excessive compressive stresses under earthquake forces in this direction. To meet the situation in future constructions, it is recommended that all specifications of IS: 4326-1976 must be adhered to. Moreover, all five storeyed buildings, if constructed, since not included in the above code, must be provided with vertical reinforcement even in Zone III as per calculations made for a seismic coefficient four times the basic coefficient specified in IS: 1893-1975.

It can, therefore, be concluded that one brick thick load bearing masonry buildings can be confidently built upto four storeys in seismic Zones III and IV following the detailed specifications contained in IS: 4326-1976 using the specified cement mortars only. The design procedure for load bearing walls may be adopted according to the National Building Code of India. In addition, for five storeyed buildings, a seismic analysis using four times the basic seismic coefficient, specified in IS: 1893-1975, is also recommended.

16.6.2 Use of Precast Joists, Planks and Channel Units:

The following precautions and modifications are suggested/for such units:

(a) The units should be transversely connected with reinforcing bars to avoid opening of joints.

(b) In seismic zones, it will be necessary to ensure their integral action to work as horizontal diaphragm by proper detailing of reinforcement in the joints between the units and around the units on the supporting walls.

(c) Reinforced topping concrete could also be done as an effective alternative to meet the point (b) above. ■

16.7 References

1. Agnihotri, V.K. (1962), "Strength of Single Storeyed Brick Shear Walls Against Earthquake Forces," M.E. Thesis, Civil Engineering Department, University of Roorkee, Roorkee.
2. Arya, A.S. (1968), "Constructions of Small Buildings in Seismic Area," Bull. Indian Society of Earthquake Technology, Vol. 5, No. 3 and 4.
3. Arya A.S., Chandra, B., Gupta, Satyendra P. and Pankaj (1982), "Evaluation of Seismic Performance of Multi-storeyed Brick Buildings Constructed Under Experimental Housing Scheme of National Buildings Organisation," Earthquake Engineering Studies, EQ 82-14, Department of Earthquake Engineering, University of Roorkee, Roorkee.
4. Chandra, B. (1963), Strength of Single Storeyed Unreinforced Brick Buildings Against Earthquake Forces, M.E. Thesis, Civil Engineering Department, University of Roorkee, Roorkee.
5. IS: 1893-1975, Indian Standard Criteria for Earthquake Resistant Design of Structures, ISI, New Delhi.
6. IS: 4326-1976, Indian Standard Code of Practice for Earthquake Resistant Design and Construction of Buildings, Indian Standards Institution, New Delhi.
7. ———————- (1980), Basic Concepts of Seismic Codes - Vol. I, Part II, International Association for Earthquake Engineering, Japan.
8. ———————- (1981), A Manual of Earthquake Resistant Non-Engineered Construction, Indian Society of Earthquake Technology, Roorkee.
9. Krishna, J. and Chandra B. (1965), "Strengthening of Brick Buildings Against Earthquake Forces," Proc. of the Third World Conference on Earthquake Engineering, New Zealand.
10. ———————- (1970), National Building Code of India, Indian Standards Institution, New Delhi.

Chapter 17

An Earthquake Resistant As Well As Economical Brick Building System[1,2,3]

17.1 Introduction

Load bearing brick masonry wall construction is probably the most common type of construction in almost all developing countries, the reasons being the availability of material, the comparative ease and low cost of construction, and its ability to insulate the weather conditions. In India also, construction in ordinary brick masonry, in various types of mortars, is still the one most used for rural, semi-urban and even urban buildings.

Study of past earthquakes shows that brick and stone masonry buildings have suffered the maximum damage in moderate to severe ground shaking. This is because of their heavy weight and large stiffness, resulting in short time periods and large amplification under short period waves in the earthquake motions. On the other hand, masonry has very low tensile and shear strengths and fails in a brittle manner. Thus, during shaking, it is easily cracked and must be suitably reinforced by a ductile material such as steel to impart to it tensile strength as well as ductility. Since one of the main factors for the popularity of this construction is its low cost, the strengthening measures adopted should not necessarily increase the cost of the construction too much. Also, at the present time, there is a need to conserve building materials like cement, brick, timber and steel because of their shortage, on the one hand, and increasing demand of dwelling units on the other. Hence, the challenge before the building engineers is to evolve ways and means of not only strengthening but also economising in brick masonry construction.

In conventional building systems of one to two storeys, which constitute the largest proportion of dwellings, both external and internal walls upto roof level are built with one brick (22.8 cm) thick walls. This thickness of walls is apparently more than sufficient from the vertical load bearing strength point of view. Therefore, it is proposed here that the thickness of walls be reduced to 11.4 cm (half brick), so as to reduce the requirements of building

---

1 *A.S.Arya, Professor & Head, Deptt. Of Earthquake Engineering, University of Roorkee, Roorkee*

2 *Ajay Kumar, Asst. Executive Engineer, ONGC, Atlanta Building, Nariman Point, Mumbai*

3 *VII Symposium on Earthquake Engineering, University of Roorkee, Nov.10-12, 1982,vol.I, pp 255-260.*

materials as well as cost of construction. To make these structures earthquake resistant, appropriate wall reinforcing arrangement has been proposed (see Fig. 1). The main aim of this paper is to study the proposed 11.4 cm thick brick wall constructions from the point of view of their earthquake resistance through half-size specimens tested on the shake table.

17.2 Description of Test Specimens

Four half-size specimens of single storeyed brick structures were constructed on the shock table using half-scale bricks. The outer dimensions of the specimens were 2.14 m x 1.8 m and height 1.74 m. The roof was a 7.6 cm thick reinforced concrete slab. The shear walls contained windows and the cross walls contained doors placed symmetrically one each in the centre. The layout plan of these models is shown in Fig. 2.

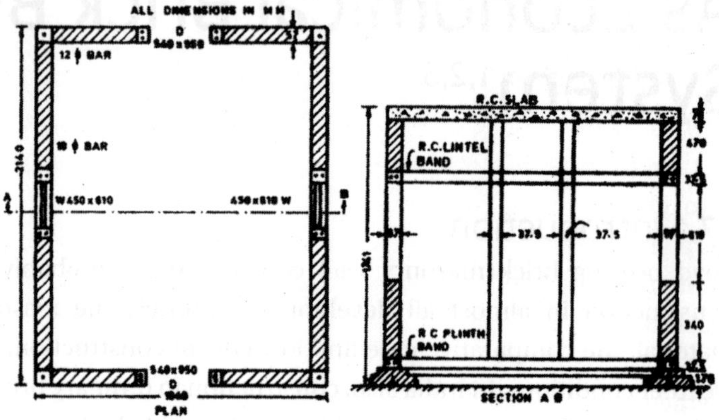

Fig.1: Brick building specimen

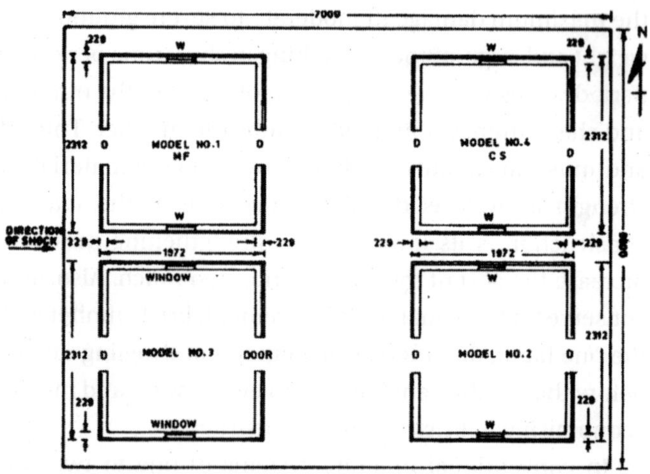

Fig.2: Layout plan of brick building models constructed on railway wagon shake table

Steel angles were welded to the table for anchoring the specimens to the table top against horizontal displacement. The lowest course of foundation bricks was laid on a bed of 1:3 cement-sand mortar. All the four models were reinforced in a similar pattern. The reinforcing arrangement was as follows:

i) *Lintel Band:* It consisted of 3.75 cm thick and 5.7 cm wide R.C. band in M 15 concrete with two 6 mm mild steel bars, one on each face.

ii) *Plinth Band :* It was similar to the lintel band.

iii) *Vertical Reinforcement :* The wall was divided into brick panels by R.C.C. members having 3.75 cm width, except for corner ones which were of square section of 5.7 cm x 5.7 cm. The corner members were provided with one 6 mm dia, mild steel bar, and the other with two 5 mm dia, bars, one on each face.

The arrangement of reinforcement and plan and elevation of specimen are shown in Fig. 1. The four specimens are further described below:

Model MF. Wall panels between the R.C. members were in mud mortar, above the plinth level. The vertical reinforcement bars were welded to the table base and taken inside the roof slab at top. This specimen is referred to as mud mortar-fixed base specimen or, MF.

Model MS: Wall panels in between R.C. members were in mud mortar. A sliding joint was provided at the plinth level. The structure above the plinth level was constructed after applying a layer of used mobil oil at the plinth. The bars of vertical reinforcement were embedded in a R.C. band just at the base of the superstructure walls above the sliding surface. This specimen is referred to as mud mortar sliding type or, MS.

Model CF: The panel walls of this structure above the plinth level were constructed in 1:6 cement-sand mortar, otherwise it was similar to model MF. For reference, it is called cement mortar fixed base or briefly, CF.

Model CS: The panel walls of this structure above the plinth level were in 1:6 cement-sand mortar, otherwise, it was, sliding type, similar to specimen MS. This is referred to as cement mortar sliding type or simply, CS.

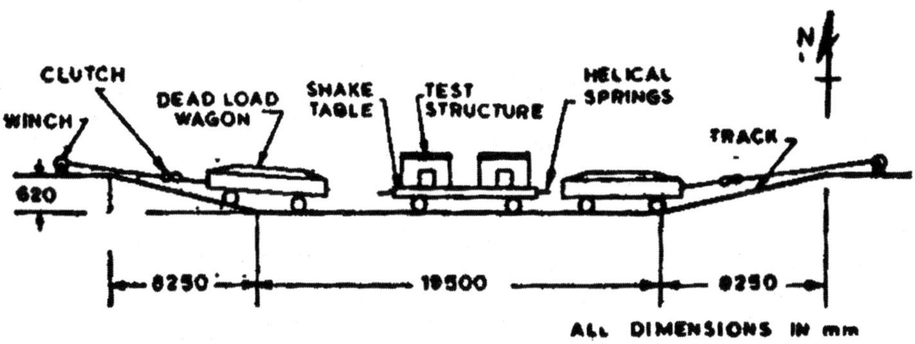

Fig.3:-Schematic view of railway wagon shake table test facility

Uniformity of workmanship was maintained in the construction of the various specimens as the same mason and labourers were employed throughout.

17.3 Testing Procedure

A general idea of the testing facility, which consists of rigid platform, 7m x 6m in plan, built on an old railway wagon frame is shown in Figs. 2 and 3. Models to be tested are built on this platform. A second wagon with sides, loaded with boulders and sand, rolls down the gentle incline (about 8% grade) and imparts an impact to the wagon platform through helical springs, driving it into collision with the third wagon, loaded with boulders and sand, which is kept at rest on the rear side, thus, giving to it a reverse impact. The roles of two dead load wagons are interchanged so that the main impact occurs from the opposite direction. The collisions impart half-sine wave type horizontal pulses to the table top in a manner somewhat like an earthquake, the time period of the pulses depending on the stiffness of the helical springs.

The test specimens were subjected to dynamic load through impacts, as stated above. Before applying the impact, the platform table was every time positioned on a fixed mark on the track. One dead load wagon was hauled up the incline, while the other was placed on the other side near the platform wagon at a fixed distance. The wagon release mechanism was then struck by a sledge hammer to release the dead load wagon instantly. The intensity of shock was increased in steps in each subsequent shock by further pulling the dead load wagon up the incline.

Accelerations were recorded for each shock given to the table. The accelerations were recorded with accelerometers, fixed at the roof of every model and to the table platform. The accelerometer signal was amplified through an universal amplifier and records obtained on direct ink-writing oscillographs.

After each shock, the damage of the structure was studied i.e. cracking deformations and displacements were observed. The cracks developed in walls were marked with black paint along with the shock numbers so as to differentiate the cracks of subsequent shocks and to study their progressive development.

The displacements of sliding type structures were measured at plinth level after each shock from a reference line marked on the four corners of the model.

17.4 Relative Competence of Test Structures

Two parameters are defined here, which were used in this study for quantitative comparison of damage to the different specimens as tested here as well as those tested earlier (Qamruddin 1978).

i) Input energy is defined for the purpose of working out cumulative input effort. It has the merit, that effect of all the shocks, rebounds, etc. can be taken into account by scalar addition. Since the base motion imparted to the table consisted of distinct half-wave shocks, the energy in each shock could be computed as

$$U_t = \frac{1}{2} m_t V_t^2 \quad (1)$$

Where,

U_t = total energy input in the shock

m_t = total mass of the table including mass of models

V_t = velocity of table attained at the end of the shock, equal to the area of the acceleration pulse.

Table 1:- Accelerations, Damage Levels And Displacements

Shock No.	Direction of shock	Acc. of Table(g)	Accelerations at the Roof and Damage Level									
			Model MF		Model MS			Model CF		Model CS		
			Acc. (g)	D.L.	Acc. (g)	D.L.	Sliding (mm)	Acc. (g)	D.L.	Acc. (g)	D.L	Sliding (mm)
1.	EW	0.35	0.31	0	0.59	0	0	0.31	0	0.36	0	0
2.	EW	0.62	0.51	I	1.07	I	1	0.65	0	0.48	0	1
3.	EW	1.33	1.46		1.60			1.30	0	0.55		
	R1WE	0.48	0.55	II	0.67	I	25	0.78		0.83	I	21
4.	WE	0.50	2.46		0.74			-	0	0.42		
	R1EW	0.17	-	II	-	II	37	-	0	-	II	30
5.	EW	1.00	0.82		1.33			2.3	0	1.75		
	R1WE	0.56	0.64	IV	-	II	47	0.46	0	-	II	52
6.	WE	0.86	3.44		1.30			-		0.69		
	R1EW	0.86	1.56	V	0.60	II	28	-	0	0.55	II	2
7.	WE	1.60	2.24		0.52			2.55		0.42		
	R1EW	1.07	1.28		0.52			1.09		0.42	II	14
	R2WE	0.80	0.64	Fail-ed	0.26	III	65	0.73	0	0.42		
8.	EW	2.32			0.77			2.95		0.53		
	R1WE	1.29			0.77			1.47		0.42		
	R2EW	0.77			-	III	70	1.05	I	0.32	II	90
9.	WE	2.98			1.33			4.00		0.95		
	R1EW	2.47			0.89			2.27		0.76		
	R2EW	1.08			0.89	III	80	1.60	II	0.48	II	0
10.	EW	6.43			1.07			3.30		0.86		
	R1WE	1.57			0.40	IV	110	1.57	III	0.57	III	180

Table 1 (Contd)

Shock No.	Direction of shock	Acc. of Table(g)	Accelerations at the Roof and Damage Level									
			Model MF		Model MS			Model CF		Model CS		
			Acc. (g)	D.L.	Acc. (g)	D.L.	Sliding (mm)	Acc. (g)	D.L.	Acc. (g)	D.L	Sliding ing(mm)
11.	WE	4.86			0.67			3.53		1.00		
	R1EW	6.60			1.13			3.05		1.60		
	R2EW	1.08			0.87			0.76		0.60		
	R3WE	0.51			0.67			0.20		0.50		
	R4WE	2.63			0.40			1.52		0.50		
	R5EW	1.37			0.40			1.33		0.40		
	R6Ew	1.02			0.53	V	190	0.48	V	0.30	III	120

Notes:
1. E : East, W : West, R1 : First Rebound, R2 : Second Rebound, D.L.= Damage Level
2. The damage level and sliding were measured at the end of each shock, hence, recorded in the last line of each shock. They represent the total effect of the shock including rebounds.

Thus, the energy per unit of mass shaken will be

$$U = \frac{1}{2} V^2_t \qquad (2)$$

ii) *Damage Level.* Damage level is defined on the basis of the crack length. That is, the length of cracks has been chosen for a comparative study of damage. Calling the damage state as 100% when part of the structure is on the point of collapse with widespread cracking all over, the length of cracks at this stage gives a basis of assigning different percentages to damage at earlier stages based on the length of cracks at that stage. Based on such calculations the following six damage states are defined here:

0 *No damage* - No cracks are seen in the walls at all.

I *Slight damage* - Extent of damage upto 5%, very fine cracks of small length.

II *Moderate damage* - Extent of damage 6% to 35%, Cracks of small length and narrow width.

III *Severe damage* - Extent of damage ranging from 36% to 65%, wide and deep cracks of large length in the walls.

IV *Destruction damage* - Extent of damage between 66% and 95%, wide gaps created, separation of walls into blocks but without collapse.

V *Total damage* - Extent of damage 96% to 100%, Some parts of the structure fall down, some on the verge of collapse.

17.5 Observed Behaviour of Test Structures

The progressive damage of specimens after each shock and the accelerations are recorded in Table 1. The shocks were applied east to west (E-W) or from W to E. Thus, the north and south side walls of the specimens acted as shear walls and those on the east and west sides acted as 'cross' walls. The significant visual observations are the following.

a) In all cases of specimens fixed at the base, the severe damage occurred in the shear walls, the cross-walls receiving only minor cracks. But in the specimens with the sliding superstructure, the critical damage occurred in 'cross' walls which were bulged due to out-of-plane bending whereas the shear walls suffered little damage.

b) The reinforced concrete corner elements showed a tendency to separate from the wall panels above the lintel level band, indicating insufficient bond with the masonry. It appears that if the wall panels are left 'toothed' before casting the concrete elements, this tendency shall be much reduced and shear strength of the bounded panels increased.

c) The sliding motion of sliding specimens was found reversible in reversed shocks, the residual displacement being, thus, reduced after each complete cycle.

The amplification of acceleration from the base to the roof of specimens is shown in Fig. 4 and the *extent of damage vs. the total input energy* curves plotted in Fig. 5. A summary of significant results of the test on the present half brick specimens and the one brick specimens tested by Qamruddin earlier (1978) is presented in Table 2. From these results, the following observations are made:

1) The results show that the panelled wall model in 1:6 cement sand mortar with the fixed or sliding base(CF/CS) withstood the maximum table accelerations, up to 6.6g, without collapse.
2) The comparison of any model CF with the corresponding model CS shows that the fixed base model withstood shocks better than the sliding base model in the initial shocks but ultimately it was the sliding model which was less damaged.

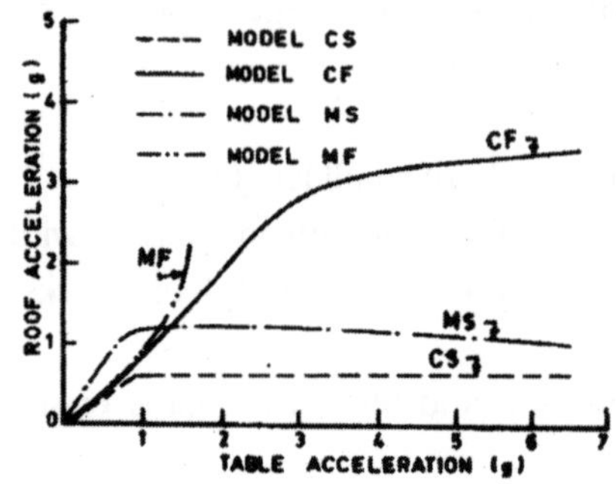

Fig.4: Base vs roof accelerations

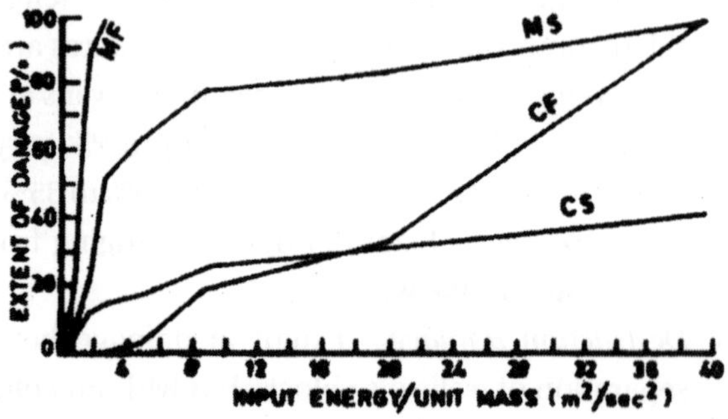

Fig.5: Progressive damage of panelled wall models

3) The model in mud mortar with sliding base (Model MS) also showed extremely good resistance and it could also withstand base shocks of 6.6g acceleration though with more severe damage than the specimen in cement mortar, but still without collapse. The specimen MF did not perform so well and reached its ultimate failure stage at base acceleration of 1.6 g.
4) The panelled wall models in 1:6 cement sand mortar (Models 3 & 4 in Table 2) resisted 7.5 times as much input energy per unit mass and 2.7 times as high table accelerations, upto ultimate failure as the one brick thick wall reinforced models (Model 11 & 12) and about 19 times more input energy and 5 times higher table accelerations than one brick thick conventional model (Model 7).
5) The panelled wall model in mud mortar with fixed base (Model 1) resisted about 1.4 times more input energy and about 1.2 times higher table acceleration than the one brick thick reinforced model in mud mortar (Model 6) and about 3.3 times more input energy and 1.5 times higher table acceleration than the unreinforced model (Model 5).
6) The panelled wall model in mud mortar with sliding base (Model 2) showed exceptionally good resistance against shock loads. It resisted about 12 times more input energy and 5 times higher accelerations as compared to the one brick thick reinforced model in mud mortar (Model 6), and about 20 times more energy and 3 times higher acceleration than the one brick thick wall reinforced model with the sliding base (Model 9).
7) The amplification of acceleration from the base to the roof decreased as the base acceleration increased and the damage progressed.

---

* *Refer(Qamaruddin,1978.)*

Table 2:- Summary Of Significant Results of Tests on Half and One Brick Thick Wall Specimens

Sl. No.	Description of specimen	Base Accelerations (g)		Cumulative Input Energy per unit mass $(m/sec)^2$	
		At First Crack a	Maximum a_{max}	At first crack U	Maximum U_{max}
A. PRESENT SERIES (HALF BRICK REINFORCED PANEL)					
1.	MF	$0.35 < a < 0.62$	1.60	$0.13<U<0.18$	2.63
2.	MS	$0.35 < a < 0.62$	6.60	$0.13<U<0.18$	39.43
3.	CS	$0.62 < a < 1.33$	6.60	$0.18<U<0.91$	39.43
4.	CF	$1.60 < a < 2.32$	6.60	$2.63<U<4.77$	39.43
B. EARLIER SERIES (ONE BRICK)*					
5.	MF-NS	$0.51 < a < 0.57$	1.05	$U < 0.35$	0.80
6.	MF-LVJ	$0.57 < a < 1.05$	1.26	$U < 0.56$	1.88
7.	CF-NS	$0.57 < a < 1.05$	1.24	$U < 1.56$	1.96
8.	CF-LVJ	$a < 0.83$	1.85	$U < 1.04$	4.75
9.	MS-LPB	$a < 0.51$	2.15	$U < 0.11$	1.90
10.	CF-L	$0.51 < a < 2.15$	2.15	$U < 0.11$	1.90
11.	CS-LPB	$0.51 < a < 2.15$	2.45	$0.11<U<1.74$	5.18
12.	CF-LPVJ	$0.51 < a < 2.15$	2.45	$0.11 < U < 1.74$	5.18

Notes:
MF = Mud Mortar, Fixed at Base
CF = Cement Mortar, Fixed at base
MS = Mud Mortar, Sliding at Plinth
CS = Cement Mortar, Sliding at Plinth
L = Lintel Band with 3 bars 6 mm ϕ
P = Plinth Band with 3 bars 6 mm ?
V = Vertical steel at corners 1 bar 6 mm ?
J = Jamb Steel 1 bar 6 mm ?
B = Bond Beam just above plinth band, with 3 bars 6 mm ϕ

17.6 Conclusions

The main conclusions of the experimental study reported above are that the half brick panelled wall system proposed herein is more resistant against earthquake forces as compared to the conventional unreinforced one brick thick wall system as well as that which is reinforced for seismic forces in accordance with IS: 4326. Therefore, brick buildings can be constructed with half-brick thick panel walls resulting in considerable economy in construction materials like bricks and cement. The saving in cost was estimated as 30% of the cost of construction using one brick thick construction reinforced as per IS: 4326-1976. ■

17.7 References

1. Arya A.S., Chandra B., Qamaruddin M., "A New System of Brick Buildings for Improved Behaviour During Earthquakes", Seventh World Conference on Earthquake Engineering, Istanbul, Turkey, September 1980, p.4/2250.

2. "IS: 4326-1976, Indian Standard Code of Practice for Earthquake Resistant Design and Construction of Buildings", Indian Standard Institution, New Delhi, 1977.
3. Qamaruddin M., "Development of Brick Building System for Improved Earthquake Performance", Ph.D. Thesis, Department of Earthquake Engineering, Univ. of Roorkee, September 1978.
4. Qamaruddin M., Arya A.S., Chandra B., "Experimental Evaluation of Aseismic Strengthening Method of Brick Buildings, Sixth Symposium on Earthquake Engineering, Roorkee, October 1978, pp 353-360.

Chapter 18

Dynamic Response of Multi-storeyed Brick Buildings[1,2,3]

18.1 Introduction

Masonry continues to be the most common method of building construction in almost all parts of the world for its many advantages, in spite of its poor record of safety during past earthquakes (1-3). Recognising that masonry construction will continue to be the most common mode of building construction for a long time in all countries except perhaps the relatively more developed countries, it is necessary that methods should be evolved for incorporating suitable features in masonry buildings so that they become adequately earthquake resistant if not earthquake proof. Already a lot of research effort has been made in this direction in the last 25 years and the strengthening methods as recommended (4-15) in some of the earthquake regulations of the countries are meant for brick buildings to improve their performance, with a small increase in the overall cost. But systematic dynamic studies do not appear to have been carried out for investigating their realistic seismic capabilities and a dynamic evaluation of the recommended specifications is still required.

Typical floor plans of one-to-four-storeyed real buildings have been chosen for carrying out their seismic response analysis. For parametric studies of such buildings, the number of storeys, the wall thickness and damping are chosen as variables representing their physical properties. Koyna and El Centro accelerograms are used here for dynamic analysis of the buildings.

The coupled equations of motion for a shear-beam-type multi-degree-of-freedom oscillator are uncoupled into modal equations and are then solved numerically, whence the timewise earthquake response is computed employing the modal superposition method. Overturning and torsional effects due to earthquake forces are worked out and superimposed. The timewise net stresses in the building elements are then computed and their capabilities examined for resisting an earthquake shock.

---

1 *M.Qamaruddin , Deptt. of Civil Engineering, University of Petroleum & Minerals, Dharan, Saudi Arabia,*

2 *A.S.Arya, Professor, B. Chandra, Professor, Deptt. of Earthquake Engineering, University of Roorkee, Roorkee*

3 *Earthquake Engineering and Structural Dynamics, 1985, Vol.13, pp 135-150.*

The present seismic analysis confirms that the tensile and shearing strengths are certainly the critical parameters in the safety of brick buildings during severe earthquakes. The requirements of reinforcing steel have been estimated for the critical sections and then compared with the provisions of the Indian Standard Code.

18.2 Mathematical Model of Brick Building

Fig. 1 shows a schematic view of a typical single-roomed three-storeyed indealised brick building. Non-structural elements as may be present in actual buildings are not shown in the figure as they are assumed to make no contribution in resisting seismic forces. For ground shaking parallel to the x-axis, walls A-1 and B-1 (Fig. 2) would behave as shear walls whereas walls C-1 and D-1 would behave as cross-walls. But when the ground motions is along the y-axis, the situation would be reversed. Therefore, each wall has to be analysed as a shear wall for the relevant direction of the earthquake shock. The following assumptions are made for arriving at the mathematical model.

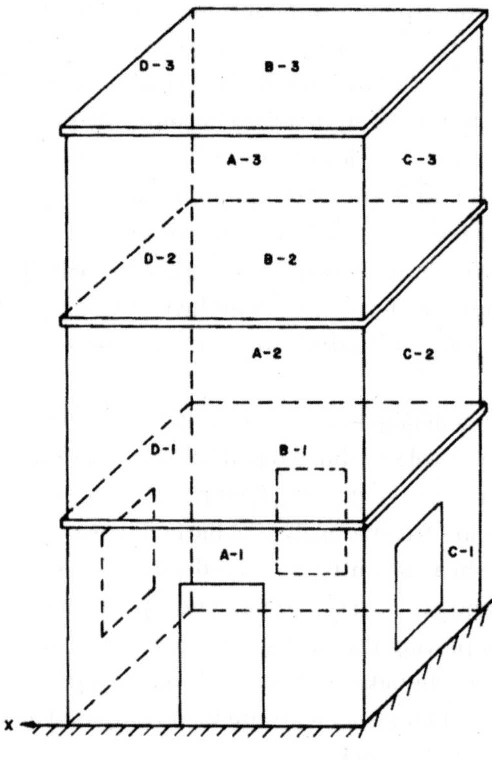

Note - Door Window Opening in Upper Storeys not Shown For Clarity of the Figure

Fig.1: Schematic view of a typical representative multi-storeyed brick building

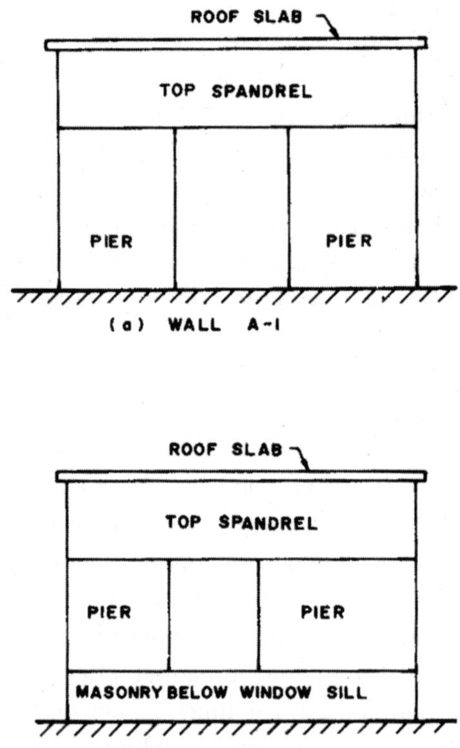

Fig. 2: First storey shear walls for ground shaking parallel to the x-axis

(1) The building material is elastic and lies within the limits of proportionality. The limitation of this assumption is that if tension arises in the building material, then its value is assumed not to be large enough to induce cracking.

(2) The buildings are subjected to only one horizontal component of ground motion at a time and the building vibrations only in the direction of ground motion.

(3) The building behaves as a shear structure while vibrating laterally. Since the spandrels of the walls are relatively much stiffer than the building elements and also axial deformations of the piers due to the overturning moment are likely to be small compared to their bending and shearing deformation, the lateral displacements at the top of the piers in a storey of the building are assumed to be equal. This is further strengthened by observation of the actual behaviour of a three-storeyed brick building after the Gediz earthquake (16) of 1970 (Turkey).

(4) The shear walls resist the earthquake force without any contribution from the cross-walls, that is, the `flange-effect' of cross-walls is neglected. Owing to the usual toothed joints adopted between walls at right angles, there remains a plane of weakness at the junction of the shear and cross-walls. As such, the joints may open out (17) very easily during ground shaking, leaving the shear walls to act separately in resisting seismic forces.

(5) Interfloor damping is represented in the form of viscous damping. The material damping values assumed in the study depend on the ratio of mortar mix and the rate of straining as well. (10,18).

(6) The building configuration is symmetrical about its principal axes. Although this assumption has been made to achieve a simple mathematical model for the building, a building having an unsymmetrical configuration could also be analysed by computing the additional shear on account of the torsional effects.

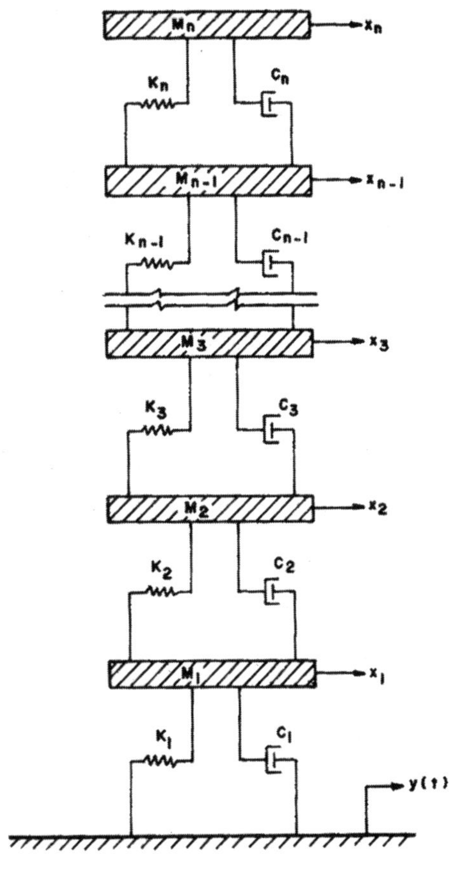

Fig.3: Mathematical model of a multi-storeyed brick building

A multi-storeyed brick building is represented by a multi-degree-of-freedom shear-beam system. The piers, which are located parallel to the direction of earthquake shaking, are assumed to provide spring action. The mass of the walls and slabs are assumed lumped at the storey levels. The lumped masses are assumed to be connected to each other through massless spring and viscous dampers. The degree of each mass in horizontal translation is one, neglecting the vertical translational and rotational degrees of freedom. Thus, the brick building system leads to a simplified mathematical model as shown in Fig. 3.

18.3 Equations of Motion

The equations of motion for a multi-degree-of-freedom system subjected to ground motion can be written as

$$[M]\{\ddot{Z}\}+[C]\{\dot{Z}\}+[K]\{Z\}=-[M]\{\ddot{Y}\} \qquad (1)$$

where [M] is a diagonal mass matrix, [C] is a viscous damping matrix, [K] is a tridiagonal stiffness matrix, {Z} is the vector of the relative displacement of masses with respect to ground and y is the horizontal component of ground displacement. Superscript dots represent differentiation with respect to time. The storey stiffness, K_{st}, in a building is evaluated as

$$K_{st}=\sum_{i=1}^{S_W}\sum_{j=1}^{b_c}k_{ij} \qquad (2)$$

where

b_e = total number of building elements in a wall

S_w= total number of shear walls in a storey

k_{ij}= lateral stiffness of a building element considering its bending and shear deformations.

Equation (1) represents simultaneous linear differential equations of second order which are transformed into a set of N uncoupled equations in normal coordinates, {β}.

The rth mode equation is written as

$$\ddot{\beta}_r+2p_r\,\xi_r\dot{\beta}_r+p_r^2\beta_r=-Q_r\,\ddot{y}(t) \qquad (3)$$

the solution of which is given by

$$\beta_r=\frac{Q_r}{p_{dr}}\int_o^t\ddot{y}(\tau)\exp\{-p_r\xi_r(t-\tau)\}\sin p_{dr}(t-\tau)\,d\tau \qquad (4)$$

where p_r and ξ_r are the natural circular frequency and fraction of critical damping, respectively, in this mode of vibration. Q_r and p_{dr} are the rth mode participation factor and damped natural frequency, respectively, and τ is the time variable for integration. The relative displacements are computed as

$$\{Z\} = \sum_{r=1}^{N} \{\phi_r\}\{\beta_r\} \qquad (5)$$

where $\{\phi_r\}$ is the rth modal column vector for a square matrix $[\phi]$ and N is the number of modes considered in the analysis. The Runge-Kutta fourth order method (19) is employed for numerical integration of equation (4).

18.4 Seismic Stress Analysis of The Building Elements

Knowing the absolute accelerations at various storey levels of the building subjected to ground motion in any mode, the storey shears are computed by multiplying the absolute acceleration by the mass lumped at the corresponding storey level. The storey shears resulting from seismic force without torsion at any instant of time are then obtained by superimposing the modal shears at that instant of time. Where the condition of no-torsion is not satisfied, the distance between the centres of mass and rigidity gives the eccentricity of the earthquake force in plan. The torsional moment is then equal to the seismic force times the eccentricity and the resulting shears in the building elements are computed using the usual simplified procedure (20). The shear force in each shear wall of the storey is then determined by distributing the storey shear among the shear walls in proportion to their stiffness (21). The timewise stresses in each building element are then computed as explained elsewhere (17). The maximum of the timewise stress in the building element thus gives its maximum stress for the ground motion. A computer code has been developed for timewise seismic response computation of a multi-storeyed brick building. Provisions are also made in the code to perform timewise seismic stress analysis of the building elements.

18.5 Data for Study of Brick Buildings

18.5.1 Buildings studied

A typical plan of a residential building (Figs. 4 and 5), as actually constructed at New Delhi has been chosen for the study. The number of storeys varies from one to four. The same plan has been adopted for different storeys. The thickness of the main walls in different storeys has two possible values. The buildings studied are referred to in the later descriptions as follows:

(a) B1 - single-storeyed building one brick thick main walls (Fig. 4);
(b) B2 - two-storeyed building having one brick thick (223 mm) main walls in both the storeys (Fig. 4);
(c) B3A- three-storeyed building with one brick thick main walls in all the storeys (Fig. 4);
(d) B3B- three-storeyed building having one and a half bricks thick (343 mm) main walls in the first storey (Fig. 5) and one brick thick walls in the upper storeys (Fig. 4);

(e) B4A- four-storeyed building with one brick thick main walls in all the storeys (Fig. 4);
(f) B4B- four-storeyed building with one and a half brick thick main walls in the first and second storeys (Fig. 5) and one brick thick walls in the upper storeys (Fig. 4).

18.5.2 Damping

Three values of viscous damping have been taken in the fundamental mode. These are: 5, 10 and 15% of critical damping. The fraction of critical damping in higher modes has been assumed in proportion to their respective modal frequencies.

18.5.3 Ground Motion Characteristics

The seismic response has been computed for two actually recorded severe accelerations, viz the El Centro shock of May 18, 1940 (N-S component) and the Koyna shock of December 11, 1967 (longitudinal component). The Koyna accelerogram was recorded close to the epicentre of the shock and had high acceleration pulses and frequency content, while the El Centro accelerogram was recorded at about 50 km from the epicentre of the shock and had relatively lower acceleration ordinates as well as frequency content. In this way, the influence of the pattern of the accelerogram on the structural response has been studied with these two different earthquakes.

The damaging potential of a shock is represented more realistically by its spectral intensity (22, 23) which gives a quantitative idea of structural response over a range of structural periods. Although Housner (22) has defined the spectral intensity (SI) as the area under velocity spectrum curves between the periods 0.1 and 2.5 s, yet SI values of the two chosen earthquakes have been computed between the periods 0.004 and 0.30 s for 5% of critical damping so as to cover the range of periods of interest in the buildings under study. It is found that the spectral intensity (17) thus computed for the Koyna shock is almost two times that of the El Centro shock. Incidentally, these two earthquakes could be considered appropriate for MM Intensity IX and VIII zones so far as short period structures are concerned. Considered in this light, the Indian Standard Specifications for strengthening brick buildings (IS: 4326) could be checked with the strengthening requirements based on dynamic response calculations.

18.5.4 Material elasticity and strength

For brickwork in various mortars, the values of compressive, tensile and shear strengths and modulus of elasticity are listed elsewhere (17). Cement mortar of 1:6 proportion is assumed for the present analysis for which 5% of critical damping is a representative value in the uncracked condition and 10% of critical damping for the cracked condition with reinforcing bars. The higher value of 15% of critical damping will indeed represent the cracked state of mud mortar construction and has been included just for comparison.

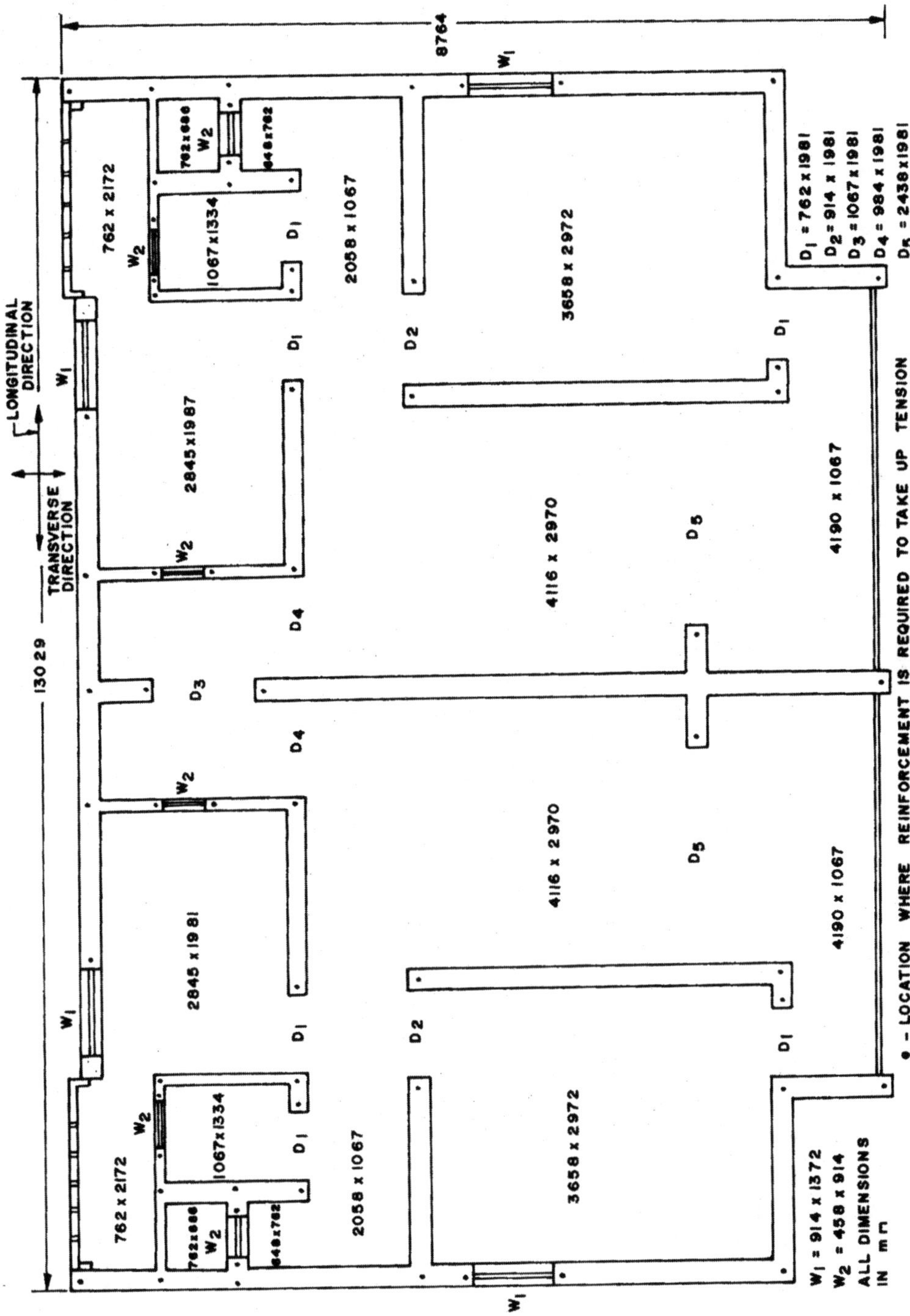

Fig.4: Typical plan at ground and other floors of a multi-storeyed brick building (main walls 1-brick thick)

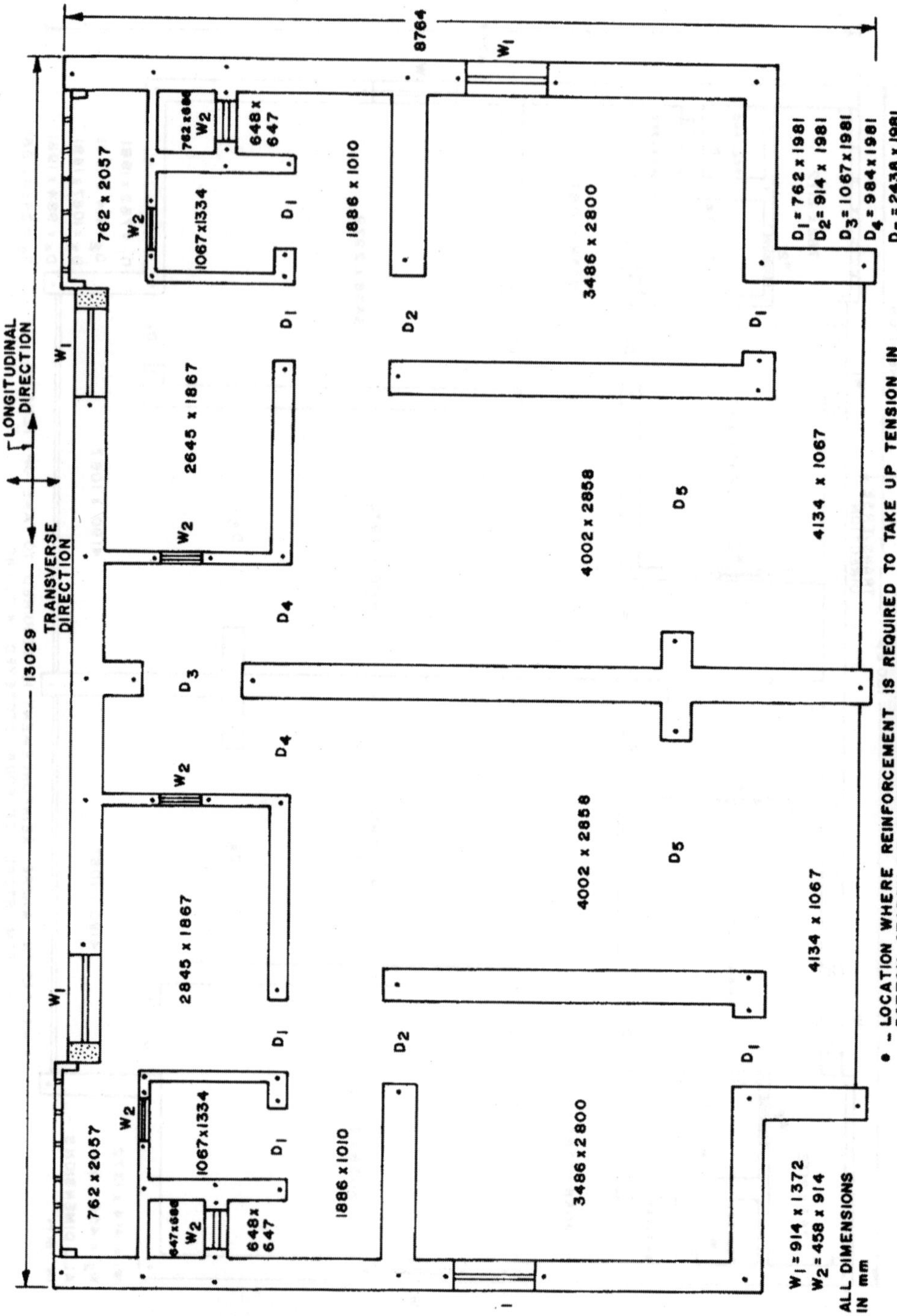

Fig.5: Typical plan at ground floor of a multi-storeyed brick building (main walls 1½ bricks thick)

18.6 Results and Discussion

18.6.1 Parametric study

The typical results of seismic analysis of unstrengthened brick buildings (B1 to B4B as defined earlier) when subjected to the Koyna and El Centro shocks are presented in Tables 1 and 2. The variations of the stress along the height of the buildings are plotted in Figs. 6 and 7. The influence of different parameters of the buildings on the stresses is discussed as follows:

18.6.2 Number of storeys

(i) In all the buildings maximum stresses go on reducing in the upper storeys as compared to that in the lower storeys, but the rate of decrease and its patter is different with the type of stress. For instance, referring to Fig. 7(b), the shape of the curve for the bending stress, which is based on the storey shears, and that for the overturning stress, which depends on the cantilever bending of the whole building, are different.

(ii) The single-storey building is similar in all respect to the top storeys of multi-storeyed buildings in cases B2, B3B and B4B. Similarly, the results for the second storey from the top in B3B and B4B are similar in all respects. From the typical results (for the Koyna earthquake and 5% of critical damping) shown in Table 3, it turns out that the earthquake stresses in the upper two storeys of multi-storeyed buildings are higher, but the maximum margin is about 28%. The higher stresses in the upper storeys occur in spite of the fact that the taller buildings have a longer time period and less spectral acceleration, but the mode shape transfers the higher shears towards the upper storeys.

18.6.3 Thickness of main walls

(i) Dissimilar patterns of stress distribution are obtained in the two cases of main wall thickness (Figs. 6 and 7) for various damping values and both the ground motions. This should be expected since there is a change of stiffness suddenly at a level in the non-uniform thickness case.

(ii) It is significantly observed from Figs. 6 and 7 that the stresses in the elements of buildings with non-uniform wall thickness are reduced in the lower thick wall storeys but increased in the upper thinner wall. This trend is seen for all damping values and both the earthquakes. The reason for stress reduction in the bottom storeys is the reduction of inertia force due to weight reduction in the upper storeys. In the upper storeys, the stresses are increased due to a larger reduction in the pier cross-section than the decrease in the inertia force.

18.6.4 Damping.

The seismic stresses in all the buildings decrease with an increase of the damping. Generally, the stresses reduce faster when the damping is increased from 5 to 10% than 10 to 15%.

Table I. Maximum Stresses in Three-Storeyed Brick Building of Uniform Thickness of Main Walls (Type - B3A)

Stresses	Damping (%)	Maximum stresses in various storeys (kg/cm^2)					
		Storey I		Storey II		Storey III	
		Koyna earthquake	El Centro earthquake	Koyna earthquake	El Centro earthquake	Koyna earthquake	El Centro earthquake
Bending	5	30.63	15.74	21.52	11.57	8.98	5.04
	10	24.91	13.13	17.13	9.48	7.06	4.07
	15	20.99	12.28	14.28	9.65	5.91	3.63
Overturning	5	7.79	4.10	4.25	2.33	1.03	0.58
	10	6.26	3.39	3.37	1.90	0.81	0.47
	15	5.23	3.13	2.81	1.71	0.68	0.41
Compressive (net)	5	40.91	22.35	27.42	15.55	10.82	6.42
	10	33.67	19.02	22.15	13.03	8.68	5.35
	15	28.72	17.91	18.74	12.02	7.39	4.85
Tensile (net)	5	36.27	17.70	24.47	12.59	9.55	5.16
	10	29.02	14.38	19.19	10.08	7.42	4.08
	15	24.07	13.26	15.79	9.07	6.13	3.59
Shear	5	18.20	9.35	12.79	6.87	5.34	2.99
	10	14.80	7.80	10.18	5.63	4.20	2.42
	15	12.48	7.30	8.48	5.14	3.51	2.16

Table 2. Maximum Stresses in Four-Storeyed Brick Building of Non-Uniform Thickness of Main Walls (Type-B4B)

Stresses	Damping (%)	Maximum stresses in various storeys (kg/cm^2)							
		Storey I		Storey II		Storey III		Storey IV	
		Koyna earthquake	El Centro earthquake	Koyna earthquake	El Centro earthquake	Koyna earthquake	El Centro earthquake	Koyna earthquake	El Centro earthquak
Bending	5	31.45	21.55	25.47	17.33	22.92	15.35	9.95	6.33
	10	26.71	15.36	20.92	12.21	18.36	10.94	7.55	4.55
	15	24.61	13.94	19.05	11.02	16.54	9.76	6.77	4.04
Overturning	5	8.86	6.05	6.48	4.37	4.60	3.02	1.14	0.72
	10	7.38	4.28	5.25	3.10	3.60	2.16	0.86	0.52
	15	6.76	3.87	4.75	2.78	3.24	1.92	0.77	0.46
Compressive (net)	5	43.03	30.06	33.90	23.48	29.18	20.02	11.90	7.86
	10	36.51	22.18	27.92	17.14	23.61	14.76	9.23	5.88
	15	33.80	20.38	25.56	15.64	21.44	13.34	8.35	5.31
Tensile (net)	5	37.59	25.23	30.08	20.09	26.22	17.06	10.64	6.60
	10	31.73	17.28	24.57	13.71	20.66	11.80	7.96	4.62
	15	29.00	15.46	22.20	12.19	18.48	10.38	7.09	4.04
Shear	5	18.69	12.80	15.13	10.29	13.62	9.12	5.91	3.76
	10	15.87	9.12	12.43	7.26	10.91	6.50	4.49	2.71
	15	14.62	8.28	11.32	6.55	9.83	5.80	4.02	2.40

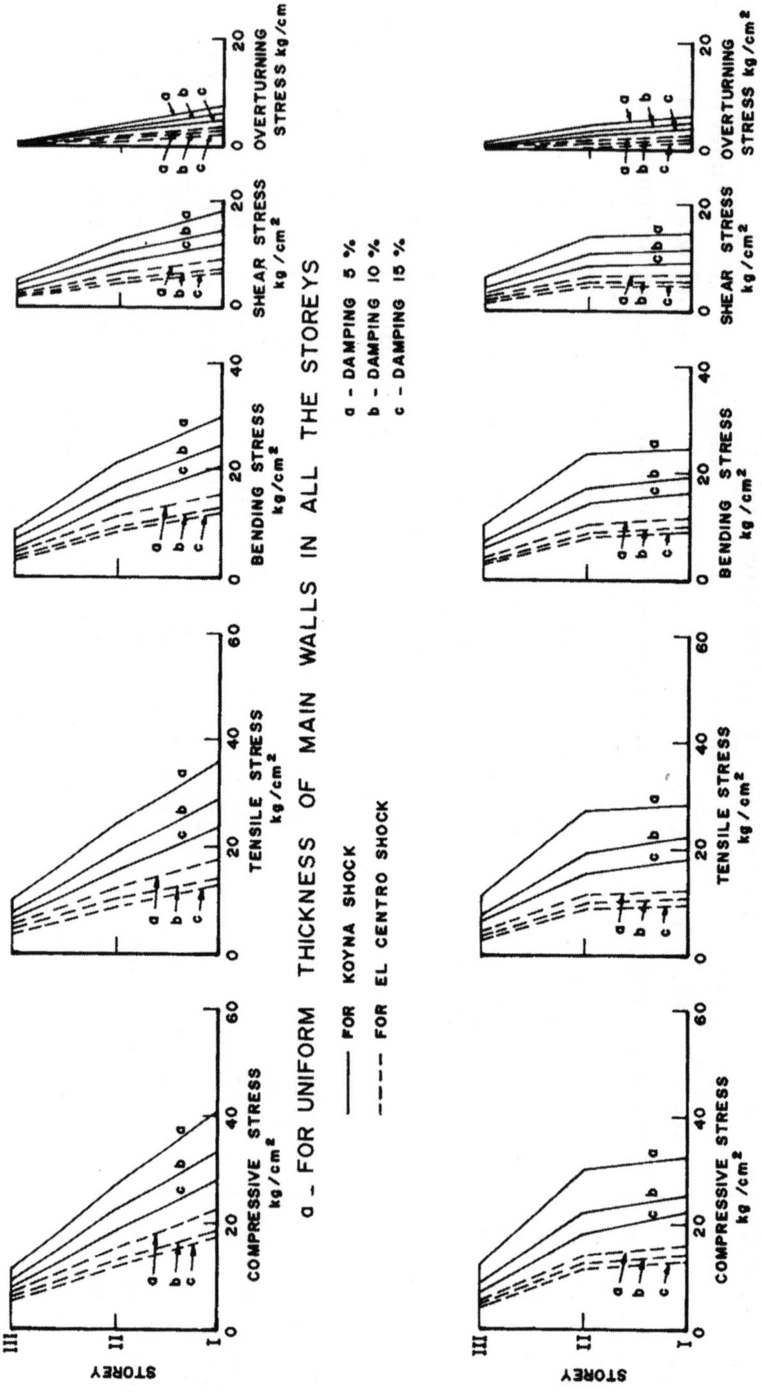

Fig.6: Maximum stresses in different storeys of three-storeyed brick building

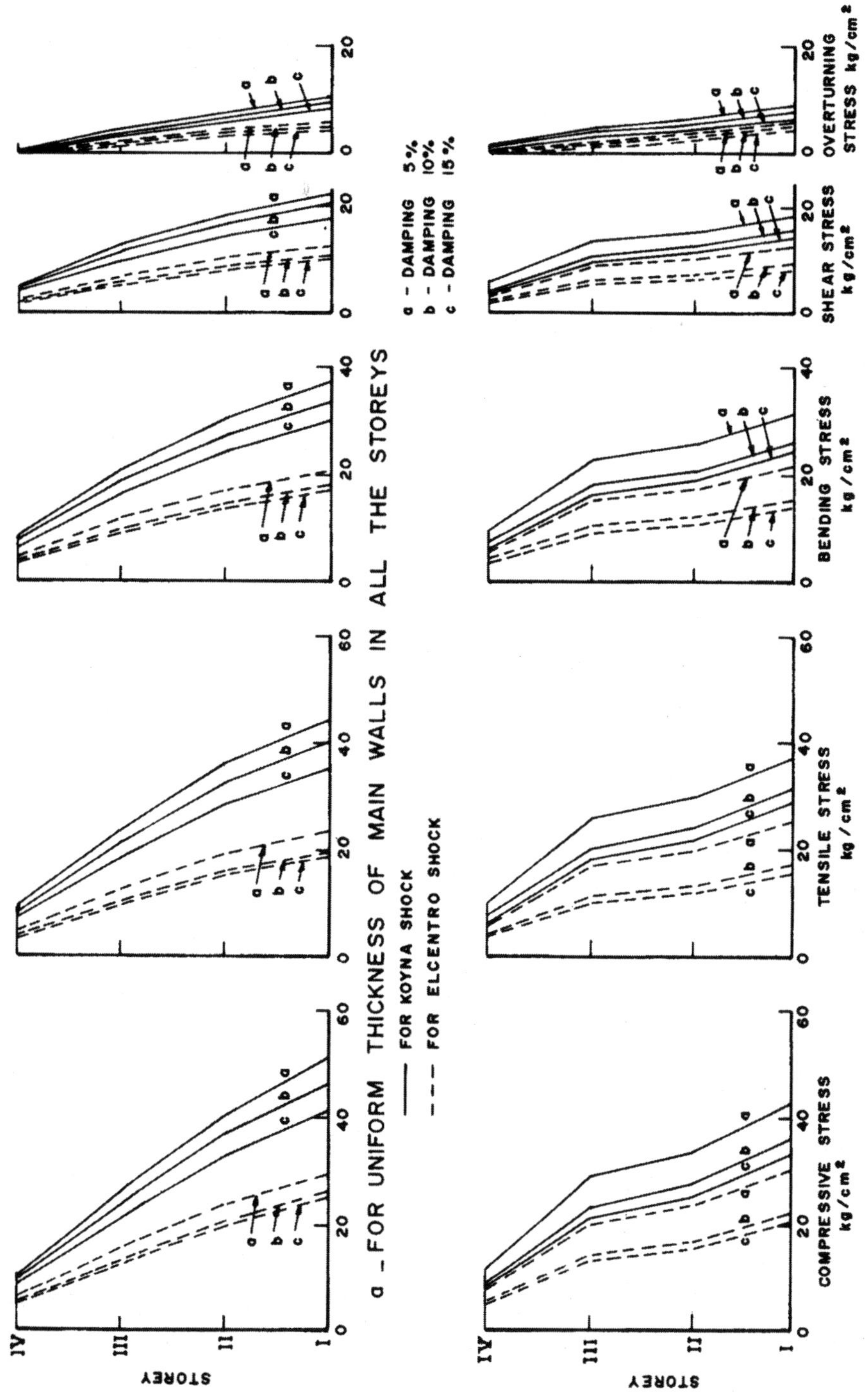

Fig. 7: Maximum stresses in different storeys of four-storeyed brick building

Table 3. Comparison of stresses in different storeys of the buildings

Storey	Building type	Bending stress (kg/cm^2)	Shearing stress (kg/cm^2)	Net tensile stress (kg/cm^2)	Ratio to base value
	B4B	9.95	5.91	10.64	1.21
	B3B	10.56	6.27	11.31	1.28
Top	B2	8.99	5.34	9.57	1.09
	B1	8.21	4.88	8.69	1.00
Second	B4B	22.92	13.62	26.22	1.20
from	B3B	24.12	14.33	27.68	1.27
top	B2	19.02	11.30	21.69	1.00
Third	B4B	25.47	15.13	30.08	1.04
from top	B3B	24.43	14.51	28.96	1.00

18.6.5 Type of earthquake

The pattern of the stresses in different building types is similar (Figs. 6 and 7) in the two shock cases for all the parameters in the analysis. But the values of the stresses computed for all types of building under the Koyna shock are about double the corresponding values for the El Centro shock. This observation corroborates well with the ratios of the spectral intensities of the two earthquakes, as discussed earlier.

18.6.6 Strength requirements

From the present seismic stresses, as presented in Tables 1 and 2, developing in the unstrengthened multi-storeyed brick buildings, it is seen that the demands on tensile and shearing strengths of material in both the Koyna and El Centro earthquakes are very severe. Thus, provision of vertical steel reinforcement at the sections developing more than cracking tension and horizontal steel reinforcement at the sections developing higher shearing stresses is essential to check the development of cracks and sliding in the joints. As regards evaluation of their compressive strength, it is observed that the maximum compressive stress is less than the ultimate compressive strength, leaving some margin of safety.

18.6.7 `Requirements of reinforcing steel

The critical sections developing tensile stresses are identified in unstrengthened brick building plans (Figs. 4 and 5). In estimating the steel section, it is assumed that the total tension developing on any building element is resisted by the steel and the stress in the steel is allowed to reach the yield stress at the most severely stressed sections in view of the earthquake severity. In the case of more severe shocks, steel at some sections may go into the inelastic yielding range and save the building from collapse by absorption of energy. Also for calculation of steel, an assumption of 10% of critical damping is justified since the cracking in brick work associated with the development of tensile stresses in steel will raise the damping value to about 10%. The vertical steel bars estimated on this basis are shown in Table 4.

Table 4. Vertical Reinforcement at the Most Critical Section of Longitudinal walls for $\xi = 0.10$

S. no.	Type	No. of Storeys	Storey	Diameter for single bar (mm): Required for Koyna earthquake	IS: 4326- for severe case+	Required for El Centro earthquake	IS: 4326 for moderate case	Thickness of main wall in the storey (mm)
1	B1	one	-	10	12	nil	nil	223
2	B2	two	top	10	12	8	nil	223
			bottom	16	16	10	nil	
3	B3A (uniform wall thickness)	three	top	10	12	8	12	223
			middle	16	16	12	12	
			bottom	20	16	16	16	
4	B3B (non-uniform wall thickness)	three	top	10	12	8	12	223
			middle	16	16	12	12	223
			bottom	20	16	12	16	343
5	B4A (uniform wall thickness)	four	top	10	12	8	12	223
			third	16	16	12	12	
			second	20	20	16	16	
			bottom	22	25	16	16	
6	B4B (non-uniform wall thickness)	four	top	10	12	8	12	223
			third	16	16	12	12	223
			second	20	20	16	16	343
			bottom	22	25	16	16	343

+Severe cases are those in seismic zones where the design seismic coefficient becomes more than 0.08 (Zones V and IV) and moderate cases where the design seismic coefficient can come between 0.06 and 0.08 (Zones IV and III).

The estimated vertical steel has been compared with the reinforcing provisions of the Indian Standard Code. For this, the results obtained for the buildings subjected to the Koyna shock are compared with the provision of steel as given in the code for areas with design seismic coefficients equal to and more than 0.08. Similarly, the case of buildings subjected to the El Centro shock is compared with the provisions for areas having design seismic coefficient of 0.06 or more but less than 0.08. This comparison shows that the reinforcing provisions of the Indian Code are surprisingly close to the calculated requirements for vertical steel.

Table 5:- Horizontal Reinforcement at the Most Critical Section of Longitudinal Transverse Walls for ξ = 0.10

S. no.	Type	No. of Storey	Storey	Diameter for single bar (mm)				Thickness of main wall in the storey (mm)
				Required - for Koyna shock		Required for El Centro shock		
				Longitu -dinal wall	Trans -verse wall	Longitu -dinal wall	Trans -verse wall	
1	B1	one	-	5	nil	nil	nil	223
2	B2	two	top	4	nil	nil	nil	223
			bottom	2	3	4	nil	
	B3A		top	5	nil	nil	nil	
	(uniform		middle	2	3	3	nil	223
3	wall	three	bottom	1	2	2	4	
	thickness)							
4	B3B (non		top	4	nil	nil	nil	
	uniform	three	middle	2	4	3	nil	223
	wall		bottom	1	3	3	nil	343
	thickness)							
	B4A		top	4	nil	nil	nil	
	(uniform		third	1	3	3	nil	
5	wall	four	second	1	2	2	4	223
	thickness)		bottom	1	1	1	3	
	B4B							
	(non-		top	2	nil	nil	nil	
6	uniform	four	third	1	3	3	nil	223
	wall		second	1	3	3	nil	
	thickness)		bottom	1	2	2	4	343

Horizontal steel at the sections developing higher shearing stresses has been estimated (24) here for the most critical section in longitudinal and transverse walls and for both the earthquakes. It is seen from Table 5 that the horizontal reinforcement is definitely required in all the storeys of all the buildings considered when subjected to the Koyna shock. But in the case of the El Centro shock, horizontal steel is not required in some of the top storeys. Therefore, recommendations for such horizontal steel are called for so as to safeguard the brick building from failure due to higher shearing stresses.

18.7 Conclusions

Analytical response studies of one-to four-storeyed brick buildings confirm that the usually low tensile and shearing strengths of brick work are the main cause of cracking and collapse during severe earthquakes. Estimations of required steel reinforced in one-to four-storeyed brick buildings show that the reinforcing provisions of the Indian Code are generally adequate in imparting the desired strength for the probable maximum earthquakes. The only deficiency is in the shearing strength of piers for which specifications for providing horizontal steel reinforcement are called for. ■

18.8 References

1. A.S.Arya, B. Chandra and P. Rani, “Influence of Natural Disasters (Earthquakes) on Educational Facilties, Part I-Collection and analysis of available information- annotated Bibliography”, Report, School of Research and Training in Earthquake Engineering, University of Roorkee, Roorkee, 1977.
2. J.E. Monge, “Seismic Behaviour and Design of Small Buildings in Chile”, Proc. 4th world conf., earthquake eng., Chile (1969).
3. K.V.Steinbrugge and R. Flores, “The Chilean earthquake of May 1960, a Structural Engineering View Point”, Bull. Seism. Soc. Am. 53, No. 2, 225-307 (1963).
4. A.A. Moinfar, “Earthquake Resistant Design of Brick Masonry Buildings’, Eur. Symp. earthquake eng., Imperial College, London, 1972.
5. Alan H. Yorkdale, “Masonry Building Systems”, State of Art Report No. 9, Technical Committee No.3, Int. conf. planning design tall buildings, Lehigh University (1970).
6. A.S. Arya, “Design and Construction of Masonry Buildings in Seismic areas’, Bull. ISET 4, No. 2 (1967).
7. A.S. Arya, “Construction of Small Buildings in Seismic Areas”,Bull. ISET 5, Nos. 3, and 4 (1968).
8. H.C. Plummer and J.A. Blume, “Reinforced Brick Masonry - Lateral Force Design,” Structural Clay Products Institute, Washington, 1953.
9. J. Krishna and A.S. Arya, “Earthquake Resistant Design of Buildings”, J. inst. engrs (India) 45, No. 7, 610-627 (1965).
10. J. Krishna and B. Chandra, “Strengthening of Brick Buildings Against Earthquake forces”, Proc. 3rd world conf. earthquake eng., New Zealand (1965).
11. J. Krishna and B. Chandra, “Strengthening of Brick Buildings in Seismic Zones”, Proc. 4th world conf. earthquake eng., Chile (1969).

12. Earthquake Resistant Regulations - A World List, International Association of Earthquake Engineering, 1973.
13. B. Chandra and K. Kumar "Earthquake Resistant Construction of Brick Buildings", Jai Krishna 60th Birthday Anniversary Commemoration Volume, Sarita Prakashan, Meerut, 1974.
14. IS: 4326-1976, Code of Practice for Earthquake Resistant Design and Construction of Buildings, Indian Standards Institution, New Delhi, 1977.
15. R.W. Clough, P. Gulkan and R.L. Mayes, "Earthquake Behaviour of One Storey Masonry Houses", Proc. ASCE 2nd specially conf. dyn. Response struct. (1981).
16. A.S. Arya, B. Chandra and S.K. Thakkar, "Influence of Natural Disasters (Earthquakes) on Educational Facilities, Part II', Final Report, Submitted to Educational Facilities Division, UNESCO, School of Research and Training in Earthquake Engineering, University of Roorkee, December 1977.
17. M. Qamaruddin, "Development of Brick Building Systems for Improved Earthquake Performance", Ph.D. Thesis, University of Roorkee, 1978.
18. D.V. Mallick, "Damping Characteristics of Brick Masonry in Different Mortars", M.E. Thesis, University of Roorkee, 1961.
19. A.A. Bennett, W.E. Miline and H. Bateman, "Numerical integration of Differential Equations," Dover Publications, New York, 1956.
20. J.A. Blume, N.M. Newmark and L.H. Corning, "Design of Multistorey Reinforced Concrete Buildings for Earthquake Motions," Portland Cement Association, Illinois, 1961.
21. J.R. Benjamin, "Statically Indeterminate Structures," McGraw-Hill, New York, 1959.
22. G.W. Housner, "Behaviour of Structures During Earthquakes", J. eng. Mech. Div., ASCE 85, No. EM4, 109-129 (1959).
23. B.Chandra, "Study of Inelastic Response of Multistory Frames During Earthquakes", Ph.D. Thesis, University of Roorkee, 1971.
24. J.E. Amrhein, Reinforced Masonry Engineering Handbook, Masonry Institute of America, California, 1978.

Chapter 19

Seismic Isolation of Masonry Buildings - An Experimental Study[1,2,3]

19.1 Introduction

Pronounced seismic vulnerability of masonry structures was confirmed in numerous past earthquakes worldwide, having as a consequence major devastation of the built environment and loss of human lives. To enhance seismic performance of masonry structures, rational strengthening arrangements, with emphasis on minimum use of reinforcing steel, have been proposed, verified and incorporate in building codes and guidelines worldwide such as IAEE (1986). As an alternative to conventional seismic strengthening methods, the base isolation concept emerges as a viable means for reducing the vulnerability of both new and existing masonry buildings to strong seismic motions. In the last few decades, base isolation was used in the design of new structures, as well as in seismic upgrading of the existing ones. Field applications of various base isolation systems were preceded by extensive experimental studies. However, only a few research investigations so far were related to base isolation of masonry structures. By and large, such studies treated masonry structures decoupled from the foundations above plinth level by means of a "sliding joint" (Li, 1984; Qamaruddin et al., 1984). This base isolation concept is known as a Pure-Friction (P-F) sliding isolation system.

It utilises friction, allowing some parts of the structure to slide relative to the others in the course of an earthquake attack. A review of the P-F base isolation concept was made recently by Nikolic-Brzev (1994). With only a few exceptions (Paulson et al., 1991; Lou et al., 1992), previous research efforts were directed mainly to base isolation systems for single-storey masonry buildings. When subjected to earthquake-type excitation, such structures demonstrate behavior characteristic of rigid mass systems. However, due to their inverted mechanical and dynamic characteristics, medium-rise multi-storey masonry buildings subjected to the same excitation respond in a different way, characterised with amplified ground vibrations. The existing large

---

1. *S. Nikolic-brzev, Sandwell Inc., 1190 Hornby Street, Vancouver, B.C., V6Z 2H6, Canada*
2. *A.S. Arya, Department of Earthquake Engineering, University of Roorkee, Roorkee, India*
3. *Paper no.1338, Eleventh World Conference on Earthquake Engineering, Canada*

inventory of multi-storey masonry buildings, as well as the need for an innovative strategy for efficient seismic protection of the new ones, necessitate a study on feasible seismic isolation schemes. The results of an extensive experimental and analytical investigation of the P-F isolation concept for multi-storey masonry buildings were reported by Nikolic-Brzev (1993), whereas the relevant findings of experimental study are presented in this paper.

19.2 Outline of The Study

19.2.1 The Test Model

To evaluate the effectiveness of proposed isolation scheme in diminishing damage to multi-storey masonry buildings during major earthquakes, the performance of both a conventional and an isolated model structure was investigated by shake-table testing. Configuration of a model was established to resemble a two room, three-storey high masonry building, with aspect ratio (height/length/width) typical for the Indian housing practice. One door opening in each cross-wall, and four window openings, two by two in each longitudinal wall, were provided in the model, as illustrated in Fig. 1. The model structure was obtained from the prototype one by scaling down all dimensions by a factor of three (1:3 scale). All materials and

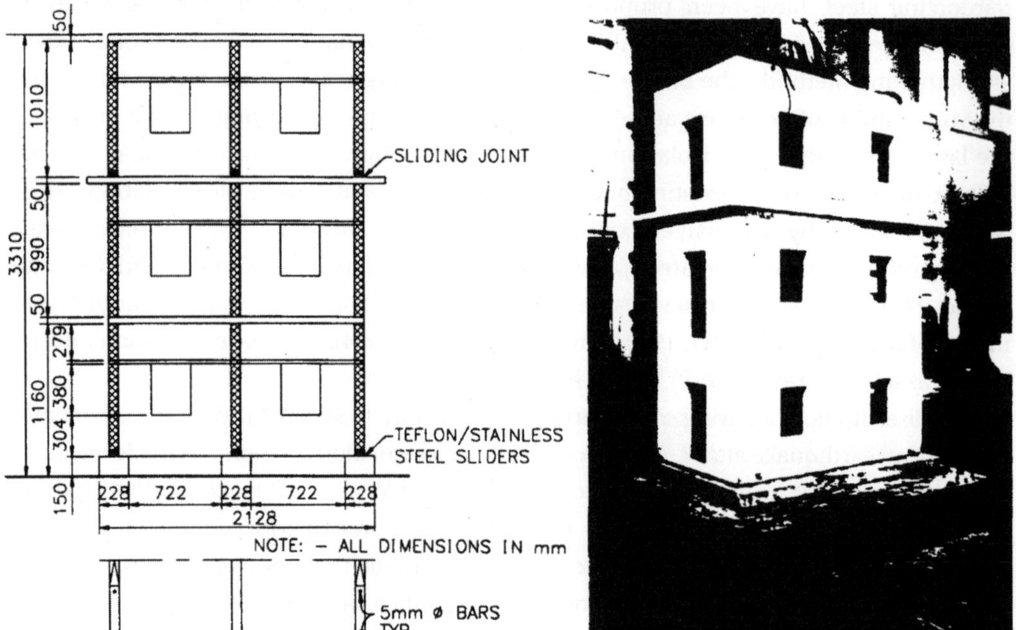

a) b)

Fig. 1: Test structure: a) Configuration, and b) Structure on shaking-table during the experiment

their characteristics were kept the same. Outside dimensions of the model structure were: length 2.13 m, width 1.65 m, height 3.31 m, and wall thickness 75 mm. Specially manufactured 1:3 scale burnt clay bricks (size 80x40x25 mm, compressive strength 25 N/mm^2), were used. Cement/sand mortar (1:6 ratio) was used for masonry construction. Vertical steel reinforcement bars (5 mm dia, mild steel f_e = 250 N/mm^2) were provided at the wall corners and junctions, as well as around the openings. One reinforced concrete lintel ring beam (cross-sectional dimension of 30x60 mm) was constructed above the openings at each storey level. Model structure was constructed atop a steel plate that was bolted to the shaking-table before the experiment. A reinforced concrete plinth beam was constructed to simulate foundations and also to provide support for the masonry superstructure. A single model structure was used for evaluating the seismic response of both conventional (fixed-base) and isolated masonry structure, by making a special arrangement for fixing the base level bond beam to the plinth structure in the initial phase of experiment.

Two different seismic isolation systems were installed at the base and the second storey floor levels respectively. The masonry superstructure was decoupled from the plinth concrete by means of a discrete isolation system comprising a set of Teflon/stainless steel sliders (Fig. 2a). In total, 18 sliding bearings were installed at the wall corners and junctions. The upper sliding surface of sliding bearing was made of a stainless steel sheet bonded to a mild steel plate, whereas a lower sliding surface comprised a steel plate-backed Teflon pad. Components of the sliding bearing were bonded to the backup structures by means of epoxy-based adhesive. A "sliding joint" at the second storey level was installed by providing a smoothly finished layer of cement/sand mortar over the floor-slab, and afterwards covering with a thin grease layer (Fig. 2b). Experimentally obtained static frictional coefficient values for Teflon/steel and grease/concrete sliding couples amounted to 0.1 and 0.4, respectively.

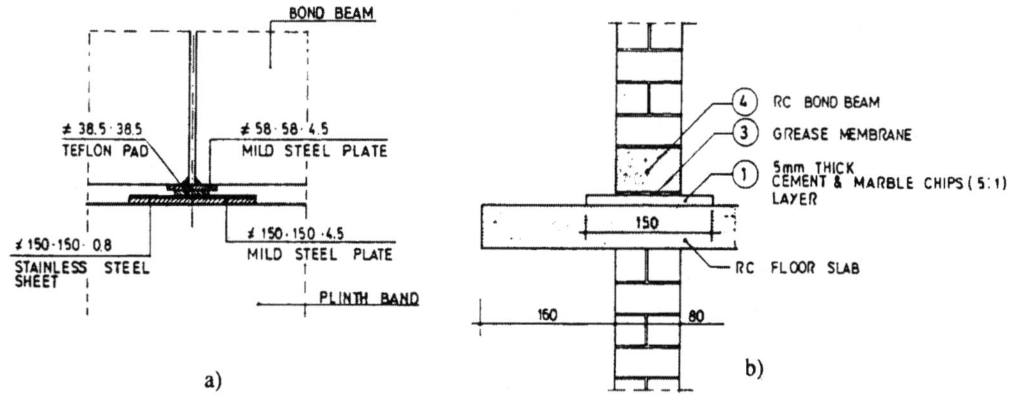

Fig.2: Seismic isolation systems: a) A typical Teflon/stainless steel sliding bearing at the base level, and b) sliding joint at the second storey floor level.

19.2.2 Base Excitation

In total, thirteen shake-table test runs were carried out on a computer-controlled shake-table facility of the Department of Earthquake Engineering, University of Roorkee, India. A single artificially generated earthquake motion of gradually increasing intensity was used throughout the experiment. The target response spectrum curve of simulated earthquake motions used in this study was determined as a smooth broadened envelope curve of the average spectra on rock and stiff soil sites. The test structure was subjected to simultaneous effects of horizontal and vertical base motions in the course of several test runs. Zero period acceleration value of the vertical excitation target spectrum curve was taken as two/thirds of the horizontal one. In the first series of six test runs, the model structure was tested in the fixed-base condition. In order to keep the model in undamaged condition for further testing, the intensity of the fixed-base model tests was purposely kept at a rather low level (peak ground acceleration was in the range of 0.2g). Input excitation level was increased during the second series of seven test runs on sliding structure, with maximum peak ground acceleration level of 0.4 g in the last run.

19.2.3 Instrumentation

The model was instrumented with ten Force Balance Accelerometers (FBA) during the experiment. For recording horizontal and vertical components of the simulated earthquake excitation, two FBAs were mounted on each floor-slab and at the shaking-table level. To evaluate response accelerations at the location of isolation systems, one horizontal FBA was mounted on the bond beam at the ground and second storey levels respectively. Data was sampled at 100 points per second, and 2048 samples per channel were stored in digital form in the course of each table run for further processing. Unfortunately, displacement transducers (LDVTs) required for measuring displacement time histories were not available; however, maximum and residual displacements at the base level were recorded after each shaking-table test run. Finely ground turmeric powder was spread over the lower sliding plate, and displacement values were obtained by measuring the steel plate area free of powder after the test.

19.3 Results of Experimental Investigation

19.3.1 Simulated Earthquake Motion

Characteristics of recorded base motions in each table run are summarised in Table 1. During the initial series of six runs, the model was tested in fixed-base condition. The recorded maximum amplitude of the horizontal table motion, i.e. Peak Ground Acceleration (PGA) varied from 0.125 g in run No. 1 to 0.207 g in run No. 5; vertical component of the base motion was introduced in runs No. 4 and 5, with recorded PGA levels of 0.07 g and 0.14 g respectively. The isolated model structure was tested in seven test runs, with horizontal table PGA values ranging from 0.201g in run No. 7 to 0.379 g in run No. 13; the vertical component of the base motion was introduced in runs No. 10 to 13, with PGAs ranging from 0.09 g to 0.186 g. Due to the feedback of the model and table interaction, it was not possible to generate exactly the same base motion in different table runs. However, peak intensities and acceleration response spectrum curves of horizontal base motions recorded in runs No. 3 (fixed-base) and No. 7 (isolated) were found to be similar. Out of all test runs with bi-directional (i.e. horizontal and vertical) excitation, similar intensities were recorded in runs No. 5 (fixed-base) and No. 10 (isolated), as illustrated in the linear response acceleration curves shown in Fig. 3. Also shown are the recorded values of roof level accelerations and corresponding fundamental periods inferred from the Fourier amplitude spectra. The results recorded in those four runs were, therefore, used as a reference for a comparison between the seismic response parameters related to fixed-base and isolated model structures respectively.

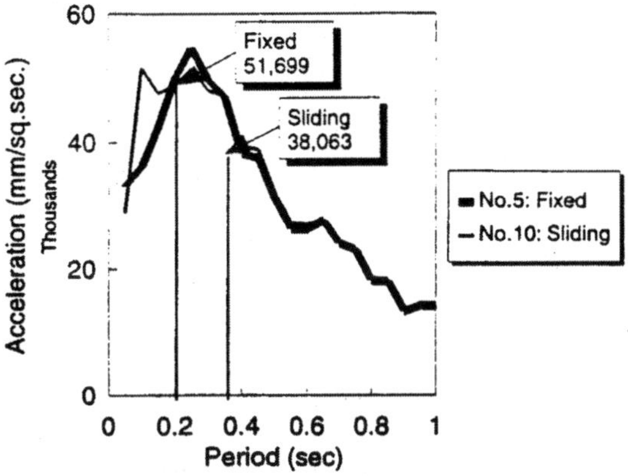

Fig.3: Acceleration response spectra for measured base motions- Runs no.5 (fixed-base) and No.10 (Sliding structure)

19.3.2 Dynamic Response Parameters

Calculated peak values of relevant seismic response parameters are summarised in Table 1. It should be noted that maximum values of certain parameters might not have necessarily occurred simultaneously. The effect of phase lag to model response has been neglected in this study. In the course of the experiment, maximum response accelerations of 0.527 g and

Table 1. Measured Dynamic Response Parameters

Run No.		1	2	3	4	5	6	7	8	9	10	11	12	13
Condition		F	F	F	F	F	F	S	S	S	S	S	S	S
Horizont.	ahg (g)	0.13	0.16	0.21	0.12	0.21	0.14	0.20	0.25	0.27	0.21	0.23	0.22	0.38
Accel.	ah3(g)	0.20	0.28	0.49	0.28	0.53	0.28	0.36	0.43	0.51	0.39	0.49	0.33	0.50
	ah3/ahg	1.56	1.76	2.39	2.31	2.51	2.04	1.81	1.73	1.90	1.87	2.14	1.48	1.31
Vertical	avg(g)	-	-	-	0.07	0.14	-	-	-	-	0.15	0.17	0.09	0.19
Accel.	av3(g)	-	-	-	0.20	0.19	-	-	-	-	0.23	0.24	0.22	0.27
	Av3/avg	-	-	-	2.90	1.34	-	-	-	-	1.56	1.42	2.45	1.43
Base Shear	Vb/W	0.91	0.68	0.65	0.49	0.79	0.53	0.59	0.62	0.65	0.50	-	1.10	1.63
Residual Displace	(mm)	-	-	-	-	-	-	10	35	82	33	N.A.	N.A.	N.A.

Note: F = fixed-base; S = sliding; ahg = peak horizontal ground acceleration; ah3 = max. top storey horizontal response acceleration; ah3/ahg = horizontal accel. amplification ratio; avg = peak vertical ground acceleration; av3 = max. top storey vertical response acceleration; av3/avg = vertical accel. mplification ratio; Vb = max. base shear; W = total model weight.

0.495 g were recorded at the roof level in the fixed-base and isolated models respectively. During the last test run, the isolated structure was subjected to excitation with a recorded horizontal PGA level of 0.4g; even at such a high excitation level, the structure had not experienced any structural distress. The fixed-base structure was subjected to peak table excitation of approximately 0.21 g without any signs of damage.

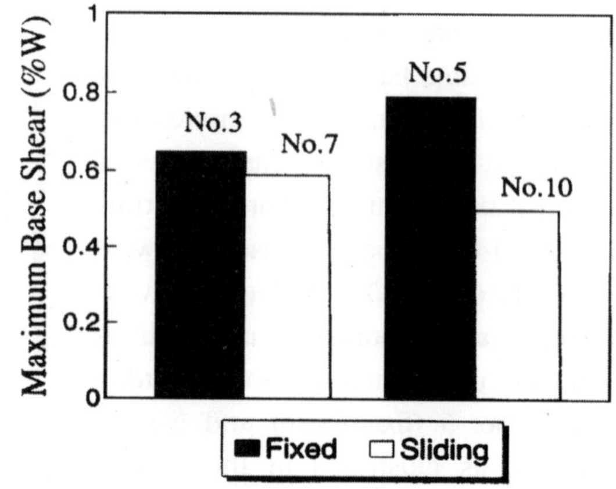

Fig.4: Comparision of maximum base shear forces in the fixed-base (runs Nos.3 & 5) and sliding structures (runs Nos.7 & 10)

Base shear forces, expressed as a fraction of the total model weight, were computed from the acceleration records. Total weights of the fixed-base and isolated model amounted to 3.65 and 4.2 tons respectively. Maximum base shear forces of 0.79 W (run No. 5) and 1.63 W (run No.13) were recorded in the fixed-base and isolated models respectively. A considerable reduction in maximum base shear forces by around 40% was reported in the isolated structure (run No.10) as compared to the fixed-base one (run No. 5). Similar results were obtained by comparing the maximum base shear forces in the runs No. 3 (fixed-base) and No. 7 (isolated structure), as illustrated in Fig. 4.

Maximum and residual sliding displacements at the plinth and second storey floor levels respectively were measured after each test run. The second storey isolation system was not initiated to slide even at a very high base excitation level of around 0.4g (run No.13). Obviously, maximum response acceleration of 0.27 g attained at the second storey level was not sufficient to induce sliding movement of the third storey superstructure.

Free vibration test of the fixed-base model was carried out before the experiment in order to determine its dynamic characteristics, particularly *natural frequency* and damping ratio. The test was performed by pulling the model in longitudinal direction with the rope attached to the roof floor-slab. The rope was suddenly released and the resulting vibrations were recorded by means of a FBA mounted on the roof floor-slab. The Fourier amplitude spectrum analysis of recorded transient vibrations indicates that the first-and second-mode frequencies were around 5.5 and 10.2 Hz respectively. On the other hand, Fourier amplitude spectra computed using the roof accelerations recorded during the first series of six runs indicate that the first-mode frequency of fixed-base model was in the range from 4.8 to 6.4 Hz. The first-mode frequencies of the isolated model structure determined from the Fourier spectrum analysis of recorded roof accelerations indicate a reduction in the first-mode frequency value by 45 to 50% as compared to that of the fixed-base model. *Structural damping ratio* of 2.37% was determined from the free vibration acceleration record by logarithmic decrement method. The obtained value is comparable with the average damping ratio value of 2.6%, that was determined from the transient vibration records of reference masonry piers constructed simultaneously with the model structure using the same building material.

Acceleration amplification is defined here as a ratio between response acceleration at a certain level of the structure (usually the uppermost level) and the PGA. Experimentally obtained values of the horizontal acceleration amplification ratio were in the range from 1.31 to 2.51. Comparison was made between the amplification ratios of the fixed-base system (run No.3) and the isolated one (run No.7) subjected to similar excitation level (PGA of approximately 0.2g). Fig. 5a shows that the amplification of horizontal accelerations was considerably higher in the fixed-base structure (run No.3) than in the isolated one (run No.7). Reduction in the acceleration response of isolated structure as compared to the fixed-base one was not as remarkable as expected. However, it was observed that the minimum value of amplification ratio (1.31) was obtained during the last test run, when the model was subjected to the highest ever excitation of approximately 0.4g. Further on, a comparative study on the system responses during test runs with increasing excitation levels was carried out. It was shown that the efficiency of the proposed isolated scheme in terms of reduction in acceleration responses was more pronounced at the higher excitations.

Effect of vertical base motion to dynamic response of isolated structure was compared in the test runs No. 7 and 10; those two runs were characterised with similar horizontal PGA levels of around 0.2g. Although vertical component of the base motion was introduced in the run No. 10, horizontal acceleration amplification ratio remained unchanged (see Table 1). Mean values of maximum vertical and horizontal acceleration amplification ratios calculated for all test runs are presented in Fig. 5b. Similar levels of acceleration

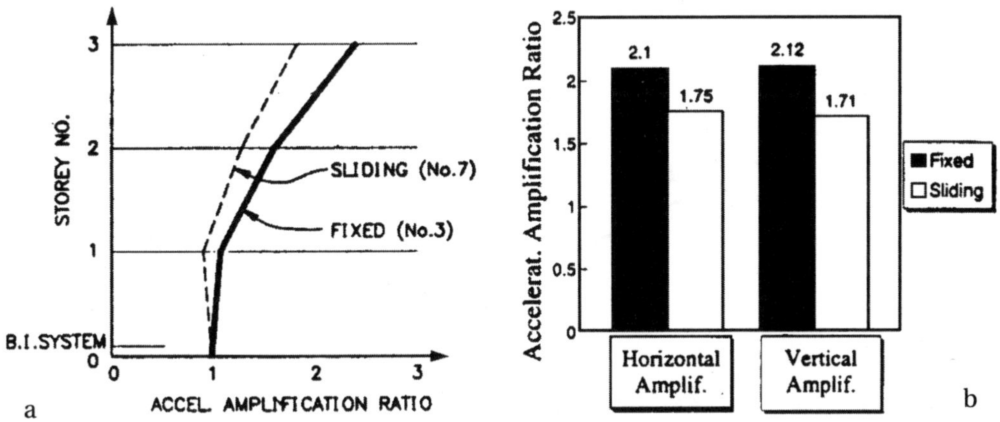

Fig. 5: Dynamic acceleration amplification ratios: a) Horizontal amplification ration in the fixed-base (run No. 3) and sliding structure (run No. 7), and b) mean values of horizontal and vertical amplification ratios

amplifications obtained in the case of models subjected to horizontal and/or vertical base motions indicate that seismic response of a sliding structure is not significantly affected by vertical base motion.

Amount of input energy transferred to the structure in the course of simulated earthquake motion is an important parameter related to the effectiveness of a seismic isolation scheme. Response accelerations recorded during the experiment served as a basis for a study on the relationship between PGA and energy input to the model structure. Relevant results of statistical regression analysis are presented in Fig. 6. The study has shown that whereas energy input increased with increasing PGA level in the fixed-base system (Fig. 6b), the reverse trend was observed in the isolated structure (Fig. 6a). This finding is in support of the previous observations concerning the pronounced efficiency of the P-F isolation scheme at higher levels of base motion.

19.4 Conclusions

As a result of the study presented in this paper, the following conclusions can be drawn:

1. An average reduction in maximum response accelerations by around 30% has been reported in the isolated model as compared to the fixed-base one. However, it was observed that the minimum acceleration amplification ratio of 1.31 was obtained in the isolated structure during the test run characterised with the highest ever excitation level of approximately 0.4g. Efficiency of the proposed isolation scheme, thus, becomes more pronounced at the higher excitation levels.
2. The model structure was exposed to simultaneous effects of horizontal and vertical base motions in several test runs. Similar levels of acceleration amplification ratios

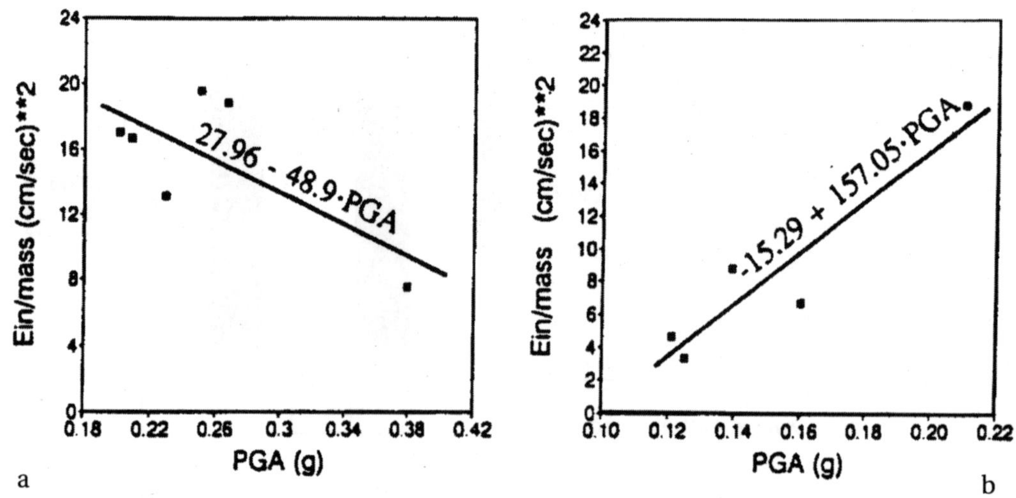

Fig. 6: Earthquake input energy verses PGA level: a) sliding structure, and b) fixed-base structure

were attained in test runs with horizontal and vertical base excitations; this indicates that seismic response of the isolated structure is not significantly affected by vertical base excitation.

3. A considerable reduction in maximum base shear forces by around 40% was reported in the isolated model structure as compared to the fixed-base one.
4. A comparative analysis of the amount of earthquake energy input to the fixed-base and sliding system was carried out. The study revealed that the amount of input energy in a fixed-base system is constantly increasing with the increase of horizontal excitation, whereas for a sliding system, an opposite trend is observed. This finding is in support of the previous notion related to the increased effectiveness of the proposed isolation scheme at higher excitation levels.

It can be finally concluded that the proposed seismic isolated scheme is capable of reducing the dynamic response of multi-storey masonry buildings to high intensity earthquakes appreciably as compared with the fixed-base structures. This scheme may be, therefore, recommended for adoption as a feasible practical means for seismic protection of low-and medium-rise masonry buildings. ■

19.5 References

1. Arya A.S. et. al. 1986, "Basic Concepts of Seismic Codes - Non-Engineered Construction," International Association for Earthquake Engineering, Tokyo.
2. Nikolic-Brzev, S. (1994), "An Innovative Seismic Protection Scheme for Masonry Buildings," Proceedings of the Tenth International Conference on Brick/Block Masonry, Vol. 1, pp. 273-282. Calgary.
3. Nikolic-Brzev, S. (1993), "Seismic Protection of Multi-storey Brick Buildings by Seismic Isolation Technique," Thesis submitted in the fulfillment of the degree of Doctor of Philosophy. Department of Earthquake Engineering, University of Roorkee, India.
4. Paulson, T.J., Abrams, D.P., and Meyes, R.L. (1991), "Shaking-Table Study of Base Isolation for Masonry Buildings," Journal of Structural Engineering ASCE 117, 3315-3336.
5. Qamaruddin, M., Chandra, B., and Arya, A.S. (1984), "Dynamic Testing of Brick Building Models," Proceedings, The Institution of Civil Engineers, U.K., 77, 353-365.
6. Li, L. (1984), "Base Isolation Measure for Aseismic Buildings in China," Proceedings of the Eight World Conference on Earthquake Engineering, Vol. 6, pp. 791-198. San Francisco.

Chapter 20

Rapid Visual Seismic Screening of Buildings[1]

20.1 RVS Procedure, Objectives and Scope

The Rapid Visual Screening method is designed to be implemented without performing any structural calculations. The procedure utilises a damageability grading system that requires the evaluator to (1) identify the primary structural lateral load-resisting system, and (2) identify building attributes that modify the seismic performance expected for this lateral load-resisting system along with non-structural components. The inspection, data collection and decision-making process typically occurs at the building site, and is expected to take a couple of hours for a building, depending on its size.

The screening is based on the Code based Seismic Intensity, Building Type and Damageability Grade as observed in past earthquakes and covered in MSK/European macro-intensity scales.

20.2 Uses of RVS Results

The main uses of this procedure in relation to seismic upgrading of existing buildings are:

i. To identify if a particular building requires further evaluation for assessment of its seismic vulnerability.
ii. To assess the seismic damageability (structural vulnerability) of the building and seismic rehabilitation needs.
iii. To identify simplified retrofitting requirements for the building (to collapse prevention level) where further evaluations are not considered necessary or not found feasible.

20.3 Seismic Hazard in India

As per IS 1893:2002 (Part 1), India has been divided into 4 seismic hazard zones (see Fig.1). The details of different seismic zones are given below:

---

1. *Dr.A.S.Arya, Professor Emeritus, IIT - Roorkee and National Seismic Advisor, GoI - UNDP, DRM Programme*

Zone II	Low seismic hazard (damage during earthquake may be of MSK Intensity VI or lower)
Zone III	Moderate seismic hazard (maximum damage during earthquake may be upto MSK Intensity VII)
Zone IV	High seismic hazard (maximum damage during earthquake may be upto MSK Intensity VIII)
Zone V	Very high seismic hazard (maximum damage during earthquake may be of MSK Intensity IX or greater)

When a particular damage intensity occurs, different building types experience different levels of damage depending on their inherent characteristics. For carrying out the Rapid Visual Screening, all four hazard zones have been considered.

20.4 Building Types Considered in RVS Procedure

A wide variety of construction types and building materials are used in the urban and rural areas of India. These include local materials such as mud, straw and wood, semi-engineered materials such as burnt brick and stone masonry, and engineered materials such as concrete and steel. The seismic vulnerability of the different building types depends on the choice of building materials and construction technology adopted. The building vulnerability is generally highest with the use of local materials without engineering inputs and lowest with the use of engineered materials and skills.

The basic vulnerability class of a building type is based on the *average expected seismic performance* for that building type. All buildings have been divided into type A to type F based on the European Macro Seismic Scale (EMS-98) recommendations. The buildings in type A have the highest seismic vulnerability while the buildings in type F have the lowest seismic vulnerability. A building of a given type, however, may have its vulnerability *different from the basic class* defined for that type depending on the condition of the building, presence of earthquake resistance features, architectural features, number of storeys etc. It is, therefore, possible to have a damageability range for each building type considering the different factors affecting its likely performance. Some variations in building type are, therefore, defined as A, B, B+ etc.

The RVS procedure presented here has considered different building types, based on the building materials and construction types that are most commonly found in India. Masonry buildings are presented in Table 1. The likely damages to buildings under future earthquake intensity occurrences have been categorised in different Grades depending on the seismic impact on the strength of the building.

20.5 Grades of Damageability

Five grades of damageability from G1 to G5 are specified in MSK and European Intensity Scale as described in Table 2:

Table 1: Masonry Load Bearing Wall Buildings

Building Type	Description
A	a) Rubble (field stone) in mud mortar or without mortar, usually with sloping wooden roof. b) Uncoursed rubble masonry without adequate 'through stones'. c) Masonry with round stones.
B	Semi-dressed, rubble, brought to courses, with through stones and long corner stones; unreinforced brick walls with country type wooden roofs; unreinforced CC block walls constructed in mud mortar or weak lime mortar.
B+	a) Unreinforced brick masonry in mud mortar with vertical wood posts or horizontal wood elements or seismic band (IS: 13828) b) Unreinforced brick masonry in lime mortar.
C	a) Unreinforced masonry walls built from fully dressed (Ashler) stone masonry or CC block or burnt brick using good cement mortar, either having RC floor/roof or sloping roof having eave level horizontal bracing system or seismic band. b) As at B with horizontal seismic bands (IS: 13828)
C+	Like C(a) type but having horizontal seismic bands at lintel level of doors & windows (IS: 4326)
D	Masonry construction as at C(a) but reinforced with bands & vertical reinforcement, etc (IS: 4326), or confined masonry using horizontal & vertical reinforcing of walls.

Table 2: Grades of Damageability of Masonry Buildings

	Classification of damage to masonry buildings
Grade 1:	*Negligible to slight damage (no structural damage, slight non-structural damage)*
Structural:	Hair-line cracks in very few walls.
Non-structural:	Fall of small pieces of plaster only. Fall of loose stones from upper parts of buildings in very few cases.
Grade 2:	*Moderate damage (Slight structural damage, moderate non-structural damage)*
Structural:	Cracks in many walls, thin cracks in RC* slabs and A.C.* sheets.
Non-structural:	Fall of fairly large pieces of plaster, partial collapse of smoke chimneys on roofs. Damage to parapets, chajjas. Roof tiles disturbed in about 10% of the area. Minor damage in under structure of sloping roofs.

Grade 3:	*Substantial to heavy damage (moderate structural damage, heavy non-structural damage)*
Structural:	Large and extensive cracks in most walls. Wide spread cracking of columns and piers.
Non-structural:	Roof tiles detach. Chimneys fracture at the roof line; failure of individual non-structural elements (partitions, gable walls).
Grade 4:	*Very heavy damage (heavy structural damage, very heavy non-structural damage)*
Structural:	Serious failure of walls (gaps in walls), inner walls collapse; partial structural failure of roofs and floors.
Grade 5:	*Destruction (very heavy structural damage)* Total or near total collapse of the building.

* RC = Reinforced Concrete; AC = Asbestos Cement

20.6 Relationship of Seismic Intensity, Building Type & Damage Grades

Table 3 provides guidance regarding the likely performance of the buildings in the event of design-level earthquake intensity postulated in the seismic zone. This information has been used in the survey forms to decide if there is necessity of further seismic evaluation of the building using higher level procedures. It can also be used to identify need for retrofitting,

Table 3: Damageability Grades of Masonry Buildings

MASONRY BUILDINGS	Type of Building	Zone II MSK VI or less	Zone III MSK VII	Zone IV MSK VIII	Zone V MSK IX or More
	A	*Many* of grade 1 *Few* of grade 2 (rest no damage)	*Most* of grade 3 *Few* of grade 4 (rest of grade2or1)	*Most* of grade 4 *Few* of grade 5 (rest of grade 3,2)	*Many* of grade 5 (rest of grade 4&3)
	B and B+	*Many* of grade 1 *Few* of grade 2 (rest no damage)	*Many* of grade 2 *Few* of grade 3 (rest of grade 1)	*Most* of grade 3 *Few* of grade 4 (rest of grade 2)	*Many* of grade 4 *Few* of grade 5 (rest of grade 3)
	C and C+	*Few* of grade 1 (rest no damage)	*Many* of grade 1 *Few* of grade 2 (rest of grade 1,0)	*Most* of grade 2 *Few* of grade 3 (rest of grade 1)	*Many* of grade 3 *Few* of grade 4 (rest of grade 2)
	D		*Few* of grade 1	*Few* of grade 2	*Many* of grade 2 *Few* of grade 3 (rest of grade 1)

Note:

1. As per MSK scale, *Few, Many* and *Most* may be taken as: *Few*: 15%, *Many*: 50% and *Most*: 75%.
2. Buildings having vertical irregularity may undergo severe damage in seismic Zones III, IV & V if not specifically designed. Hence, they will require special evaluation. Also buildings sited in liquefiable or landslide prone areas will require special evaluation for seismic safety.
3. Buildings having plan irregularity may undergo a damage of one grade higher in Zones III, IV & V. The surveyor may recommend reevaluation.

and to recommend simple retrofitting techniques for ordinary buildings where more detailed evaluation is not feasible.

The Indicative quantities Few, Many and Most as defined in European Intensity Scales are as follows:

Few: Less than (15±5) %; *Many*: Between (15±5) to (55±5) %;
Most: Between (55±5) to100%

As per MSK Intensity scale the average values of these terms may be taken as

Few: 5-15% *Many*: 50% *Most*: 75%

Table 3 is generally based on MSK descriptions.

20.7 RVS Survey Forms - Special Points

The RVS survey forms are developed here for all the seismic Zones II to V, based on the probable earthquake Intensities, building types and damageability grades as described above. Some special cases included therein are described below:

20.7.1 Importance of Building/Structure

As per IS: 1893-2002, an important factor I is defined for enhancing the seismic strength of buildings & structures, as follows:

Important buildings*: Hospitals, schools, monumental structures; emergency buildings like telephone exchange, television, radio stations, railway stations, fire stations, large community halls like cinemas, assembly halls and subway stations, power stations, Important Industrial establishments, VIP residences and residences of important emergency persons.

For these important buildings, the value of I is specified as 1.5, by which the design seismic force is increased by a factor of 1.5. Now the seismic zone factors for Zones II to V are as follows.

Zone	II	III	IV	V
Zone Factor	0.10	0.16	0.24	0.36

It is seen that one Unit change in Seismic Zone Intensity increases the Zone Factor 1.5 times.

Hence, to deal with the damageability of important buildings in any zone, they should be checked for one unit higher zone. The assessment forms are designed accordingly.

20.7.2 Special Hazards

There are some special hazardous conditions to be considered:

i) *Liquefiable condition*: Normal loose sands submerged under the high water table are susceptible to liquefaction under moderate to high ground accelerations; a building founded on such soils will require special evaluation and treatment.
ii) *Land Slide Prone Area*: If the building is situated on a hill slope which is prone to land

* *Any building having more than 100 Occupants may be treated as Important for purpose of RVS.*

slide/ land slip or rock-fall under monsoon and/or earthquake, special geological and geotechnical evaluation of the site and treatment of the building will be needed.

iii) *Irregular Buildings*:

Irregularities in buildings are defined in Cl.7.1 of IS: 1893 - 2002 under the following sub-heads:

i. Plan Irregularities: These are defined in Table 4 of the above Code as follows:
   a) Torsion Irregularity
   b) Re-entrant Corners
   c) Diaphragm Discontinuity
   d) Out of Plane Offsets
   e) Non - Parallel Systems

The Geometric Irregularities in building plans which can be easily identified are shown in Fig.2

These irregularities enhance the overall damage (increased grade of damage e.g. at re-entrant corners). Such a building may be recommended for detailed evaluation.

ii. Vertical Irregularities: The following vertical irregularities may be seen in masonry buildings (see Fig. 3).
   a) Mass Irregularity
   b) Vertical Geometric Irregularity
   c) In-Plane Discontinuity in vertical Elements Resisting Lateral Forces.

If any of these irregularities are noticed, the building should be recommended for detailed evaluation.

iv) *Falling Hazard*: Where such hazards are present, particularly in Zones IV & V, recommendations should make reference to these in the survey report as indicated.

v) *Type of Foundation Soil*: IS 1893-2002 defines three soil types: hard/stiff, medium & soft. No effect of these is seen in the design spectra of short period buildings, with T< 0.4 second, covering all masonry buildings, hence, the effect may be considered not so significant.

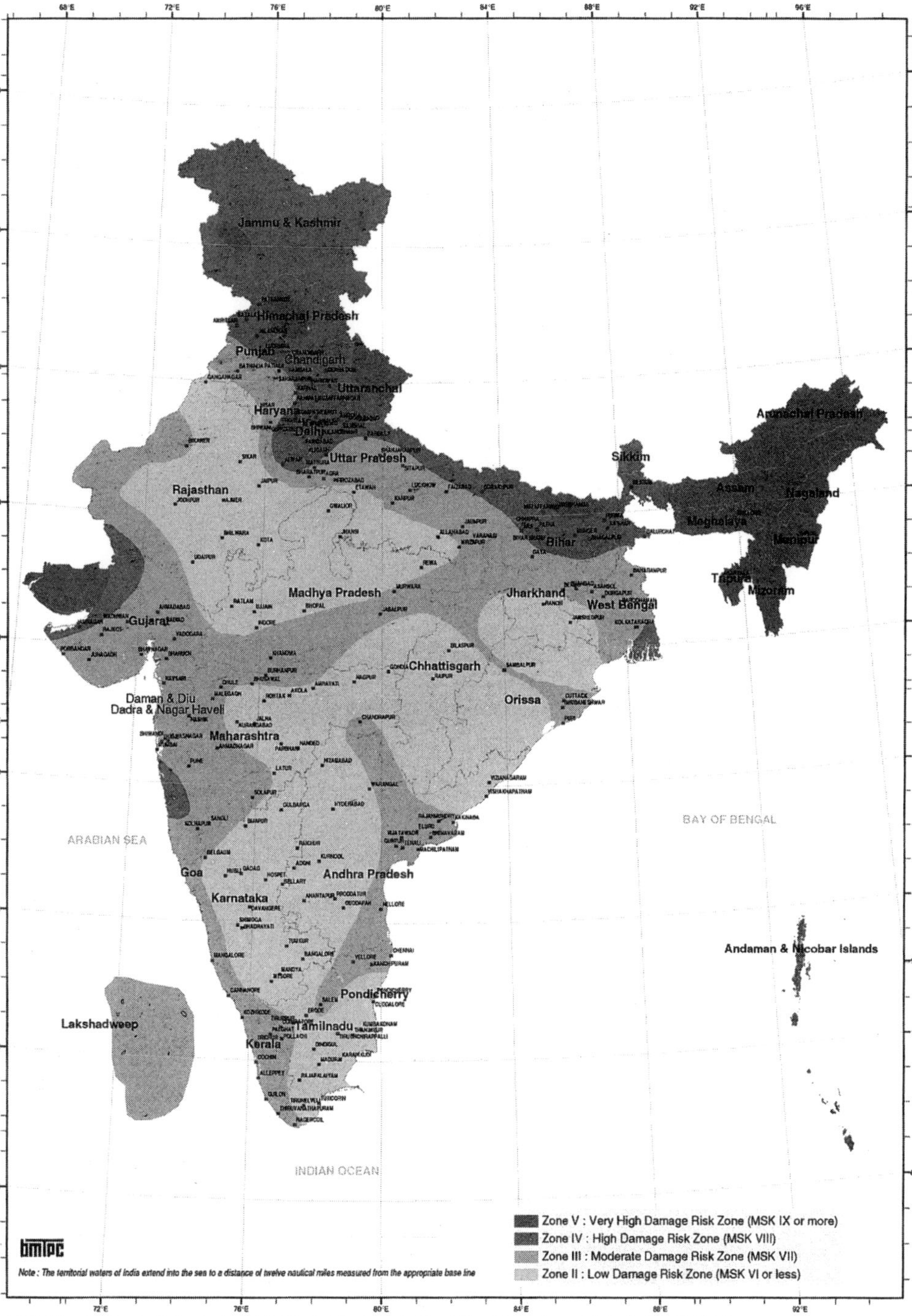

BMTPC : Vulnerability Atlas - 2nd Edition; Peer Group, MoH&UPA; Map is Based on digitised data of SOI, GOI; Seismic Zones of India Map IS:1893 - 2002, BIS, GOI, Seismotectonic Atlas of India and its Environs, GSI, GOI

Fig.1 Earthquake Hazard Zones 2002

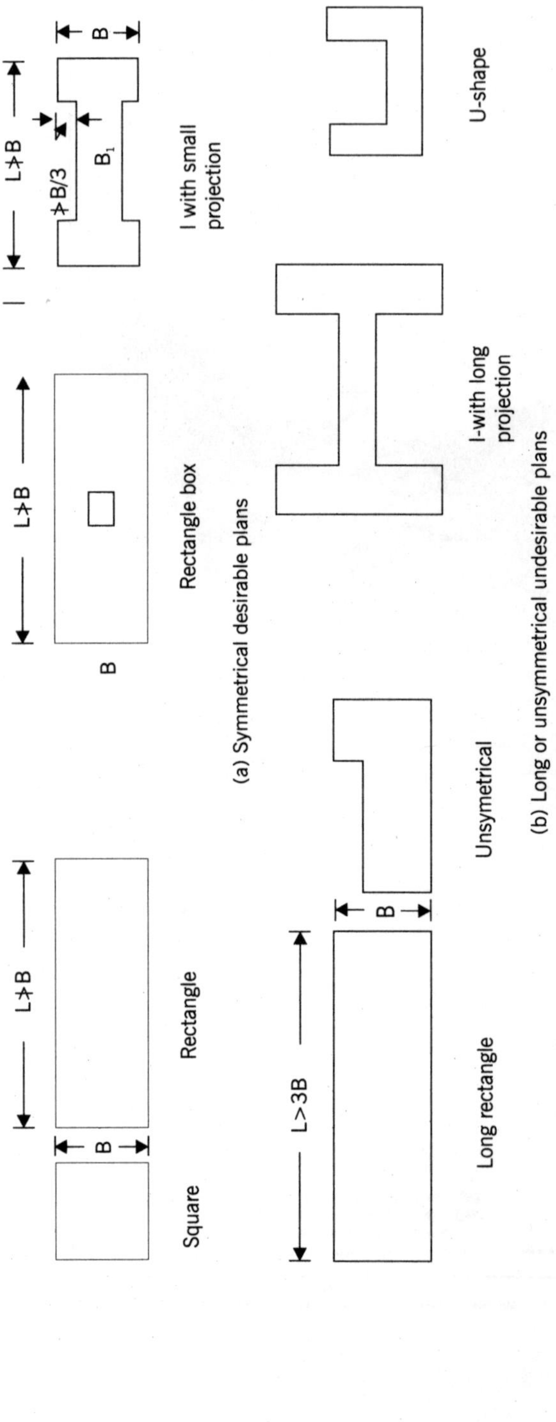

Fig. 2: Horizontal Irregularities

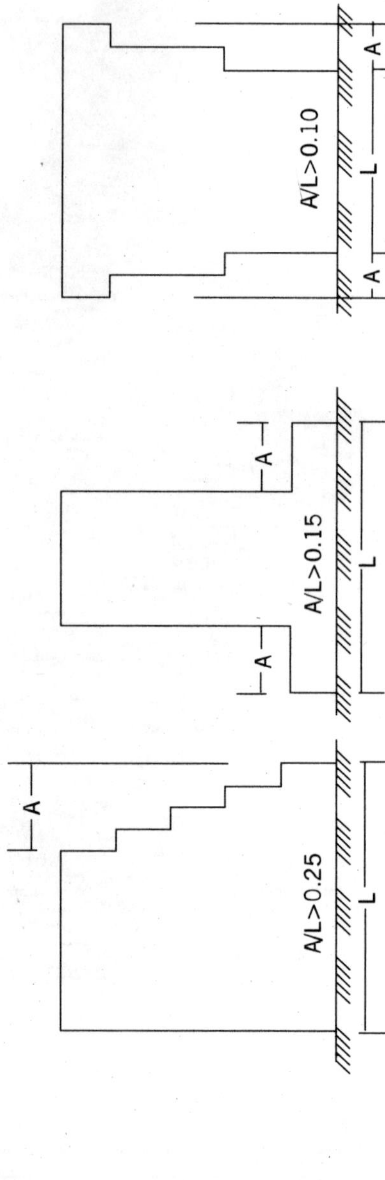

Fig. 3: Vertical Irregularities

Rapid Visual Screening of Masonry Buildings for Seismic Hazards – Form 1

Seismic Zone II Ordinary Building

Photograph

1.1 Building Name ____

1.2 Use ____

1.3 Address ____

____ Pin ____

1.4 Other Identifiers ____

1.5 No. of Stories ____ 1.6 Year Built ____

1.7 Total Covered Area; all floors (sq.m) ____

1.8 Ground coverage (sq. m) ____

1.9 Soil Type ____ 1.10 Foundation Type ____

1.11 Roof Type ____ 1.12 Floor Type ____

1.12 Structural Components:

1.12.1 Wall Type: BB* ☐ Earthen ☐ UCR* ☐ CCB* ☐

1.12.2 Thickness of Wall ____ 1.12.3 Slab Thickness ____

1.12.4 Mortar Type : Mud ☐ Lime ☐ Cement ☐

1.12.5 Vert. R/F bars: Corners ☐ T-Junctions ☐ Jambs ☐

1.12.6 Seismic bands : Plinth ☐ Lintel ☐ Eaves ☐ Gable ☐

* BB - Burnt Brick, * UCR - Uncoursed Random Rubble
* CCB: Cement Concrete Block

Sketch Plan with Length & Breadth

2.0 Occupancy	3.0 Special Hazard	4.0 Falling Hazard
*2.1 **Important buildings:** Hospitals, Schools, monumental structures; emergency buildings like telephone exchange, television, radio stations, railway stations, fire stations, large community halls like cinemas, assembly hall and subway station, power stations, important industrial establishments, VIP residences and Residences of Important Emergency person.* *• Any building having more than 100 Occupants may be treated as **Important.*** ***2.2 Ordinary buildings:** Other buil ding having occupants<100*	3.1 High Water Table (within 3m) & if sandy soil, then liquefiable site indicated Yes ☐ No ☐ 3.2 Land Slide Prone Site Yes ☐ No ☐ 3.3 Servere Vertical Inregularity Yes ☐ No ☐ 3.4 Severe Plan Irregularity Yes ☐ No ☐ **5.0 SPECIAL OBSERVATION** 5.1 Length of wall between two cross walls are as per IS:4326 or IS:13828 Yes ☐ No ☐ 5.2 Percentage of openings in walls is as per IS:4326 or IS:13828 Yes ☐ No ☐ 5.1 Ratio of height & width of wall is as per IS:4326 or IS:13828 Yes ☐ No ☐	4.1 Chimneys ☐ 4.2 Parapets ☐ 4.3 Cladding ☐ 4.4 Others ☐

RECOMMENDED ACTION

- ☐ Ensure adequate maintenance.
- ☐ If any Special Hazard 3.0 found, re-evaluate for possible retrofitting.
- ☐ If any of the falling hazard is present, either remove it or strengthen against falling.
- ☐ Special observation if not compliant may lead to more severe damage and will call for retrofitting.

5.0 Probable Damageability in Few/Many Building

Building type	**5.1 Masonry Building**			
Damage-ability in Zone II	A	B/B+	C/C+	D
	G2	G2/G1	G1/G1	-

Note: + sign indicates higher strength hence somewhat lower damage expected as stated. Also average damage in one building type in the area may be lower by one grade point than the probable damageability indicated.

Surveyor will identify the Building Type; encircle it, also the corresponding damage grade.

Surveyor's
Sign : ____
Name : ____
Executive
Engineer's Sign: ____
Sign ____
Date of Survey: ____

Rapid Visual Screening of Masonry Buildings for Seismic Hazards – Form 2

Seismic Zone III Ordinary Building also Seismic Zone II Important Building

Photograph

1.1 Building Name ______

1.2 Use ______

1.3 Address ______

______ Pin ______

1.4 Other Identifiers ______

1.5 No. of Stories ______ 1.6 Year Built______

1.7 Total Covered Area; all floors (sq.m) ______

1.8 Ground coverage (sq. m) ______

1.9 Soil Type ______ 1.10 Foundation Type ______

1.11 Roof Type ______ 1.12 Floor Type ______

1.12 Structural Components:

1.12.1 Wall Type: BB* ☐ Earthen ☐ UCR* ☐ CCB* ☐

1.12.2 Thickness of Wall ______ 1.12.3 Slab Thickness ______

1.12.4 Mortar Type : Mud ☐ Lime ☐ Cement ☐

1.12.5 Vert. R/F bars: Corners ☐ T-Junctions ☐ Jambs ☐

1.12.6 Seismic bands : Plinth ☐ Lintel ☐ Eaves ☐ Gable ☐

* BB - Burnt Brick, * UCR - Uncoursed Random Rubble
* CCB: Cement Concrete Block

Sketch Plan with Length & Breadth

2.0 Occupancy

*2.1 **Important buildings:** Hospitals, Schools, monumental structures; emergency buildings like telephone exchange, television, radio stations, railway stations, fire stations, large community halls like cinemas, assembly hall and subway station, power stations, important industrial establishments, VIP residences and Residences of Important Emergency persons.*

Any building having more than 100 Occupants may be treated as **Important.*

***2.2 Ordinary buildings:** Other buil ding having occupants<100*

3.0 Special Hazard

3.1 High Water Table (within 3m) & if sandy soil, then liquefiable site indicated Yes ☐ No ☐
3.2 Land Slide Prone Site Yes ☐ No ☐
3.3 Servere Vertical Inregularity Yes ☐ No ☐
3.4 Severe Plan Irregularity Yes ☐ No ☐

4.0 Falling Hazard

4.1 Chimneys ☐
4.2 Parapets ☐
4.3 Cladding ☐
4.4 Others ☐

5.0 SPECIAL OBSERVATION

5.1 Length of wall between two cross walls are as per IS:4326 or IS:13828 Yes ☐ No ☐
5.2 Percentage of openings in walls is as per IS:4326 or IS:13828 Yes ☐ No ☐
5.1 Ratio of height & width of wall is as per IS:4326 or IS:13828 Yes ☐ No ☐

RECOMMENDED ACTION

☐ Ensure adequate maintenance.
☐ Detailed evaluation of B type for need for retrofitting.
☐ Detailed evaluation of A types for need for reconstruction or possible retrofitting.
☐ If any Special Hazard 3.0 found, re-evaluate for possible prevention / retrofitting.
☐ If any of the falling hazard is present, either remove it or strengthen against falling.
☐ Special observation if not compliant may lead to more severe damage and will call for retrofitting.

5.0 Probable Damageability in Few/Many Buildings

Building type	5.1 Masonry Building			
Damage-ability in Zone II	A	B/B+	C/C+	D
	G4	G3/G2	G2/G1	G1

Note: + sign indicates higher strength hence somewhat lower damage expected as stated. Also average damage in one building type in the area may be lower by one grade point than the probable damageability indicated.

Surveyor will identify the Building Type; encircle it, also the corresponding damage grade.

Surveyor's
Sign : ______
Name : ______
Executive
Engineer's Sign: ______
Sign ______
Date of Survey: ______

Rapid Visual Screening of Masonry Buildings for Seismic Hazards – Form 3

Seismic Zone IV Ordinary Building (Also for zone III Important Building)

Photograph

1.1 Building Name ______

1.2 Use ______

1.3 Address ______

______ Pin ______

1.4 Other Identifiers ______

1.5 No. of Stories ______ 1.6 Year Built______

1.7 Total Covered Area; all floors (sq.m) ______

1.8 Ground coverage (sq. m) ______

1.9 Soil Type ______ 1.10 Foundation Type ______

1.11 Roof Type ______ 1.12 Floor Type ______

1.12 Structural Components:

1.12.1 Wall Type: BB* ☐ Earthen ☐ UCR* ☐ CCB* ☐

1.12.2 Thickness of Wall ______ 1.12.3 Slab Thickness ______

1.12.4 Mortar Type : Mud ☐ Lime ☐ Cement ☐

1.12.5 Vert. R/F bars: Corners ☐ T-Junctions ☐ Jambs ☐

1.12.6 Seismic bands : Plinth ☐ Lintel ☐ Eaves ☐ Gable ☐

* BB - Burnt Brick, * UCR - Uncoursed Random Rubble
* CCB: Cement Concrete Block

Sketch Plan with Length & Breadth

2.0 Occupancy	3.0 Special Hazard	4.0 Falling Hazard
*2.1 **Important buildings:** Hospitals, Schools, monumental structures; emergency buildings like telephone exchange, television, radio stations, railway stations, fire stations, large community halls like cinemas, assembly hall and subway station, power stations, important industrial establishments, VIP residences and Residences of Important Emergency person.* ** Any building having more than 100 Occupants may be treated as **Important.*** ***2.2 Ordinary buildings:** Other buil ding having occupants<100*	3.1 High Water Table (within 3m) & if sandy soil, then liquefiable site indicated Yes ☐ No ☐ 3.2 Land Slide Prone Site Yes ☐ No ☐ 3.3 Servere Vertical Inregularity Yes ☐ No ☐ 3.4 Severe Plan Irregularity Yes ☐ No ☐ **5.0 SPECIAL OBSERVATION** 5.1 Length of wall between two cross walls are as per IS:4326 or IS:13828 Yes ☐ No ☐ 5.2 Percentage of openings in walls is as per IS:4326 or IS:13828 Yes ☐ No ☐ 5.1 Ratio of height & width of wall is as per IS:4326 or IS:13828 Yes ☐ No ☐	4.1 Chimneys ☐ 4.2 Parapets ☐ 4.3 Cladding ☐ 4.4 Others ☐

RECOMMENDED ACTION

- ☐ A or B: evaluate in detail for need of reconstructions or possible retrofitting to achieve type C & D
- ☐ B + C: evaluate in detail for need for retrofitting
- ☐ If any Special Hazard 3.0 found, re-evaluate for possible prevention/retrofitting.
- ☐ If any of the falling hazard is present, either remove it or strengthen against falling.
- ☐ Special observation if not compliant may lead to more severe damage and will call for retrofitting.

5.0 Probable Damageability in Few/Many Building

Building type	**5.1 Masonry Building**			
Damage-ability in Zone II	A	B/B+	C/C+	D
	G5	G4/G3	G3/G2	G2

Note: + sign indicates higher strength hence somewhat lower damage expected as stated. Also average damage in one building type in the area may be lower by one grade point than the probable damageability indicated.

Surveyor will identify the Building Type; encircle it, also the corresponding damage grade.

Surveyor's

Sign : ______

Name : ______

Executive

Engineer's Sign: ______

Sign ______

Date of Survey: ______

Rapid Visual Screening of Masonry Buildings for Seismic Hazards – Form 4

Seismic Zone V All Building (Also for Zone IV Important Building)

Photograph

1.1 Building Name ____________________

1.2 Use ____________________

1.3 Address ____________________

____________________ Pin ____________

1.4 Other Identifiers ____________________

1.5 No. of Stories ____________ 1.6 Year Built ____________

1.7 Total Covered Area; all floors (sq.m) ____________

1.8 Ground coverage (sq. m) ____________

1.9 Soil Type ____________ 1.10 Foundation Type ____________

1.11 Roof Type ____________ 1.12 Floor Type ____________

1.12 Structural Components:

1.12.1 Wall Type: BB* ☐ Earthen ☐ UCR* ☐ CCB* ☐

1.12.2 Thickness of Wall ________ 1.12.3 Slab Thickness ________

1.12.4 Mortar Type : Mud ☐ Lime ☐ Cement ☐

1.12.5 Vert. R/F bars: Corners ☐ T-Junctions ☐ Jambs ☐

1.12.6 Seismic bands : Plinth ☐ Lintel ☐ Eaves ☐ Gable ☐

* BB - Burnt Brick, * UCR - Uncoursed Random Rubble
* CCB: Cement Concrete Block

Sketch Plan with Length & Breadth

2.0 Occupancy	3.0 Special Hazard	4.0 Falling Hazard
*2.1 **Important buildings:** Hospitals, Schools, monumental structures; emergency buildings like telephone exchange, television, radio stations, railway stations, fire stations, large community halls like cinemas, assembly hall and subway station, power stations, important industrial establishments, VIP residences and Residences of Important Emergency person.* ** Any building having more than 100 Occupants may be treated as **Important.*** ***2.2 Ordinary buildings:** Other buil ding having occupants<100*	3.1 High Water Table (within 3m) & if sandy soil, then liquefiable site indicated Yes ☐ No ☐ 3.2 Land Slide Prone Site Yes ☐ No ☐ 3.3 Servere Vertical Inregularity Yes ☐ No ☐ 3.4 Severe Plan Irregularity Yes ☐ No ☐ **5.0 SPECIAL OBSERVATION** 5.1 Length of wall between two cross walls are as per IS:4326 or IS:13828 Yes ☐ No ☐ 5.2 Percentage of openings in walls is as per IS:4326 or IS:13828 Yes ☐ No ☐ 5.1 Ratio of height & width of wall is as per IS:4326 or IS:13828 Yes ☐ No ☐	4.1 Chimneys ☐ 4.2 Parapets ☐ 4.3 Cladding ☐ 4.4 Others ☐

RECOMMENDED ACTION

☐ A or B,B+: evaluate in detail for need of reconstructions or possible retrofitting to achieve type C+ or D

☐ C: evaluate in details for need for retrofitting to achieve type C+, D.

☐ Wood: evaluate in detail for Retrofitting

☐ If any Special Hazard 3.0 found, re-evaluate for possible prevention/retrofitting.

☐ If any of the falling hazard is present, either remove it or strengthen against fall.

☐ Special observation if not compliant may lead to more severe damage and will call for retrofitting.

5.0 Probable Damageability in Few/Many Building

Building type	5.1 Masonry Building			
Damage-ability in Zone II	A	B/B+	C/C+	D
	G5	G5/G4	G4/G3	G3

Note: + sign indicates higher strength hence somewhat lower damage expected as stated. Also average damage in one building type in the area may be lower by one grade point than the probable damageability indicated.

Surveyor will identify the Building Type; encircle it, also the corresponding damage grade.

Surveyor's

Sign : ____________

Name : ____________

Executive

Engineer's Sign: ____________

Sign ____________

Date of Survey: ____________

Chapter 21

Seismic Assessment of Masonry Buildings[1,2]

21.1 Introduction

In the assessment of expected seismic performance of existing load bearing masonry wall buildings, the most direct approach will be to compare the safety provisions specified in IS: 4326-1993 for such buildings with the situation actually prevailing in the existing building. Where the existing condition is found *compliant* with IS: 4326, it will be considered safe and acceptable. But where the existing condition is found deficient, hence, *non-compliant,* it will be considered as weak, requiring upgradation or strengthening or retrofitting.

21.2 Building Categories

For specifying the seismic safety requirements and the reinforcing details in load bearing masonry wall buildings, the buildings are categorised according to the seismic coefficient specified for design in the seismic zone and the important factors specified for the building type. Taking these into account, the following building categories have been specified in IS:4326-1993 as amended in 2005.

Table1: Building Categories (for use with IS 4326 &IS 13928)

Building Use	Building Category in Seismic Zone			
	II	III	IV	V
Ordinary	B	C	D	E
Important (I=1.5)	C	D	E	E+

21.3 Factors Considered in Seismic Safety as Per IS: 4326-1993

The most important factors considered in IS: 4326-1993 for ensuring seismic safety of various category buildings are the following:

1. *Dr. A.S.Arya, Professor Emeritus, IIT - Roorkee and National Seismic Advisor,GoI - UNDP, DRM Programme.*
2. *This method of assessment by the author has been included in IS: 13935 - 2006 (First Revision).*

Table 1A: The Buildings Considered with Importance Factor I =1.5

S.No.	Structure	Importance factor
1	Important service and community buildings, such as hospitals; schools; monumental structures; emergency buildings like telephone exchange, television stations, fire station buildings; large community halls like cinemas, assembly halls and subway stations, power stations	1.5
2	All other buildings	1.0

Notes:-

1) The design engineer may choose values of importance factor/greater than those mentioned above.
2) Buildings not covered in S.No. 1 & 2 above may be designed for higher value of I, depending on economy, strategy considerations like multi-storey buildings having several residential units.
3) This does not apply to temporary structures like excavations, scaffolding etc. of short duration.

I) Walls
- (i) Mortar
- (ii) Door, window openings in walls
- (iii) Length of wall between cross walls
- (iv) Height of wall above floor to ceiling
- (v) Unconnected perpendicular walls, parapets, cantilever balconies, etc.
- (vi) Horizontal Seismic Bands at
  - (a) plinth level
  - (b) door-window lintel level
  - (c) ceiling of flat floor/roof or eave level of sloping roofs
  - (d) gable ends and top of ridge wall
  - (e) window sill level
- (vii) Vertical steel bar
  - (a) at each corner/junction of walls
  - (b) at door & window jambs

II) Roofs or Floors
- (i) Roofs/floors with prefabricated or precast elements
- (ii) Cantilever balconies
- (iii) Roofs/floors with wooden joists with various covering elements
- (iv) Sloping roofs with sheets or tile covering
- (v) Jack arch roof or floors
- (vi) Sloping raftered roofs

The safety requirements of the various elements as per IS: 4326-1993 are given in Tables 2, 3 and 4 for ready reference.

21.4 Assessment of a Building

For the building under assessment for seismic safety, first its building Category should be determined. For example, for a residential building in Delhi, the seismic zone is IV and Importance factor will be 1.0. Therefore, from Table 1, the Category of the building is D. Now compile the values for various elements for the actual building in col (2) of Table 5. The corresponding Codal values may be picked from Table 2 & 4 and entered in col (3) of Table 5. Comparing the values in col (2) with those in col (3) in Table 5, one could mark the appropriate elements as compliant or non-compliant in col (4) of Table 5. For the elements found non-compliant, improvement by retrofitting needs to be done for which suitable action is indicated in the last column of Tables 2, 3 & 4.

Table 2: Main Provisions in IS 4326:1993 about Masonry Walls

S.No	Item of Masonry	Requirement as per IS 4326: 1993 for Building Category				Action for Retrofitting if Code Requirement not found satisfied.
		B	C	D	E, E^{+}	
1	Mortar	CLS-1:2:9 or CS-1:6		CLS-1:1:6 or CS-1:4		See cl.8.4. Alternatively, walls may be strengthened by ferro-cement plating or injection grouting
2	Door, Window openings -. See figure below b_5 minimum	0.0	230 mm	450 mm	450 mm	Increase by building up or strengthening by ferro-cement plating.
	$(b_1+b_2+b_3)/l_1$ max.$(b_6+b_7)/l_2$					Attain the limit by closing/ narrowing an opening Or reinforce the opening by seismic belting.
	one storey	0.6	0.55	0.50	0.5	
	two storey	0.50	0.46	0.42	0.42	
	three storey	0.42	0.37	0.33	0.33	
	four storey	0.42	0.37	0.33	Not allowed	

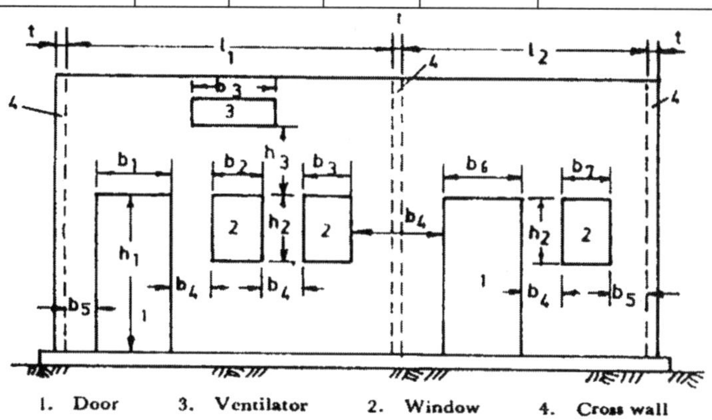

Fig.1: Door 2. Window 3. Ventilator 4. Cross Wall

	b_4, minimum	340 mm	450 mm	560 mm	560 mm	Increase by build-up or reinforce with belt.
3	Length of wall between cross walls	---	Max length = 35 x thickness or 8m whichever less			If length more, provide pilaster or buttress to reduce effective length.
4	Height of wall from floor to ceiling	---	Maximum = 15 times thickness or 4 m whichever less			If height more, add pilaster to increase effective thickness
5	Random-Rubble walls	• 'Through' or Header stones, one each in 0.72sq.m surface area of wall. • Long stones at corners of walls, in each wall in every alternate course.				If not provided, install RC Headers in holes made by removing stones.

Table 2A:Improvements Against Global Deficiencies

S.No	Item	B	C	D	E, E$^+$	Retrofitting Action if code provision not satisfied
1.	Sloping raftered roofs		Preferably use full trusses			Convert rafters into A-frames or full trusses to reduce thrust on walls
2.	Unsymmetrical Plans		Symmetrical plans are suggested			Inserting new walls to reduce dissymmetry
3.	Perpendicular Walls not connected at corners and T-junctions		Perpendicular walls should be integrally constructed			Stitch the perpendicular walls using tie rods in drilled holes fully grouted or box them with seismic belts

Table 3: Provisions for Seismic Elements in Masonry Buildings (IS:4326 - 1993)

S.No	Item of Masonry	Requirement as per IS 4326:1993 for Building Category				Action for Retrofitting if Code req. not found satisfied.
		B	C	D	E, E$^+$	
1	*Horizontal Seismic Bands-* • Plinth level	Needed if soft (type III) soil at base				Provide seismic belt if plinth height = 90 cm
	• Door window lintel level	Needed in all cases with varying reinforcement and thickness specified in each case				Provide seismic belt of equivalent strength on both sides of walls
	• Ceiling or eave level	Needed in sloping roofs or floors and roofs of prefabricated elements.				-do- -do-

	• Gable or ridge wall	Needed in case of pitched roofs				
	• Window sill level or dowels	Not requ-ired	Not requ-ired	Requ-ired in 3 & 4 storeyed	Requ-ired in all buil-dings	-do-
2.	Vertical bar at each corner and T-junction of wall	Needed in only 4 storey building	Needed in 3 and 4 storey building	Needed in all building	Needed in all building	Install equivalent bars or vertical belts at corners and T-junctions
3.	Vertical bar at jambs of window &door	Not nee-ded	-do-	-do-	-do-	Install equivalent seismic belts around the opening

Table 4 - Provisions for Roofs & Floors in IS 4326:1993

S.No	Item of Roof/ Floor	Requirement as per IS 4326:1993 for Building Category				Retrofitting Action if Code provision not found satisfied.
		B	C	D	E, E^+	
1.	Roof/floor with prefabricated/ pre-cast elements	Tie beam all round		All round tie beam and RC screed		Provide RC screed and seismic belt or band around
2.	Roof/floor with wooden joists, various covering elements and earth fill	___	All round seismic band and integration of units as a rigid horizontal diaphragm			Provide seismic belt around , inter-connect beam ends through wooden planks and diagonal x-ties
3.	Jack arch roof/floor	___	Connect the steel joists by horizontal ties to prevent spreading and cracking of arches. Provide seismic band all round			Install steel flats as ties by welding them to the steel joists and provide seismic belt
4.	Sloping roofs with sheeting or tile coverings	___	(i) Horizontal x-bracing at level of ties of the trusses. (ii) X-bracing in the planes of the rafters and purlins			Install the x-bracings as per the code IS:4326. Anchor trusses into walls and rafters into seismic belt at eave level.
5.	Sloping rafter roof	___	Preferably use full trusses			Convert rafters into A frames.
6.	Cantilever balcony	___	To be taken out from RC floor or anchored back for $A_v = 3 A_h$			Upgrade the anchorage

Table 5 - Seismic Design Compliance Assessment of Building

Building Category (tick mark as applicable) - C ☐, D ☐, E ☐, E+☐

S.No (1)	Data of Building under Assessment (2)	Required as per Code(3)	Compliant Yes/no
1	Number of storeys	≤ 4	
2.	Wall building unit BB/CCB (solid)/CCB (Hollow)	Compressive strength ≥ 50 mpa	
3.	Wall thickness G.F. I.F. II. F. III.F	BB = 230 mm CCB = 200 mm	
4.	Largest size room ....m X ...... m	8 m X 8 m	
5.	Mortar used C:S =		
6.	Door, Window openings (Based on building height) (i) Overall $(b_1 + b_2 + \ldots)/l^1$, max = (ii) B_4 *min.* = (iii) B_5 min.=		
7.	Wall length/thickness ratio t = l = l/t =		
8.	Wall height/thickness ratio t = h = h/t =		
9.	Soil at base Soft/ hard/medium		
10.	Floor type (tick mark) RC slab/RB slab/ Precast beams or slabs		
11.	Roof type (tick mark) Horizontal flat/sloping/RC or RB slab/ trusses or rafters		
12.	Seismic Bands (yes/no)		
	(i) at plinth (ii) at lintel level (iii) at ceiling or eave level (iv) at window sill level (v) at gable ends (vi) at ridge top		
13.	Vertical bar (yes/no) (i) at external corners (ii) at external T-junctions (iii) at internal corners (iv) at internal T-junctions (v) at jambs of door (vi) at jambs of windows		
14.	Sloping Roofs (yes/no) (i) rafters (any x-bracing ?) (ii) trusses (x-bracing in plan?)(x-bracing in slopes?) (iii) tile covering (with holding down systems?)		
	BB - Burned Brick, CCB - Cement Concrete Block		

21.5 Testing of Materials

Sometimes the strength of the soil, the building units such as bricks or blocks and the mortar will need to be tested, particularly for old (heritage type) buildings. Particular attention in this regard may be paid to important and/or critically significant buildings so as to ensure their durability over many years of future life, besides required strength performance.

21.6 Summary of Results of Comparative Assessment

The major deficiencies indicating non-compliance with codal provisions should be noted. These will need to be considered for upgrading by retrofitting suitably to prevent the total or partial collapse of the building in future, if and when the probable maximum earthquake intensity will strike the area. ■

21.7 References

1. IS:4326-1993 "Earthquake Resistant Design and Construction of Buildings - Code of Practice (Second Revision)".
2. IS:13935.2007 "Repair and Seismic Strengthening of Masonry Buildings - Guidelines" (First Revision 2007).

Chapter 22

Damage Scenario of A Hypothetical 8.0 Magnitude Earthquake in Kangra Region[1,2]

22.1 Introduction

The state of Himachal Pradesh (HP) is located between 30.3° - 33.0°N latitude and 75.6° - 79.0°E longitude in the western Himalayas. Seismically it lies in the great Alpide-Himalayan seismic belt running from the Alps mountains through Yugoslavia, Turkey, Iran, Afghanistan, Pakistan, India, Nepal, Bhutan and Burma. The terrain is hilly all through the state of Himachal Pradesh, the ranges varying from the Shivaliks in the south to the tall snow clad Pir Panjals in the north. These are traversed by major rivers, Sutlej, Beas and their tributaries. The state has not only been shaken by earthquakes occurring in its territory but also in the neighbouring areas of Jammu and Kashmir in the north, Tibet in the east and U.P. Hills in the south-east. A number of damaging earthquakes have occurred in the H.P. territory during this century for which information is well recorded. Information about earthquake occurrence before the famous 1905 Kangra earthquake is not, however, available and is a matter of research through historical and archival records. Herein, there is presented a profile of the seismic hazard in the state, the types of prevalent construction, their vulnerability to the earthquake hazards and the resulting seismic risk from a hypothetical earthquake of Magnitude 8.0 on Richter scale, if it were to occur in 1991 in the same area in which as large an earthquake had occurred in 1905.

22.2 The Earthquake Occurrence in H.P.

The earthquake activity in H.P. is attributed to the Himalayan orogeny. Based on the latest concept of plate tectonic model of the earth, the Himalayan mountains have formed due to continuous thrusting of the Indian plate with Eurasian plate since Cretaceous times. The present geological structure and the tectonics of the Himalayas have been formed as a result of this continued collision. There are regional tectonic features in Himalayas like the Main

---

1. *A.S. Arya, Professor Emeritus, University of Roorkee.*
2. *Chapter 5, by A.S. Arya Techno-Legal Aspects, Report of the Expert group on Natural Disaster Prevention, Preparedness and Mitigation having hearing on Housing and related Infrastructure, pub. By BMTPC, 1998*

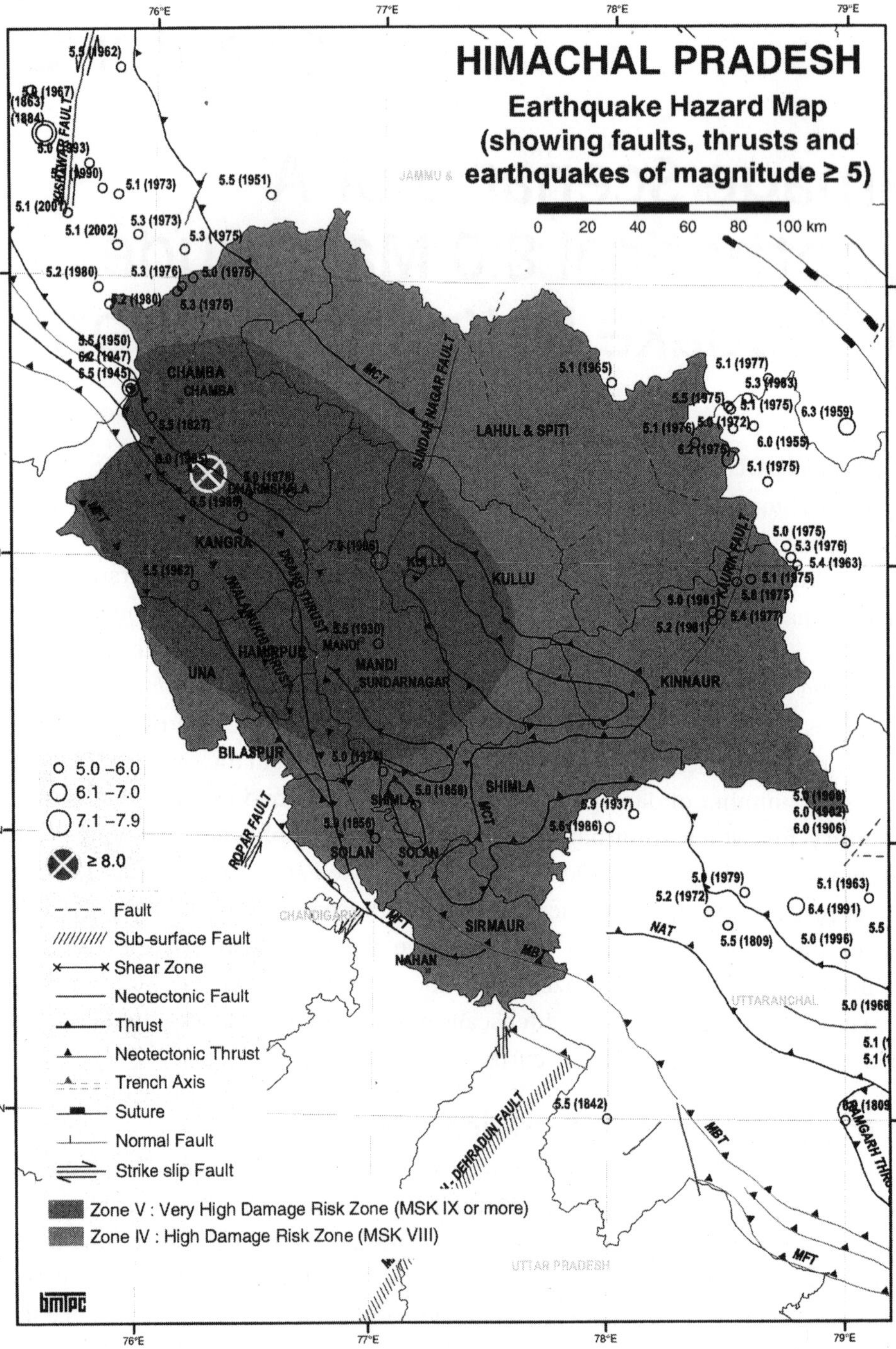

BMTPC : Vulnerability Atlas - 2nd Edition; Peer Group, MoH&UPA; Map is Based on digitised data of SOI, GOI; Seismic Zones of India Map IS:1893: 2002; Seismotectonic Atlas of India, GSI, GOI

Fig.1: Seismic zoning map of Himachal Pradesh

Boundary Fault (MBF) and Main Central Thrust (MCT) running parallel to the strike length of the Himalayas. Apart from these regional tectonic features, there are lineaments running transverse to the Himalayan Trend. Slow movements result in the elastic strain build up and the sudden release of tectonic strain energy along any of these tectonic features causes the observed earthquake activity.

The region surrounding Dharmsala and Kangra is the seat of the great Kangra earthquake of 1905 (M=8.0 on Ricther's Magnitude scale) and historically has high seismicity. The region falls in seismic Zone V of the seismic zoning map of India which is based on probable Intensity 'IX and more' on Modified Mercalli 'MM' or MSK, Intensity Scale. Fig. 1 shows the Seismic Zoning Map of Himachal Pradesh. There are 250 earthquakes of Magnitude 4.0 and more, including 81 earthquakes with magnitude above 5.0 which have rocked the state of H.P. and adjoining regions of Jammu and Kashmir in the last ninety years.

Among these earthquakes, three earthquakes have caused considerable damage and loss of life. These are briefly described below:

22.2.1 Great Kangra Earthquake of 1905

This earthquake having a magnitude 8.0 on the Richter scale had its epicentre at 32o 15'N, 76o 15'E and occurred at 00h 50m GMT or 06h 20m IST. It had taken a toll of 20,000 lives, caused MM intensity X and more in the epicentral region and was felt over an area of 416000 sq.km. It was concluded that the earthquake may have been caused due to a displacement taking place along a low angle fault at a depth of 34 to 64 km. The Isoseismal of this earthquake on the Rossi-Forrel Intensity scale as observed and MSK scale as interpolated are shown in Fig. 2.

Fig.2: Isosesimal map of Himachal Pradesh

The first violent effect of this earthquake was observed at Shahpur at a distance of 55 km from Pathankot towards Dharmsala. The shops at the road side built in sun-dried bricks with heavy slate roofs were partly ruined (roof collapsed, but walls standing). At Shahpur, half the buildings were ruined and the rest were rent apart. All buildings were built of roughly shaped sun-dried bricks and sometimes with stone foundations raised abut 15 cm above ground. Roofs were normally of slates but sometimes thatch was also used.

Beyond Shahpur, in the direction of Dharamsala and Kangra, the damage was of ascending scale. The route between Shahpur and Dharmsala showed intense damage to the small and big villages and lines of communication. The road was broken at many places. At village Cheri, the effect was maximum and the whole village was destroyed, with the exception of only one or two stronger buildings which were also half ruined. Generally speaking, the houses had become mere heaps of sun-dried bricks, mixed with slates and rafters.

The Dharamsala township suffered severe damage amounting to total destruction at many places and casualties reached a very high figure. The military and civil staff was reduced to about one-half by the deaths. The barracks at the cantonment consisted of long single storey buildings, the principal walls being built oo sun-dried bricks and the two end walls of cut stones. The roofs were of thick slates. In all cases, the two long walls had rocked over, generally as a whole, in the downhill direction, with the exception of the portions immediately connected with the stone at the gable ends. Buildings constructed throughout of dressed stone were badly cracked and rent and the portions like gable ends had been flung out as a whole flat on to the ground. The iron framed roof with two rods across from wall to wall preserved the Armoury building from complete collapse. The Magazine consisting of a central square room, very strongly built of well dressed stones and lime mortar with arched roof, showed only cracks.

At Forsythganj bazaar, buildings were constructed of sun-dried bricks, especially in the lower storeys and partly of wood mostly in the upper storeys and verandahs. All shops to the east of the road were ruined while on the west many of them survived total collapse.

Mcleodganj bazaar was levelled to the ground with no building standing even partially. The situation was the same at Kotwali bazaar which was reduced to a thick deposit of house rubbish about 1.5m high through which roads and lanes were cut for rescue and salvage operation. At Kangra, the devastation was total. Not a single house was standing. The horror of the actual calamity was beyond imagination. There was no one left alive for directing rescue operations. *All the subordinate officials were killed* and panic stricken people fled fearing that yet worse tremors would envelop them.

After the earthquake, the district headquarters were shifted to Dharamsala which was rebuilt using *wood-framed brick-nogged buildings* locally called *Dajji,* the present office and residence of the deputy commissioner there being the existing examples of this construction. This appears to have become the typical construction then, even being used by private people for their shops and houses. But later on, gradually and more particularly after the independence of the country, when the pace of development became fast and wood started becoming scarce, this very highly earthquake resistant construction method, namely Dajji, was superseded by the plains-type brick and hill-type stone constructions. Their unsuitability for meeting the earthquake shaking became clearly manifest in the April 26, 1986 earthquake of M=5.7 and slightly obvious during the earlier quake of June 15, 1978 with M=5.0 occurring in the same general epicentral region.

22.2.2 Kinnaur Earthquake of January 19, 1975

The Richter magnitude was estimated as 6.7 and maximum observed intensity in the region was IX on the MM scale. The epicentral distance was about 25 km from Kinnaur town. The earthquake caused the death of sixty people and several hundred were severely injured. Traditional construction in the area was non-seismic and had little resistance against the lateral forces. Nearly 2000 dwellings are reported to have suffered heavy damage, even in that sparsely populated area.

Recent random rubble masonry and dressed stone masonry construction suffered extensive damage. Heavy flat roofs suffered greater damage. Buildings constructed in hollow concrete blocks or dressed stone masonry in cement-mortar developed small cracks in the walls. *Light structures made of corrugated iron sheets nailed to timber frames and arches did not suffer any damage.* The temples (monastries) and monuments also suffered badly.

22.2.3 Dharamsala Earthquake of April 26, 1986

The epicentre of this earthquake of M=5.7 was at Lat. 32.1oN and Long. 76.3oE, that is, very close to that of the 1905 earthquake. The origin time was 13h 5m 17s Indian Standard Time and focal depth was shallow, about 10km. The loss of life was fortunately small, only about 6 persons, since the earthquake did not last long, although a very large number of dwellings got cracked, including many government buildings at Dharamsala and other towns. The most significant damage, requiring reconstruction of houses, was to the adobe and stone houses in the villages near Dharamsala, such as Narghota, Naddi, Kaned, Sukar and Khanyara. The total financial loss was estimated as Rs.65 crores (1986 prices) according to newspaper reports.

In comparison to the Kangra earthquake of 1905, the Dharamsala earthquake had a destructive energy of only about 1/28000. The maximum Modified Mercalli intensities in the two earthquakes were X and VII+ respectively, showing again that the 1986 quake was a rather minor one as compared to the great earthquake of 1905.

22.3 Building Construction In H.P.

The building construction in Himachal Pradesh, except for the newer R.C. framed buildings in larger towns, follows the traditional pattern of load bearing walls in burnt bricks, stone or unburnt clay blocks (called 'Adobe' construction in modern writings) with sloping roofs in

Table: 1 House Types in the Districts of Himachal Pradesh

(based on Census of Housing 1991)

S. No	Name of District	Total Houses	Percent Area in (Kangra E.Q. 1905)			Number of House Types				Remarks
			MSK ≥IX	MSK VIII	MSK VII	A	B	C	X	
9.01	Bilaspur	111,107	-	-	100	98,410	10,373	1,373	951	House Type: A-Walls of mud, unburnt bricks, random-rubble masonry in mud, field stone (any roof) B- walls of burnt bricks laid in any mortar with any roof C- Walls of concrete, R.C. frame construction, wooden walls or frame, Ekra construction X- Walls of galvanised iron metal sheets, bamboo, thatch, grass leaves, etc.
9.02	Chamba	145,344	2	7	40	134,087	4,723	4,406	2,128	
9.03	Hamirpur	144,408	-	17	83	123,775	19,351	477	805	
9.04	Kangra	397,728	33	17	50	323,614	64,387	2,873	6,854	
9.05	Kinnaur	39,351	-	-	20	29,029	516	8,908	898	
9.06	Kullu	123,455	1.0	29	70	110,992	3,700	8,398	365	
9.07	Lahul &Spiti	14,521	-	-	-	13,951	20	148	402	
9.08	Mandi	313,362	19	33	48	291,492	10,875	9,845	1,150	
9.09	Shimla	266,871	-	-	100	209,866	34,582	18,355	4,068	
9.10	Sirmour	150,990	-	-	100	124,930	20,326	1,546	4,188	
9.11	Solan	139,261	-	-	100	96,185	39,319	1,742	2.015	
9.12	Una	129,538	-	-	100	68,239	58,112	767	2,420	
9.0	Total	1975,936				1624,570	266,284	58,838	26,244	

high rainfall areas and flat roofs in the dry areas. A very good idea of the house construction in the whole state may be obtained with reference to the 1991 Census of Houses, the main data from which is presented in Table 1. The Census houses include dwellings, residences, shop sum residences, workshops-cum-residences, hotels, sarais, dharamsalas, tourist houses, inspection houses, business houses and offices, restaurant, sweet shops and eating places; cinemas and panchayat ghars, places of worship and those lying vacant.

From Table 1, it is clearly seen that the predominant materials used in walls are: A-type (earthern and stone walls) 82.2%, B-type (ordinary burnt-brick walls) 13.5%, C-type (concrete or wood-framed walls) only 3%, and X-type (rest including metallic sheets, thatch etc.) 1.3%. Among these, the A-type are highly vulnerable to severe damage in MSK IX and VIII, which form the basis of seismic Zones V and IV respectively. Wood and concrete walls which are rather safe in moderate earthquakes constitute only about 3.4% and the X-type which are rather insensitive to seismic effects are only 1.3% of the total housing.

The predominant materials used in the roofs (not shown in Table 1) are thatch and reeds, etc. which are about 30%, and tile, slate and shingle comprise 55.6% . Thatch when wet and the tiled roofs tend to be heavy and cause large earthquake forces on the structure without providing the binding effect on the walls, therefore, are unsuitable for seismic Zones IV and V. Metal and A.C. sheet roofs which are light in weight and concrete slab roofs which have a binding effect on the walls, are both suitable from the seismic angle.

Five districts of Himachal Pradesh, namely, Chamba, Kangra, Mandi, Kullu and Hamirpur which have parts of their areas in seismic Zone V and Parts in Zone IV have higher seismic hazard than the other districts which have only seismic Zone IV. The M.S.K. intensities VII, VIII and IX as related to building damage are stated in Table 2.

Table 2 Seismic Intensity and Maximum Damage to Buildings

Building Type	Intensity VII	Intensity VIII	Intensity IX
A)Mud and Adobe houses,random-stone constructions	*Most* have large deep cracks. *Few* suffer partial collapse	*Most* suffer partial collapse. *Few* suffer complete collapse	*Most* suffer complete collapse
B)Ordinary brick buildings, building of large blocks and prefab type, poor half timbered houses	*Many* have small cracks in walls	*Most* have large and deep cracks. *Few* partial collapse	*Many* show partial collapse. *Few* completely collapse *Few* minor cracks
C)Reinforced buildings, well built wooden buildings	*Many* have fine plaster cracks.	*Most* may have small cracks in walls. *Few* may have large deep cracks	*Many* may have large and deep cracks. *Few* may have partial collapse.

* *Most* = about 75%, *Many* = about 50%, *Few* = about 5%

Table 3 Number of Houses of Various Types in Different Seismic Intensity Areas of Magnitude 8.0 Hypothetical Earthquake in 1991

S. No	Name of District	No. of A-type in MSK			No. of B-type in MSK			No. of C-type in MSK		
		≥IX	VIII	VII	≥IX	VIII	VII	≥IX	VIII	VII
9.01	Bilaspur	-	-	98,410	-	-	10,373	-	-	1,373
9.02	Chamba	2,682	9386	53,634	94	331	1889	88	308	851
9.03	Hamirpur	-	21042	102,733	-	3290	16061	-	81	396
9.04	Kangra	106,793	55014	161,807	21248	10946	17193	948	488	1,437
9.05	Kinnaur	-	-	5,806	-	-	103	-	-	1,782
9.06	Kullu	1110	32188	77,694	37	1073	2590	84	2,435	5,879
9.07	Lahul & Spiti	-	-	-	-	-	-	-	-	-
9.08	Mandi	55383	96193	139,916	2066	3589	5,220	1,870	3,249	4,726
9.09	Shimla	-	-	209,866	-	-	34,582	-	-	18,355
9.10	Sirmaur	-	-	124,930	-	-	20,326	-	-	1,546
9.11	Solan	-	-	96,185	-	-	39,319	-	-	1,742
9.12	Una	-	-	68,239	-	-	58,112	-	-	767
9.0	Total	165,968	213,823	1,139,220	23,445	19,229	205,768	2,990	6,561	38,854

22.4 Damage Potential of Magnitude 8.0 Earthquake

The damage potential of an earthquake of magnitude 8.0 on the Richter scale to the housing in Kangra region can best be estimated by referring to what happened during the 1905 earthquake in that region (see Fig..2 for Intensity Isoseismals of this quake) and the type and numbers of houses today. To arrive at a reasonable estimate of the potential damage, a number of plausible assumptions have to be made. These are listed here below:

1) It will be reasonable to assume that the size of various intensity zones will be the same as that in the 1905 earthquake.
2) Since the building pattern has not substantially changed since 1905, as evidenced by the materials of construction in Table 1 as per the 1991 Census, the damage pattern can be assumed as in the 1905 earthquake which could be taken as globally averaged in the MSK Intensity scale (see Table 2).

For using the average figures of Table 2 in quantitative estimate of damage to buildings, the intensities on the Rossi-Forrel scale shown in Fig 2 have to be interpreted in terms of the MSK scale. For this purpose, the following equivalence is used.

R.F.	1	1-II	III	IV-V	V-VI	VI-VII	VIII-	VIII+ to IX-	IX+	X
MSK	I	II	III	IV	V	VI	VII	VIII	X	XI-XII

3) It is further assumed that the population in any affected district has uniform distribution in that district. Hence, the number of total census houses lying within Isoseismal VIII can approximately be estimated by proportioning the areas under the isoseismal in the district. Thus, the number of houses of A to C type under Intensities IX, VIII and VII is worked out in Table 3.

The potential damage to buildings is estimated in Table 4 with the help of Table 2 based on MSK Scale.

In Table 4, the last line sums up, number of buildings with different degrees of damage to the nearest thousand. The following scenario emerges:-

a) Completely collapsed houses 136000

b) Those partly destroyed and partly having large and deep cracks which will require reconstruction 263000

(c) Those without total or partial collapse but with large cracks, some of these may have to be reconstructed but most are repairable with 15 to 20% of the cost of reconstruction (included strengthening) 916000

Table 4 Estimate of Potential Building Damage

Building Type	No. of Houses in each Intensity Zone		No. of Houses to sustain damage				No Damage or fine cracks
			(a) Complete Collapse	(b) Destroyed with Part collapse	(c) Large Cracks	(d) Small Cracks	
A.Mud, Adobe, FieldStone Wall	≥IX	165968	Most 122476	Bal 33194	Few 8298	-	-
	VIII	213823	Few 10691	Most 160367	Bal 32073	Few 10691	-
	VII	1139220	- -	Few 56961	Most 854415	Bal 227844	-
B.Ordinary Brick Wall	≥IX	23445	Few 1172	Many 11723	Bal 9378	Few 1172	-
	VIII	19229	- -	Few 961	Many 9615	Bal 8653	-
	VII	205768	- -	- -	- -	Many 102884	Bal 10288
C. Concrete and wood Frame/Wall	>IX	2990	- -	Few 150	Many 1495	Bal 1345	-
	VIII	6561	-	- -	Few 328	Most 4921	Bal 131
	VII	38854	- -	- -	- -	- -	Bal 3885
Total		**1815858**	**136339**	**263356**	**915602**	**357510**	**14305**

Few = 5%, Many=50%, Most=75%. Bal = balance out of total

c) Those with small cracks, repairable without evacuation 358000
d) Those with only fine cracks or no damage 143000

Note:The number of houses under (d) and (e) are for Himachal Pradesh only. Such damage could occur in adjoining areas of J&K, Punjab and U.P. also in this hypothetical earthquake.

22.5 Potential of Life and Economic Losses

22.5.1 Estimate for loss of lives

The 1991 population figures show a total population of 5170877 for the State of Himachal Pradesh, giving an average occupancy of 2.62 persons per house.

Apparently, it is too low. Two reasons are cited for it, the state being hilly, many people leave for settling in the plains. The other is the manner of counting the households (perhaps by counting the external doors of the house). For the five districts of Chamba, Hamirpur, Kangra, Kullu and Mandi which have severe Intensities IX and VIII, the average per house is 2.68.

For estimating potential loss of lives, a number of assumption have to be made as follows:

a) The family size per collapsing house is assumed as 2.68.
b) The average death rate has varied from 6 to 18% of the occupants of collapsed houses as per the average figure of earthquakes in India and China, including the recent Killari earthquake wherein the number of deaths per 'collapsed' house came to about 10%. For the buildings in H.P., assumption of 10% would appear reasonable.
c) In partially collapsed houses, the death rate is assumed as 4% of the occupants. Based on the above assumptions, the number of person likely to get killed works out as followed:

136339 x 2.68 x 0.10 = 36539
263356 x 2.68 x 0.04 = 28232

64770

say 65000 persons.

The number of potential deaths would also vary with the month of the year in which potential earthquake may occur, being more for winter and less for summer. Collapse of houses will be comparatively much more if the earthquake occurs in the monsoon season when the mud used in houses becomes soft and weak. Time of occurrence during the day or night will also be critical, the deaths being more if the quake occurs at night when people are sleeping indoors and less during the day when people are awake and working.

22.5.2 Economic loss due to the house damage (estimate as in 1995)

Assuming the cost of reconstruction of a house as Rs. 60000 and that of repair and retrofitting as Rs.15000/=, the total cost of reconstruction and repair will be around Rs.5000 crores for houses in H.P. alone. That for the social buildings, institutional buildings, roads, bridges, etc. could be another Rs. 1000 crores. Besides, the large scale misery, the relief and rehabilitation costs, and the trauma to the state and the country has also to be estimated.

22.6 Conclusion

The scenario of building damage and potential of loss of lives and the economic losses worked out above under the hypothetical recurrence of Mag. 8.0 earthquake in Himachal Pradesh in the year 1991 in the Kangra area brings out a rather dreadful situation : 136000 fully collapsed, 263000 structurally destroyed and partially collapsed and 916000 badly cracked houses; 65000 persons killed; and Rs.5000 crores in economic losses (1995 prices), involving 11 of the 12 districts of the state, but five of them viz. Chamba, Hamirpur, Kangra, Kullu and Mandi being the worst affected . Post earthquake emergency response, even if fully prepared beforehand, can save some lives, but will not be able to reduce property losses and reconstruction and rehabilitation costs. To reduce the disastrous effects on buildings, infrastructure, life and economy, the earthquake resistant design codes must be implemented for all new constructions and seismic-retrofit measures incorporated in existing weak buildings through a concerted and well thought out time-bound programme. ■

22.7 References

1. Arya A.S., "Damage Scenario of M 8.0 Hypothetical Earthquake Occurence in Himachal Pradesh", Bulletin Indian Society of Earthquake Technology Vo. No. 19...
2. Arya A.S. et. al., "Report on Dharamsala, Himachal Pradesh Earthquake, April 26, 1986-Seismicity, Building Damage and Recommendations for Strengthening and Reconstruction", Deptt of E.Q. Engineering, University of Roorkee, July 1986.
3. "IS:1893-1984, Criteria for Earthquake Resistant Design of Structures", BIS, New Delhi.
4. "IS:4326-1993, Recommendations for Earthquake Resistant Construction of Buildings", BIS, New Delhi.
5. "The Kangra Earthquake of April 4, 1905", Memoirs of the Geological Survey of India, Vol. 38, GOI, 1910 (reprinted 1981).
6. "Vulnerbility Atlas of India", Expert Group, Ministry of Urban Affairs & Employment, Government of India, Building Materials & Technology Promotion Council, 1997

Chapter 23

Repair, Restoration and Retrofitting of Masonry Buildings[1,2]

23.1 Introduction

The Kachchh earthquake in Gujarat occurred on the January 26, 2001 and caused massive destruction to property and loss of life in the towns of Bhachau, Anjar, Rapar, Bhuj and Gandhidham as well as thousands of villages. This earthquake had a Richter Magnitude of 6.9 and surface wave magnitude Mw = 7.7 (USGS) and struck the region at 8.46 am local time, with the shaking lasting for a few minutes. Other major cities in Gujarat e.g., Ahmedabad and Jamnagar, which are hundreds of kilometres away, were also affected by the earthquake.

Thousands of people were killed or badly injured in stone masonry buildings in villages as well as in towns because of:

a) poorly constructed buildings, collapsing either totally or partially
b) walls collapsing within narrow streets, burying people escaping into them
c) untied roofs and cantilevers falling onto people
d) free standing high boundary walls, parapets and balconies falling due to the sever shaking
e) gable walls falling over after cracking under lateral thrust of the roof.

Surprisingly, reinforced concrete frame buildings of very recent construction in cities like Bhuj, Gandhidham and Ahmedabad also behaved badly, many of them collapsing totally, killing the inmates in hundreds. This happened because of

a) large open spaces at ground to first floor level, in buildings on stilts for parking or shops (soft storey collapses)
b) foundations restings on soft soils, absence of interconnecting plinth level beams.
c) bad quality of design as well as construction.

Those buildings which were partly or totally destroyed have naturally to be rebuilt, and for safety in future, must be constructed using adequate earthquake resisting measures according to the various Indian standards and appropriate guidelines (see list of References.)

---

1 Anand S. Arya, Seismic Advisor, Gujarat State Disaster Management Authority (GSDMA), Gandhinagar.
2 GSDMA Guidelines, prepared by A. S. Arya March 2002

There are, however, many more buildings which are standing in a damaged condition with cracks of minor to major width and extent. They need restoration of lost strength as well as cosmetic repairs. Even when this is done, these buildings will remain vulnerable in future earthquakes.

The purpose of these guidelines is to address the issues of repair, restoration and seismic retrofitting of various types of masonry buildings. R C buildings are not covered in this guideline.

23.2 Object and Scope

These guidelines have the following objectives;

(i) To indicate appropriate methods of repair and restoration, taking into account the building type and the type of damage.

(ii) To recommend methods of seismic strengthening to upgrade the strength of the building in line with the requirements of the seismic-zoning map of India (See IS : 1893-1984) and Earthquake Resistance Codes (IS : 4326-1993) and (IS:13828-1993).

The entire district Kachchh is placed in zone V with anticipated maximum seismic intensity of IX on the MSK Intensity Scale. The districts of Banaskantha, Patan, Surendranagar, Rajkot and Jamnagar have large parts in Zone IV with probable maximum Intensity VIII, and the remaining parts of the state lie in Zone III with maximum probable Intensity VII.

The retrofitting measures are worked out here for safety of existing damaged or undamaged buildings in future MSK Intensity occurrences in the area as per the seismic zoning map.

The masonry buildings will include: walls of brick work, random rubble stone masonry and cut stone masonry whether used for housing or community buildings like Aanganwadis, primary health centres, schools or panchaytghars; and roofs with Mangalore pattern (Morbi) tiles on wooden structures, CGI or AC sheet roofs on steel under-structure, or RC slabs.

23.3 Categorisation of Damage

As specified usually in the MSK Intensity scale, five categories of damage are recognised and named as G1 to G5; G1 referring to very slight damage without loss of structural strength and G5 referring to complete collapse of the building. Description of these categories of damage as applicable to masonry buildings is presented in Table 1. So far as repair, restoration of structural strength and seismic strengthening to meet the codal requirements are concerned, Categories G1 to G3 are most relevant, since buildings or parts thereof subjected to category G4 in most cases have to be demolished and rebuilt.

23.4 Concepts of Repair, Restoration and Retrofitting

There is a need to distinguish between the terms repair, restoration and strengthening as described below:

Table 1:- Damage Categories*			
	Category	**Walls ***	**Roof / Floors**
0	No Damage	No Damage	No Damage
G 1	Slight Non-Structural Damage	Thin cracks in plaster, falling of plaster bits in limited parts. Small	Thin cracks in small areas, tiles only slightly disturbed Small cracks
G2	Slight Structural Damage	cracks in walls, falling of plaster in large areas: damage to non-structural parts like *chhajjas,* parapets.	in slabs/ A.C. sheets; tiles disturbed in about 10% area; minor damage in under-structure of sloping roof.
G3	Moderate Structural Damage	Large and deep cracks in walls; widespread cracking of walls, columns and piers; or collaps of one wall. The load carrying capacity of structure is partially reduced.	Large cracks in slabs; some AC sheets, broken; upto 25% tiles disturbed/ fallen moderate damage to understructure of sloping roofs.
G4	Severe Structural Damage	Gaps occur in walls; two or more inner or outer walls collapse; Approximately 50% of the main structural elements fail. The building takes a dangerous state.	Floors badly cracked, part may fall; under-structure of sloping roof heavily damaged, part may fall; tiles badly affected & fallen.
G5	Collapse	A large part or whole of the building collapses.	A large part or whole floor and roof collapse or hang precariously.

* Based on I A E E Guidelines, further developed through observations in earthquakes in India, by Dr. A.S. Arya, Seismic Advisor, G S D M A.

23.4.1 Repair

This consists of *actions taken for patching up superficial defects,* re-plastering walls, repairing doors and windows and services such as the following:

i) Patching up of defects such as cracks and fall of plaster and re-plastering if needed.
ii) Repairing doors, windows and replacement of glass panes.
iii) Checking and repairing electrical connections, gas connections, plumbing, heating, ventilation
iv) Rebuilding non-structural walls, chimneys, boundary walls.

v) Relaying cracked flooring at ground level and roofing sheets or tiles.
vi) Redecoration work (white or colour washing, etc.)

It would be seen that the *repair work carried out as above does not add any strength to the structure.*

23.4.2 Restoration

This includes *actions taken for restoring the lost strength of structural elements of the building.* This is done by making the columns, piers, beams and walls at least as strong as originally provided as follows:

i) Removal of portions of cracked masonry walls and piers, and rebuilding them in richer mortar. Use of non-shrinking mortar will be preferable.
ii) Addition of reinforcing mesh on both faces of the cracked wall, holding it to the wall through spikes or bolts and then covering it suitably with micro-concrete or 1:3 cement -coarse sand plaster.
iii) Injecting neat cement slurry or epoxy like material, which is strong in tension, into the cracks in walls, columns, beams, etc.

If the structural restoration is properly executed, the structure will be as strong as before the earthquake. It is also possible to strengthen a structure to take increased vertical loading, if required.

23.4.3 Seismic Strengthening (Retrofitting) :

This will involve *actions for upgrading the seismic resistance of an existing building so that it becomes safer under the occurrence of probable future earthquakes.*

The seismic behaviour of existing buildings is affected by their original structural inadequacies, material degradation due to ageing and alterations carried out during use over time. The complete replacement of such buildings in a given area is just not possible due to a number of social, cultural and financial problems. Therefore, seismic strengthening of existing undamaged or damaged buildings is a definite requirement. Seismic strengthening including structural restoration and cosmetic repairs, may sometimes cost upto 25 to 30% of the cost of rebuilding although usually it may not exceed 12 to 15%. Hence justification of strengthening work must be fully considered from the cost point of view. The main items of seismic strengthening could be some or all of the following actions:

i) Modification of roofs.
ii) Substitution or strengthening of floors.
iii) Modification in the building plan.
iv) Strengthening of walls including provision of horizontal and vertical bands or belts, introduction of 'through' or header stones in thick stone walls, and injection grouting, etc.

v) Adding to the sections of beams and columns by casing or jacketing, etc.
vi) Adding shear walls or diagonal bracings.
vii) Strengthening of foundations if found necessary (but very difficult and expensive).

23.5 Assessment of Damage

The buildings to the restored and repaired should be thoroughly surveyed and the various damage types should be recorded on scaled drawings. The width and length of each damage needs to the recorded to as to estimate the required materials and labour for restoration and repair properly.

It should also be assessed if during the process of restoration, some of the service lines will need to be disturbed, and their temporary bypassing needed. The expenses should be included in the estimates.

23.6 Usual Damage Types in Masonry Buildings

The types of damage generally observed in various masonry buildings during an earthquake are listed in Table 2. Alongside, the actions to be taken for restoration of the lost strength are also suggested. Details of each such action are described in the following paragraphs.

Table - 2: Types of Damage in Masonry Buildings

Damage Observed	Action for Restoration
a) Different types of cracks seen in masonry walls i. Vertical cracks ii. Inclined cracks iii. Cracks at the corners or T junctions, and separation of the cross-walls	a) i,ii. Cracks to be fully filled using appropriate grout or mortar. iii. Cracks at the corners or T-junctions to be filled as above but before that, the walls at right angles to be connected using ferro-cement corner plates
b) At some places, occurrence of many cracks close together in the walls, OR tilting of some wall portions out of plumb after separation OR bulging of stone wall after delamination, OR falling of some wall portions.	b) This type of cracked, fallen, tilted or bulged wall portion to be reconstructed using 1:6 or richer cement mortar after partial demo lition of wall as required.
c) Shifting of roofing tiles or rafters OR falling down and being broken	c) The roofing tiles to be removed for further work and the rafters to be properly positioned. The opposite rafters to be tied together by horizontal braces; the purlins to be adjusted and the tiles to be placed back properly.
d) Breaking of A C Sheets	d) Replacement of the sheets, increasing number of fastenings.

23.7 Methodology for Grouting of Cracks

23.7.1 Minor and medium cracks (crack width 0.5 mm to 5.0 mm)

Material/equipment required

(i) Plastic / aluminium nipples of 12 mm dia (30 to 40 mm long).

(ii) Non-shrink cement (shrinkomp of ACC or equivalent).

(iii)Polyster putty or 1:3 cement sand mortar for sealing of the cracks.

(iv) Compressor for injecting the slurry

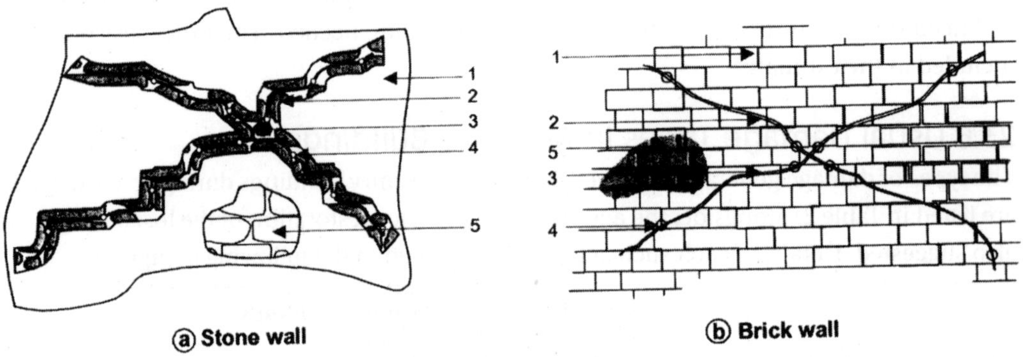

1- Plaster
2- Plaster removed and cracks cleaned
3- Cracks sealed with mortar or putty
4- Grout ports
5- Plaster fallen to be done again

Fig.1: Filling grout in cracks

Procedure:- See Fig.1

Step-1 Remove the plaster in the vicinity of crack exposing the cracked bare masonry.

Step-2 Make the shape of crack in the V-shape by chiselling out.

Step-3 Fix the grouting nipples in the V-groove on the faces of the wall at spacing of 150-200 mm c/c.

Step-4 Clean the crack with the compressed air through nipples to ensure that the fine and loose material inside the cracked masonry has been removed.

Step-5 Seal the crack on both faces of the wall with polyster putty or cement mortar 1:3 (1-cement: 3-coarse sand) and allowed to gain strength.

Step-6 Inject water, starting with the nipple fixed at higher level and moving down so that the dust inside the cracks is washed off and the masonry is saturated with water.

Step-7 Make cement slurry with 1 : 1 (1-non shrink cement: 1-water) and start injecting from lower most nipple till the cement slurry comes out from the next higher nipple and then move to the next higher nipple.

Step-8 After injection grouting through all the nipples is completed, replaster the surface and finish the same.

23.7.2 Major crack (crack width more than 5.0 mm)

Material/equipment required

(i) Plastic/aluminium nipples of 12 mm dia (30 to 40 mm long)

(ii) Polyester putty or 1:3 cement-sand mortar for sealing of cracks.

(iii) Non-shrink cement (shrinkomp of ACC or equivalent).

(iv) Compressor for injecting the slurry.

(v) Galvanised steel wire fabric (16 to 14 gauge i.e. 1.5 to 2.03mm dia wire) with 25 mm x 25 mm mesh size.

(vi) Galvanised steel clamping rod of 3.15 mm dia, or 5 mm dia 150 mm long wire nails.

Procedure:-

Step-1 Remove the plaster in the vicinity of the crack, exposing the cracked bare masonry.

Step-2 Make the shape of crack in the V-shape by chiseling out.

Step-3 Clean the crack with compressed air.

Step-4 Fix the grouting nipples in the V-groove in both faces of the wall at spacing of 150-200 mm c/c.

Step-5 Clean the crack with the compressed air through nipples to ensure that the fine and loose material inside the cracked masonry has been removed.

Step-6 Seal the crack on both the faces of the wall with polyester putty or cement mortar 1:3 (1-cement:3-coarse sand) and allowed to gain strength.

Sept-7 Inject water starting with the nipples fixed at the higher level and moving down so that the dust inside the crack is washed off and the masonry is saturated with water.

Step-8 Make cement slurry with 1:2: W (1-non shrink cement : 2-fine sand : just enough water) and start injecting from the lower most nipple till the slurry comes out from the next higher nipple and then move to the next higher nipple.

Step-9 After injection grouting through all the nipples is completed, replaster the surface and finish the same.

Alternative Procedure:- See Fig.2

Step-1 Remove the plaster in the vicinity of crack exposing the cracked bare masonry.

Step-2 Make the shape of crack in the V-shape by chiseling out.

Step-3 Clean the crack with compressed air.

Step-4 Fill the crack with cement mortar 1:3:W (1-non shrink cement : 3-fine sand : necessary water) from both sides as deep as feasible.

Step-5 Provide wire mesh on both the faces of the wall after removal of plaster in the region of repair to a width of 150 mm on each side of the crack.

Step-6 Clamp the mesh with the wall using clamps or wire nails at the spacing of 300 mm c/c.

Step-7 Plaster the meshed area with cement sand mortar of 1:3, covering the mesh by a minimum of 12 mm.

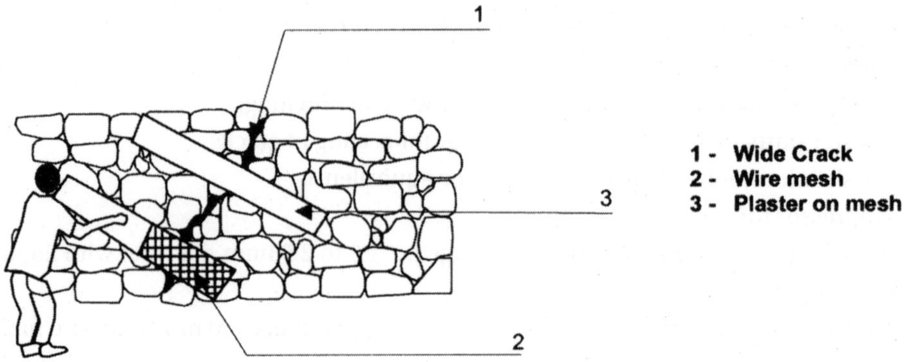

Fig.2: Fixing mesh acroos wide cracks

23.8 Installing Ferro-Cement Plates at the Corners

Before filling the cracks as in Para 7, use galvanised weld-mesh 'g14' (2.0 mm wires @ 25 x 25 mm mash) over a length of 500-600 mm on each side of the crack both inside and outside of the room in a depth of 300 mm at window sills on about 900 mm height above the floor (Fig.3) and another one at lintel level or about 2 m above the floor. But if a horizontal seismic belt is to be provided at the lintel level, the second mesh is not required.

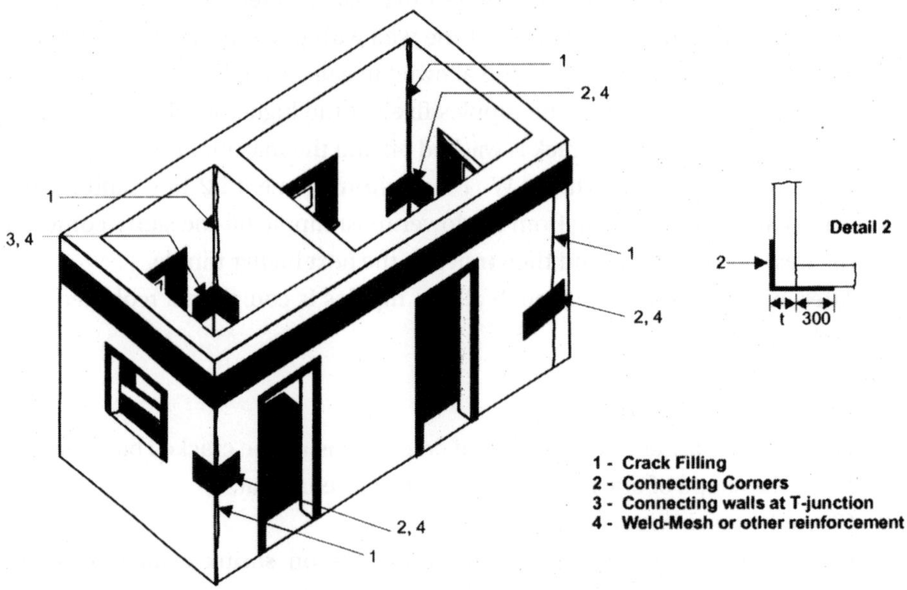

Fig.3: Connection of cracked walls at corners and junctions

23.9 Rebuilding Portions of the Wall

(i) Generally, the random stone walls are seen to be 450-600 mm thick, built by two wythes vertically (Fig.4.1). During an earthquake, the wythes get separated and either one or both get bulged (Fig.4.2) which even fall away under further vibrations (Fig.4.3). For preventing such delamination, it is necessary to use 'through' stones or RCC elements. These should be installed while rebuilding the wall (Fig.4.4).

(ii) Where portions of wall require rebuilding, the roof resting on the wall should first be supported by wooden struts, (Fig-4a,b). Then the damaged portion of the wall should be dismantled. The new portion of the wall should be constructed using cement-sand mortar of 1:6 cement-sand mortar in walls built originally in weak mortar, but using 1:4 mix for walls originally built in cement mortar.

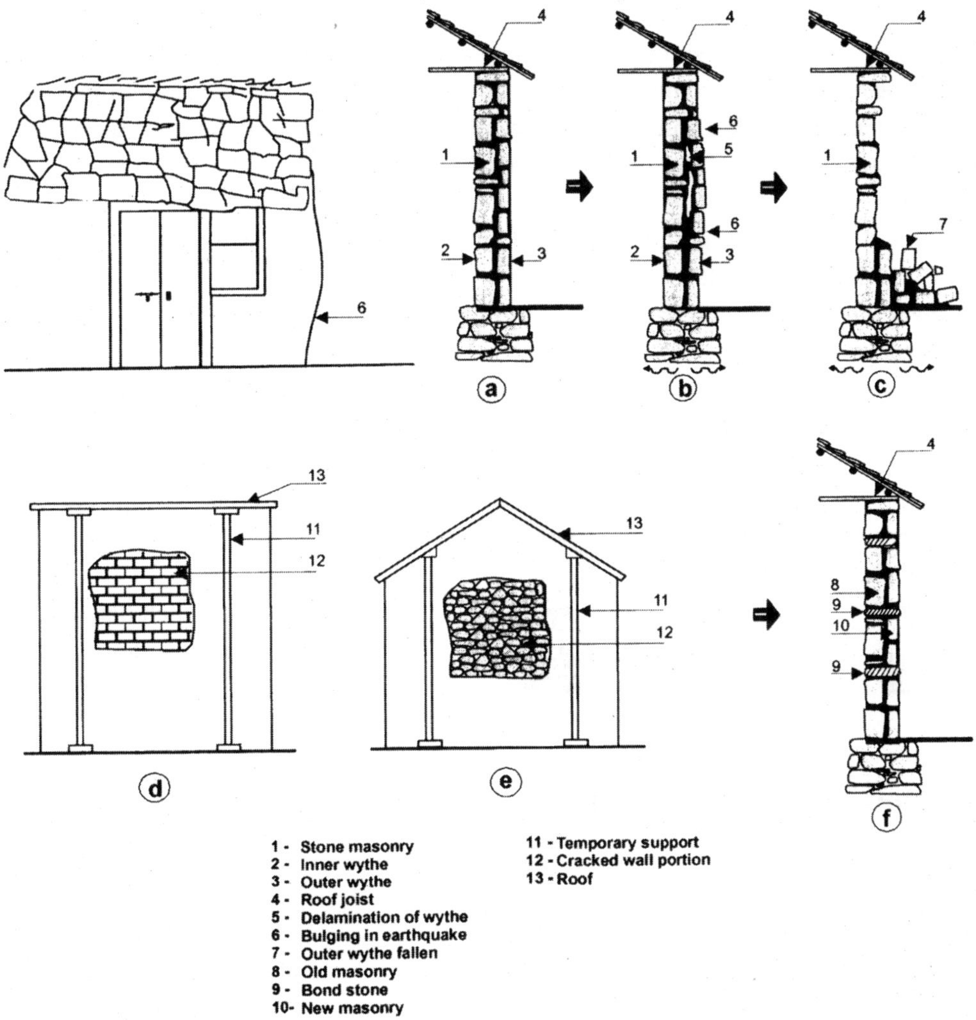

Fig.4: Rebuilding part of wall

Table 3- Simplified Building Categories in the Seismic Zones

Seismic Zone	Ordinary Buildings	Important Buildings
V	D	E
IV	C	D
III	B	C

1. This categorisation is in line with IS : 4326 - 1993 and IS: 1893 - 1984.
2. Housing falls under ordinary buildings. The community building may be considered under important buildings.

23.10 Earthquake Resistant Retrofitting of Buildings

For achieving safety of buildings against collapse in a future severe earthquake, the following retrofitting actions are recommended. The amount and placing of the retrofitting element depends upon the seismic zone, the importance of the building and the stiffness of the base soil. The categorisation of buildings is given in Table 3.

i) Check the length, height and thickness of walls and modify to conform to the Code: IS:4326 and IS:13828-1993 (see Table 3).

ii) Check the positions and sizes of openings in walls and modify as required, or provide reinforcement.

iii) If there are no 'through' stones in thick stone walls, then provide RC headers by making 'through' hole by removing the stones in opposite wythes, inserting an iron link and filling the hole with concrete.

iv) Provide a seismic belt below the roof and above the door/window lintel level. For this, use weld mesh reinforcement.

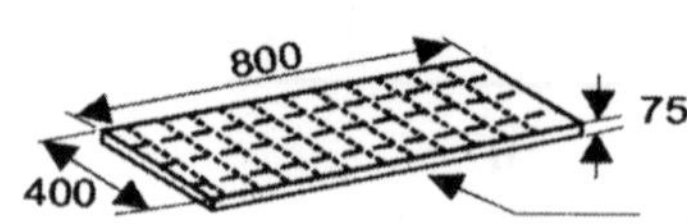

Detail 5

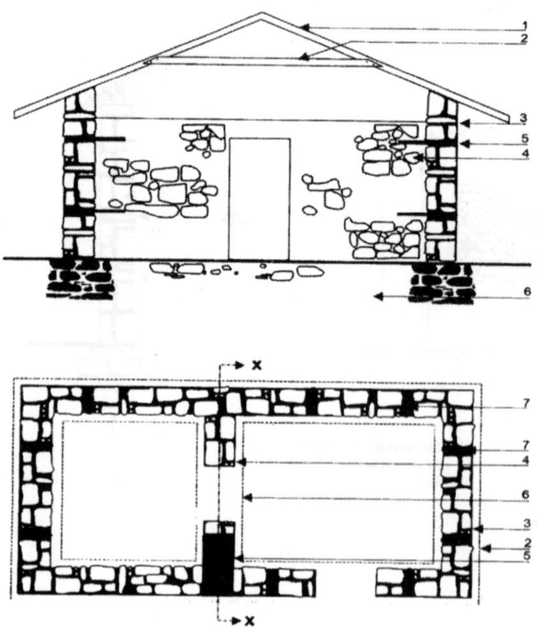

1 - Roof with rafters
2 - New tie
3 - Original Wall
4 - New cross wall
5 - Connecting R.C.elements
6 - New wall footing
7 - Through element

Fig.5: Stregthening of long walls by cross wall

v) Provide vertical reinforcement at the corners and T-junction of walls, either using bars or ferro-cement with weld-mesh reinforcement.

vi) Modify the roof structure by providing additional bracing elements and fix it to the seismic band/belt.

23.11 Control on Length, Height, Thickness of Walls

a) R.R Stone Masonry.

The wall length should not exceed 5m between cross walls in the case of mud mortar and 6m in the cement mortar case. If the length exceeds these, provide an internal wall at a spacing not farther than 4m (see Fig.5). The thickness of the new wall should not exceed 400mm. The wall height should not exceed 2.7m in mud mortar and 2.9m in cement mortar (see Table 4).

Table 4- Control on Length, Height & Thickness of Walls

Type of Masonry	Maximum Length of Walls in Room	Maximum Height of Storey
(a) R R Stone Masonry		
- in Mud Mortar	5 m	2.7 m
- in Cement Mortar	6 m	2.9 m
(b) Rectangular Unit Walls	35 t but $\leq$ 7.0 m	15 t but $\leq$ 3.5m
- in Cement Mortar	t = thickness of wall	t = thickness of wall

b) Rectangular unit masonry in cement mortar.

The wall length should not exceed 35 t and the height should exceed 15 t where t = thickness of wall. See Table 4.

23.12 Control on Door and Window Openings in Masonry

i) Door and window opening should satisfy the following: Distance of jamb from internal corner not less than 450mm and distance between two consecutive openings should be 600mm or more in the case of RR masonry and 560mm in rectangular unit masonry. In case of RR

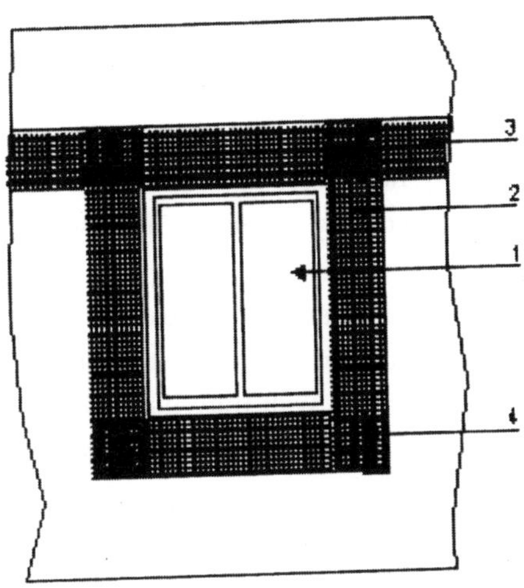

Fig.6: Reinforcing around openings

masonry in mud mortar, there should preferably be only one door or window in one wall not exceeding one-third of the wall length in the room. The combined length of openings in a wall of a rectangular unit masonry building in cement mortar should be restricted to 0.5 L in one storey, 0.42 L in two storeys and 0.33 L in three storey building where L is the length of the wall.

ii) If the above conditions are not satisfied, action should be taken to close an opening or reduce its size. Otherwise, provide strengthening around the opening (Fig.6). For details of reinforcement, see Para 16.

23.13 Making 'Through' Bond Elements in R.R. Stone Wall

a) Select points where 'through' stones will be installed at horizontal and vertical distance of about one metre apart, with 50cm horizontal stagger.

b) Remove the plaster from the surface, exposing the stones.

c) Remove the mortar around the stone to sufficient depth gently, not violently, so as to expose the stone on all sides.

d) Loosen the stone by means of gentle pushes sideways and up and down by means of a small crowbar, so that the other stones of the walls are not disturbed. Pull out the stone slowly, holding it by both hands.

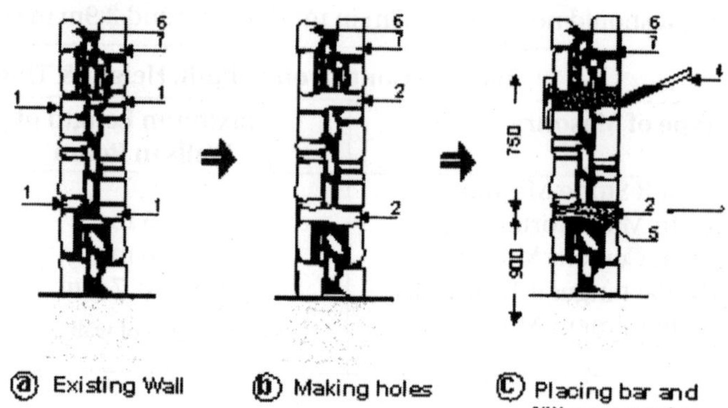

1 – Stones removed to make through holes
2 – Holes
3 – Hooked Bar
4 – Chute for pouring concrete
5 – Filled Concrete
6 – Internal Wythe
7 – External Wythe

Fig.7: Providing RC 'through' elements for 'stitching' stone wythes

e) Remove inner material gradually so that a 75mm size hole can be made in the wall. Bigger hole is not needed.

f) Locate position of the opposite stone on the other face of the wall by gentle tapping in the hole. Remove the identified stone slowly by the same gentle process.

g) The hole so made through the wall may be bigger in size on both faces and narrower inside, resembling a dumbbell shape. This is good. It does not matter if the hole is inclined instead of being horizontal.

h) Place concrete of 1:2:4 mix to fill half the depth of the hole from both sides and place 8mm dia hooked mild steel bar in the hole and fill the hole completely.

i) Cure for a minimum 10 days by sprinkling water on the exposed surfaces on both sides.

23.14 Providing Horizontal Seismic Belts

23.14.1 Seismic Belt Locations

i) Seismic belts are to be provided on all walls on both the faces just above the lintels of door and window openings and below the floor or roof.

Note: On small wall lengths in a room (less than 5m), seismic band only on the outside face will suffice. In this case, these should be connected by ties going across the rooms (see Fig.8).

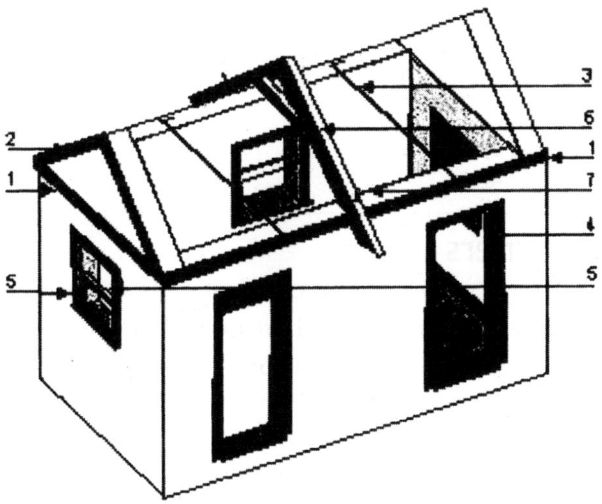

1 – Seismic belt above openings and below roof at eave level
2 – Seismic belt on gable wall
3 – Tie at eave - bend level
4 – Door
5 – Window
6 – Rafter with collar tie
7 – Tying of rafter with band

Fig.8: Overall arrangement of seismic belts (roofing removed)

ii) The roof belt may be omitted if the roof or floor is of RCC slab.

iii) The seismic belt is not necessary at plinth level, unless the plinth height is more than 900 mm.

iv) Install the similar seismic belt at the eave level of the sloping roof and near the top of gable wall, below the roof.

Note:-. If the height of eave level above the top of door in less than 900 mm, only the eave level belt may be provided and lintel level band may be omitted.

23.14.2 Description of reinforcement in belt.

The reinforcement may be of mesh types as suggested in Table 4 or any other mesh of equivalent longitudinal wires. For example in Cat. D building with room length of 6 m, MW 21 weld mesh (with long wires 5 of 4.5 mm dia spaced at 75 mm apart; cross wires of 3.15 mm dia placed at 300 mm apart) can be used, the height of the belt being kept as 375 mm.

Note: Weld mesh has to be provided continuously. If splicing is required, there should be minimum overlap of 300mm.

Table 5- Mesh Reinforcement in Seismic Belts in Various Building Categories

Length of wall	Cat. B			Cat. C			Cat. D			Cat. E		
m	Gauge	N	H	Gauge	N	H	Gauge	N	H	Gauge	N	H
≤ 5.0	g 14	9	250	g 13	9	250	g 12	9	250	g 10	10	280
6.0	g 13	9	250	g 12	9	250	g 10	10	280	g 10	14	380
7.0	g 12	9	250	g 10	10	280	g 10	14	380	g 10	18	460
8.0	g 10	9	250	g 10	14	380	g 10	18	460	g 10	23	580

1. Gauges: g10=3.25 mm, g11=2.95 mm, g12=2.64 mm, g13=2.34 mm, g14=2.03 mm.
2. N = Number of longitudinal wires in the belt at spacing of 25 mm.
3. H = Height of belt on wall in micro-concrete, mm.
4. The transverse wires in the mesh could be spaced upto 150 mm.
5. The mesh should be galvanised to save from corrosion.

23.15 Vertical Seismic Belt at Corners

Vertical reinforcing is required at the corners of rooms and junctions of walls as per Table 6. Alternatively, MW 21 weld mesh of equivalent longitudinal area could also be used. The width of this belt on each side of the corner has to be kept 25mm extra to the width of the mesh.

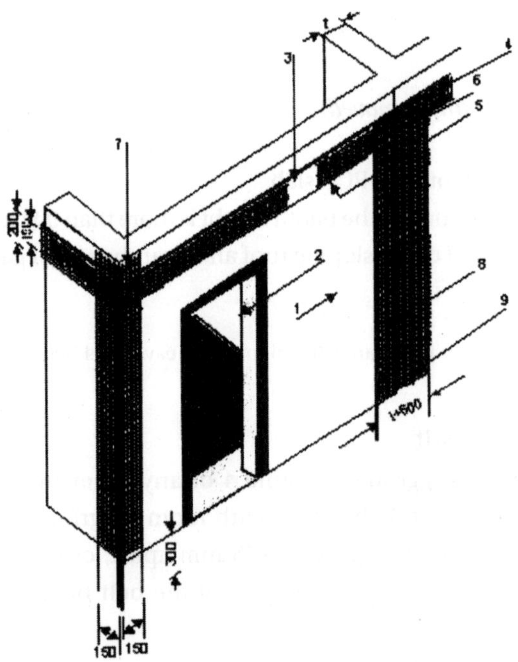

This reinforcement should be started 300mm below the plinth level and continued into the roof/eave level horizontal band. (See Fig.9).

- Masonry wall
- Door
- Plaster first layer
- Steel mesh in belt
- Wide head nails 150 long
- Plaster second layer
- Overlap in steel mesh, 200 mm each le of corner
- Vertical seismic belt at junction ɔing across horizontal belt)
- Spikes for fixing mesh

Fig.9: Vertical seismic belts at corner and junction

Table 6-Vertical Bar or Mesh Reinforcement in Vertical Belt at Corners of Rooms

No. of	Storey	Cat.C			Cat.D			Cat. E		
		Single Bar, mm	Mesh (g10)		Single Bar, mm	Mesh (g10)		Single Bar, mm	Mesh (g10)	
			N	B		N	B		N	B
One	One	-	-	-	10	10	300	12	14	400
Two	Top	-	-	-	10	10	300	12	14	400
	Bottom	-	-	-	12	14	400	16	14	400
									With 1 bars of 12 Ø tor	
Three	Top	10	10	300	10	10	300	12	14	400
	Middle	10	10	300	12	14	400	16	14	400
									With 1 bars of 12 Ø tor	
	Bottom	12	14	400	12	14	400	16	14	400
									With 1 bars of 12 Ø tor	

1. Gauge 10 (3.25 mm dia) galvanised mesh with 25 mm spacing of wires shall be used.
2. Single bar, if used, shall be HSD or TOR type. If two bars are used at a T-junction, the diameter can be taken as follows. For one of 10 or 12 mm take 2 of 8 mm, and for one of 16 mm take 2 of 12 mm.
3. N = Number of longitudinal wires in the mesh.
4. B = Width of the micro concrete belt, half on each wall meeting at the corner or T-junction.
5. The transverse wires in the mesh could be at spacing up to 150 mm.

23.16 Seismic Belt around Door/Window Openings

The jambs and piers between window and door openings require vertical reinforcement in the following situations:

i) In category **D** and **E** buildings for resistance against earthquake forces.
ii) For restoring the strength of the piers in any building category when badly damaged in an earthquake.

The following mesh reinforcement is recommended to be used for covering the jamb area on both sides of an opening or for covering the pier between the consecutive openings.

i) In category D & E buildings

Mesh of gauge 10 with 10 wires in a vertical direction spaced at 25 mm in a belt width of 280 mm.

ii) In category C buildings

Mesh of gauge 12 with 9 wires in a vertical direction spaced at 25 mm in a belt width of 250 mm.

23.17 Method of Fixing Seismic Belts

The reinforcement specified in Paras 8, 14, 15 and 16 is to be finally attached to the stone wall by nails or connectors and cement mortar. For this purpose, either 1:3 cement-coarse sand mortar or micro-concrete 1:1.5:3 is used. It is applied in two layers like plaster as described below:

Steps to construct the belt

Step-1 Remove plaster in the height of the belt.

Step-2 Rake out mortar joints to 12-15 mm depth.

Step-3 Clean the surface and wet it with water.

Step-4 Apply neat cement slurry and apply first coat of 12 mm thickness. Roughen its surface after initial set.

Step-5 Fix the mesh with 150 mm long nails at about 300 mm apart while plaster is still green.

Step-6 Apply second coat of plaster of 16 mm thickness.

Note:

1. The mesh should be continuous with 200mm overlap at the corner or elsehere.
2. Using galvanised binding wire, tie up the roof rafters with the nails of the eave level belt before applying the plaster over the mesh.
3. In brick and Bela stone walls, it will be easy to drill or chisel out holes of 75 mm dia. In that case, instead of the nails, use 3 mm galvanised mild steel wires through the holes to hold and clamp the longitudinal wires every 450 mm c/c.

23.18 Providing Vertical Reinforcement at Corners and Junctions of Walls.

The vertical reinforcement consisting of TOR bar as per Table 6 or equivalent shall be provided on the inside corner of room starting from 750 mm below the ground floor going upto the roof slab, passing through each middle floor through holes made in the slabs. (See Fig.10) The reinforcement will be connected to the walls by using L shape dowels of 8 mm TOR bar, the vertical leg of 400 mm length firmly tied to the vertical reinforcement bars and the horizontal leg of minimum 150 mm length embedded in the walls through 75 mm dia. holes drilled in the wall into which the 8 mm dia. leg of the dowel will be grouted using non-shrink cement cum polymer grout. Such dowels will be provided, first one just above plinth level and then at about every 1 m distance apart. The corner reinforcement will be covered with 1:3 cement mortar or 1:11/2:3 micro concrete fully bonded with the walls giving a minimum cover of 15 mm on the bar.

23.19 Stiffening the Flat Wooden Floor/Roof

Many of the damaged houses have flat floor or roof made of wood joists covered with wooden planks and earth. For making such roof/floor rigid, long planks 100mm wide and 25 mm thick should be nailed at both ends of the joists from below. Additionally, similar planks or galvanised metal strips 1.5 mm thick 50 mm wide should be nailed diagonally also. See fig 11:

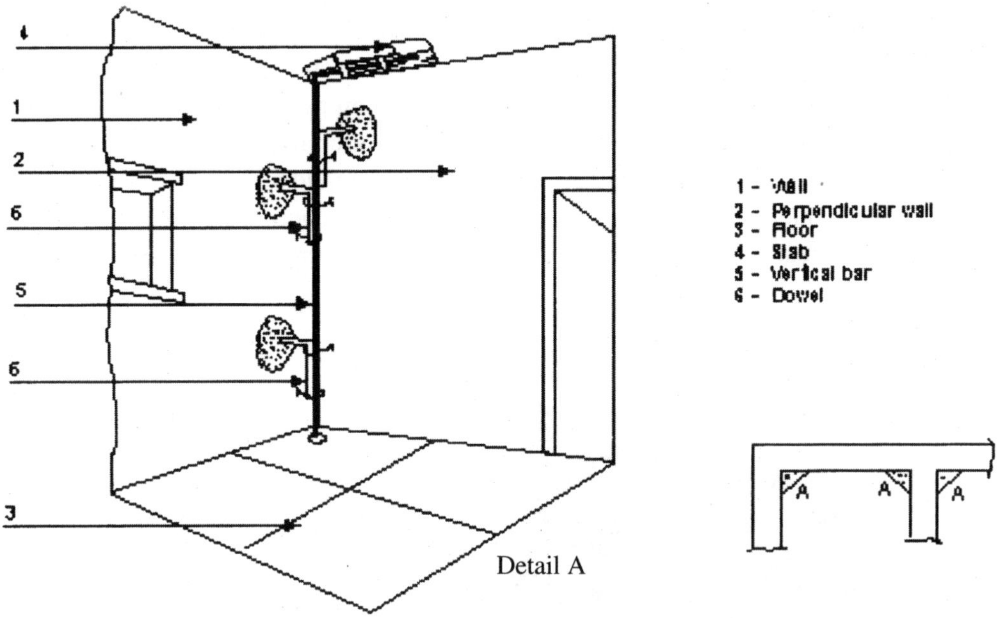

Fig.10: Vertical bar at inside corner RC 'through' elements for 'stitching' stone wythes

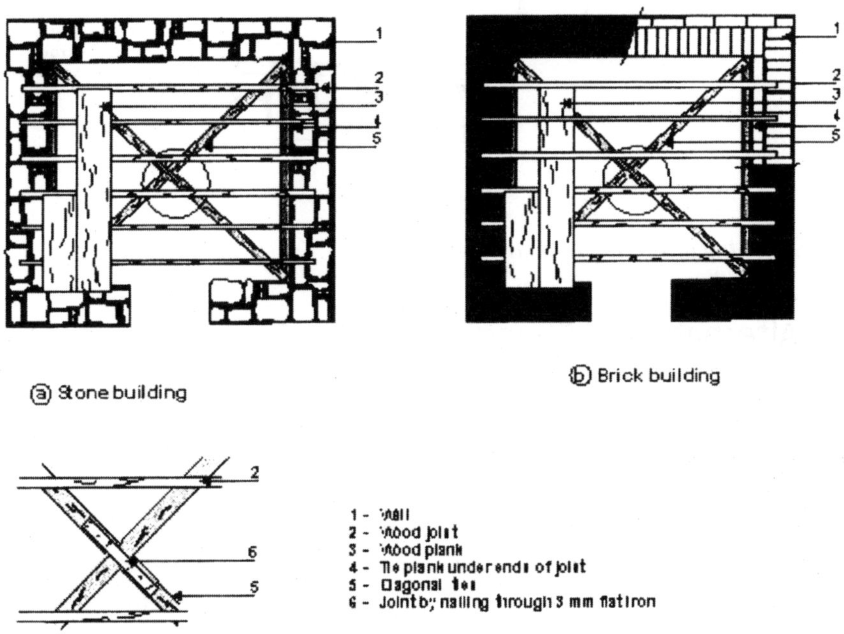

Fig.11: Stiffening flat wooden floor/roof

23.20 Stiffening the Sloping Roof Structure

Most of the sloping roofs are made of rafters, purlins, burnt clay tiles on top. Similarly, AC or CGI sheet roofs are made using wooden purlins resting on gable walls or main rafters. But trusses were not formed which require the use of ties. Such roofs push the walls outward during earthquakes. For stiffening such roofs, the rafters should be tied with the seismic belt as in the Note under para 23.17, and the opposite rafters, on both sides of the ridge need to be connected near about mid-height of the roof through cross ties nailed to the rafters (see Fig. 12).s

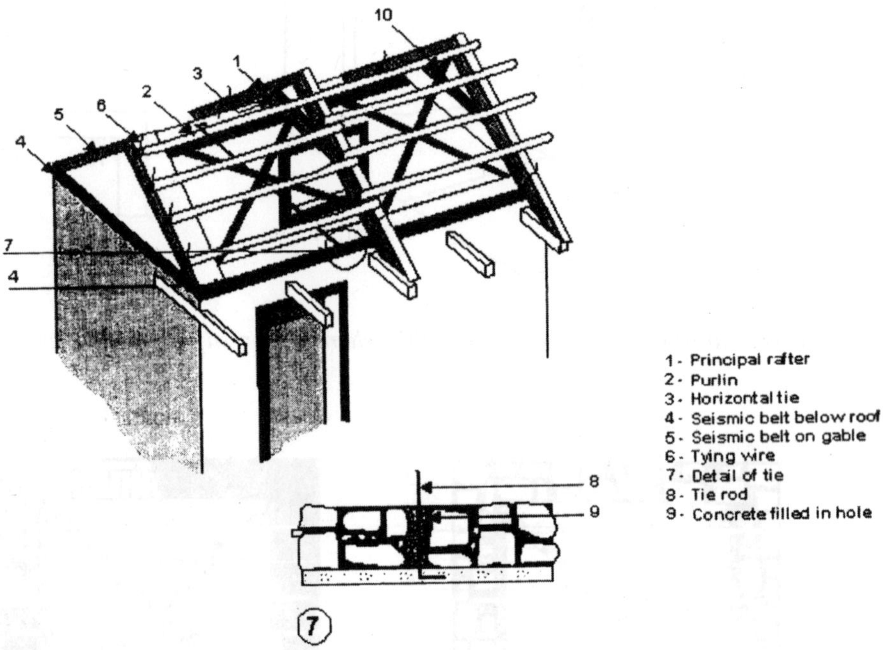

Fig.12: Stiffening of sloping roof structure

23.21 Alternative of Changing the Roof

While repairing and retrofitting the house, if it is desired to replace the roof structure or tiles with AC or CGI roof, take the following steps:-

i) Complete the repair and retrofitting work of the first storey, including provision of 'through' elements.
ii) Stiffen the first floor wooden deck.
iii) Complete the horizontal and vertical seismic belts in the first storey, keep the vertical mesh reinforcement extending beyond the first storey by 300mm and leave uncovered.
iv) Complete the repair and retrofitting of the second storey.
v) Now open the roof structure and remove gable portion upto eave level.

vi) It will be preferable to use seismic bands, instead of belts in this case. Therefore, construct the vertical seismic belts and the eave level and gable bands together, taking the vertical steel in the bands. Anchor steel wires in the bands and extend out for tying down the rafters and purlins. (Fig.12).

vii) Now erect the rafters, tie them with bands and fix the tie to make A-frames.

viii) Instal diagonal bracing in the plane of the rafters.

ix) Erect the purlins over the rafters, tie them with rafters and gable bands. Bolt down the AC or CGI sheets to the purlins using J or U bolts with iron and bitumen washers.■

23.22 References

1. Arya, A.S. et. al. (1986), *Guidelines for Earthquake Resistant Non-Engineered Construction,* IAEE Committee, The International Association for Earthquake Engineering, Tokyo.
2. Arya, A.S. (1988), " Repair and Strengthening of Damaged Stone Houses after Dhamar Earthquake, North Yemen ", *IX World Conference on Earthquake Engineering,* Tokyo-Kyoto, Aug. 2-9, Vol. VIII, pp. 1141-1146.
3. Arya, A.S. (1994), *Retrofitting of Stone Houses in Marathwada* Area of Maharashtra, Building Materials and Technology Promotion Council (BMTPC), Ministry of Urban Development, Govt.of India.
4. Arya, A.S. (1996), "Seismic Retrofitting of Stone Houses in Marathwada Area, India", Paper No. 2174, XI *World Conference on Earthquake Engineering, Acupulco, Maxico.*
5. Arya A.S. (1997), *Guidelines for Damage Assessment and Post Earthquake Action, Part-2, Brick Houses with Clay Tile Roofing, pub.* Building Materials and Technology Promotion Council, Ministry of Urban Development, Govt.of India.
6. Arya A.S. (1999), *Repair and Retrofitting of Buildings in the Chamoli Rarthquake Affected Areas,* Building Materials and Technology Promotion Council, Ministry of Urban Development. Govt. of India.
7. I:S:4326-1993, *Earthquake Resistant Design and Construction of Buildings* - Code of Practice, Bureau of Indian Standards, New Delhi.
8. I:S:13935-1993, Repair and *Seismic Strengthening of Building - Guidelines,* Bureau of Indian Standards, New Delhi.
9. *Reconstruction and New Construction of Houses in Kachchh Earthquake Affected Area of Gujarat,* (II printing), (GSDMA) Gujarat State Disaster Management Authority, Govt. of Gujarat, January, 2002.

Chapter 24

Repair and Strengthening of Damaged Stone Houses After Dhamar Earthquake of 1982[1]

24.1 Introduction

This brief paper is concerned with the repair and strengthening of those stone masonry buildings in Dhamar Province of the Yemen Arab Republic (YAR) which were damaged by varying degrees during the Dhamar earthquake of Dec. 13, 1982. Although this earthquake had a moderate magnitude of 5.8 to 6.0 on the Richter scale, the damaging effect on the stone buildings was rather destructive due to the shallow focus of the earthquake assessed as 10 km and the peculiar site and construction features of the masonry houses. According to the survey carried out by Y.A.R. Government, 1176 settlements in Dhamar Province, representing 82.2% of the total, were affected, 25045 houses (46.1% of total) were destroyed and 18458 homes (33.9% of the total) were cracked. Besides, 19 government buildings, 79 schools and 651 mosques were also destroyed. As a result, about 1600 lives were lost in spite of the fact that the earthquake accurred at noon (12.13 PM local time). Had the earthquake occurred at night, the loss of life would have been dreadfully large.

The total rehabilitation programme of the large population rendered homeless (265300 persons or 58.5% of the total population of the province) prepared by the Y.A.R. Government with the bilateral and international assistance consisted of construction of about 15000 new houses through contractors, 10000 new houses through self-help and repairing of about 13500 cracked homes. Actually, the earlier repairable estimate of about 18000 had later shown that about 25% of these houses will in fact be unfit for repairing and may have to be demolished.

Thus, a total of 13500 houses, varying from one to three storeys and having different plans, built according to the choice of generations of the population, needed to be repaired as well as strengthened to meet the future earthquake demands. A general scheme of repairing and seismic strengthening was worked out by the author, as presented in this paper, which can generally be applied to the masonry houses of one and two storey in height. It was hoped that

1. *Anand S. Arya, CSIR Emeritus Scientist and Professor, University of Roorkee, Roorkee (Govt. of India expert in Yemen Arab Republic under Itach program)*

after the proper repair and strengthening measures were applied to the damaged buildings, they would be upgraded to an extent that they would be able to sustain without collapse or serious damage an earthquake of the type which accurred in Dhamar Province on Dec. 13, 1982.

24.2 Typical Homes in Stone Masonry Construction

The stone masonry had two general variations. In the rural areas the most common construction was in random rubble field stone, locally called as 'Goum', set in mud mortar and chip filling. Sometimes, the stones were dry packed. The less common and more costly variation was the use of half-cut one-face-dressed stones on the outer face and random rubble on the inside face of the outer walls of the houses; the outside face was left unplastered, the inside walls plastered with mud or gypsum. Gypsum pointing was used on the outside face in some cases in the courses of the half cut stones. Fig. 1 shows a typical Yemeni home. Fig. 2 shows a typical cross-section of the first storey.

The stone wall thickness usually varied from 50 to 75 cm, and sometimes even 90 cm. Larger stones, from 20 to 30 cm in width and height were provided to make the outer and inner wythes of a wall, and the space in between was filled with loosely broken stone chips along with clay mud. In some old homes, the mud was washed out by rain.

The roofs in Yemen are invariably of the flat type since rainfall is scanty. The traditional roof consists of wooden beams in round, half round or rectangular shapes, placed parallel to each other at a spacing of 50 to 60 cm. The clear span of the joists would usually be 2.0 to 3.0 m and the bearing length will seldom exceed 20 cm. Thus, the room widths are limited by the span of the joists. But the length of rooms used as Diwan can be as long as the width or length of the building, say 10 to 12 m. Resting on the wooden beams, there are laid thin tree branches, to from a deck for receiving the clay topping. The clay roofing thickness usually varies from 15 to 30 cm and could even be thicker, up to half a metre. So also the thickness of the floor is quite massive with a concrete screed sometimes. In newer constructions, ply wood sheets 1.2x2.44 m in plan and about 12 mm thick are used to form the deck. The climate of the region is dry, with scanty rainfall and large diurnal temperature variations. The winters are cold at the prevailing elevations of 2400 m above mean sea level. These factors must have led to thick walls with small openings and flat heavy roofs.

24.3 Main Causes of Damage and Collapse

The stone masonry houses suffered from the following inherent weaknesses and construction defects during earthquakes:

(a) Their rigidity was high, leading to high acceleration response to the earthquake motion. The amplification of ground acceleration could be 2-3 times. Moreover, the large mass of walls as well as roofs would create large inertia forces.

(b) The tensile and shearing strength was very small, almost zero for masonry in mud mortar, thus, incapable of resisting such imposed stresses during earthquakes. Also no 'through' or 'bond' stones were used in walls which could have provided interconnection of the outer and inner wythes. Thus, they were free to separate out during shaking. This may be called 'delamination' of the wythes (see Fig. 3). The half cut stones generally become pyramidal in shape, their stability under the vertical load depending on the filler stone pieces used to keep their outer face vertical and once they become loose during shaking, the stability of the outer wythe gets disturbed. The vertical load then tends to buckle the outer wythe outwards.

(c) Due to absence of bonding between perpendicular walls, they were quite free to separate from each other at the junctions, resulting in loss of the box-like action of the enclosures, leaving each wall to resist the inertia forces on its own mass and the roof by itself as a bending wall perpendicular to its plane, unaided by the shear-wall action of perpendicular walls.

(d) The roofs had no diaphragm action to be able to distribute the inertia force to all four walls or transfer it to the two end walls which would act as shear walls.

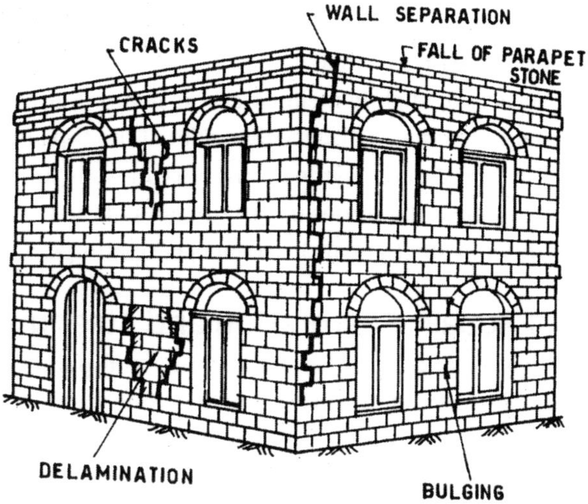

Fig.1:- Damaged Yemen home

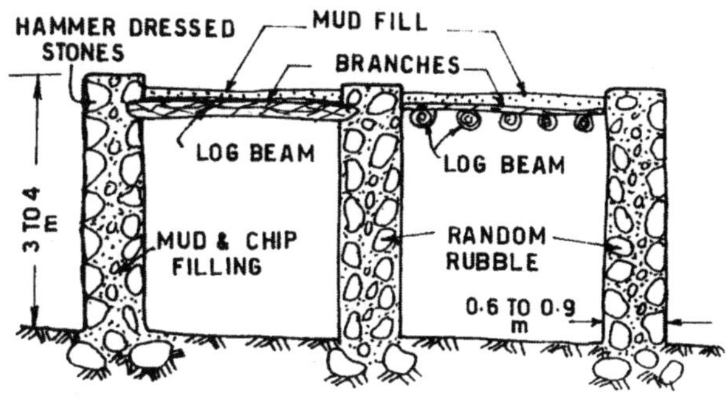

Fig.2: Typical cross-section

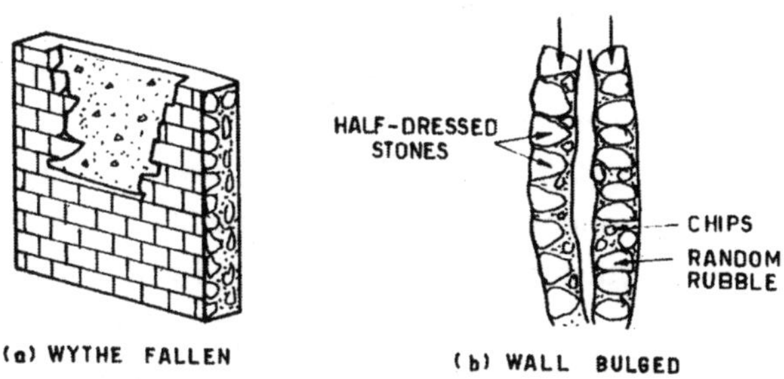

Fig.3: Delaminated wall

As a result, damage of varying degrees occurred. The damage to most repairable houses may be classified as follows:

(a) Vertical or slightly inclined cracks near outer corners indicating separating between the perpendicular walls, usually wider at the top and finer towards the base.

(b) Bulging of outer wythe of masonry in walls indicating its delamination and falling away of outer wythe stones in certain parts of the walls.

(c) Random cracks near the corners of openings, and disturbed parapet stones.

24.4 Repair, Restoration and Strengthening

By superficial repairs, like replastering and painting, the building is neither restored to its original strength nor will it be made strong for future earthquakes. Such a 'repaired' building will be very illusory, since all the structural weaknesses will be hidden and if it was shaken again, it will be damaged more seriously. 'Restoration' involves repairs to the building structure so that at least the original strength of the building is restored. It involves removal of portions of cracked masonry and rebuilding it or injecting grout into the cracks. In the present context of damaged stone buildings, it will be necessary to carry out strengthening measures also besides restoration since even the original strength of these buildings was very inadequate for resisting earthquakes of MSK intensity VIII, and simple measures of repair and restoration will be grossly insufficient for their safety in future ground motions. Now it is neither structurally possible nor economically feasible to improve all the weaknesses of stone houses. However, it is fairly possible at a reasonable cost, to remove many of the constructional defects and introduce enough strength to meet the future earthquake resistance requirements. As a first safety measure, all buildings should be reduced to two storeys only by removing the third storey. The other measures are described below:

(a) *Grouting in cracks:* The cracks in the masonry must be filled with gypsum or neat cement grout if they are fine or with cement-sand mortar if they are wide. Epoxy materials which can generate high tensile strength will be too expensive and unnecessary in stone walls in mud mortar, hence ruled out.

(b) *Stitching of stone-wythes:* It will be necessary to install 'through' bonding elements in the stone walls, particularly the outer perimeter wall in all storeys. If certain parts of outer wythes have fallen, these will have to be rebuilt using cement mortar and installing steel bar elements in the masonry itself.

(c) *Treating Bulged Walls:* Bulging of walls will indicate hollowness between the inner and outer stone wythes. This should be filled with cement-sand mortar after stitching of the two wythes is completed, otherwise the hydrostatic pressure of mortar may cause bursting of the walls. Treatment of 'seriously' bulged walls may require dismantling the bulged portion and rebuilding it. This operation may require shoring the adjoining walls or the roof and must be taken up carefully.

(d) *Installing Reinforced Concrete 'Band'* This 'band' is variously termed as 'bond beam',

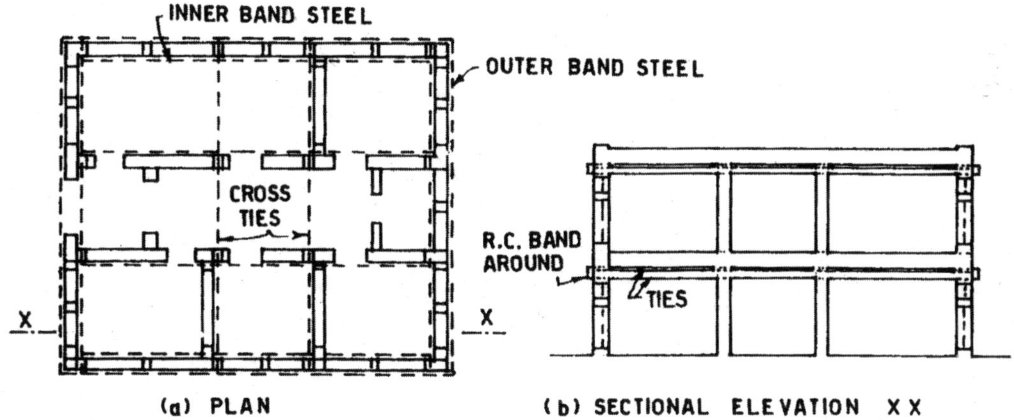

Fig.4: Belt and Cross-Ties

'perimeter beam' or 'chaining', etc. and may be installed in new constructions at various levels of a building such as the plinth level, the lintel level and just below floor and roof levels within the thicknesses. In the old buildings the best location will be just below the roof and floor beams and it will have to be installed outside the wall thickness like a *'belt'*. It will serve two important functions:

(i) To bind all the walls together and prevent their separation at the corners. Tying the perpendicular walls together at the intersections is essential to prevent 'Kinematic' instability of walls, that is, their overturning out of plane after separation at the corners.

(ii) To act as a horizontal bending beam for resisting the horizontal inertia force of the roof or floor as well as that of the mass of the outer walls, and transfer these loads to the shear walls. This action is particularly important in the present case where floors have no diaphragm action.

Ideally it will be desirable to run this *belt* (for 'band') on all internal and external walls as in new constructions. But strengthening work of the existing house can be simplified by using cross-ties instead, running by the sides of the internal walls reaching to outside of the external walls to hold the opposite external walls together. The cross-ties will also reduce the horizontal span of the belts on the external walls, thus ensuring integral box like action of all the walls. Fig. 4 shows the proposal schematically.

(e) In some old buildings, there exist horizontal wood runners which have held the masonry walls together to the extent they could. But their utility was restricted because of rather weak joints in their lengths where separation occurred and the masonry cracked. After grouting the wall cracks, the value of the wood runners will be enhanced by nailing or screwing steel straps to them across their joints. The straps should be strong enough to be equal to the full tensile strength of the wood members. In such house, the provision of reinforced concrete belt may be omitted.

24.5 Details of Repair and Strengthening Scheme

(a) *Grouting in cracks:* For fine cracks, the grout should be gypsum-water fluid like mixture or neat cement and water in the ratio about 1:1. For wide cracks, it may consist of cement and fine sand in the ratio of 1:1 or 1:2 depending upon the crack width and water added to obtain free flowing consistency. For the grouting operation, a low pressure mobile grouting pump run on diesel will be suitable. But it could be carried out by using a container kept about 1.5 to 2 m higher than the cracked area and grout allowed to flow freely through a plastic pipe into the cracks. The cracks to be grouted will be sealed with mortar first along their length leaving a few ports open to receive the nozzle of the grout pipe. The grouting is to be done starting with the lowest part and proceeding upwards after the lower part of the crack gets filled up (see Fig. 5). It may be mentioned that it will be too expensive to fill the normal joints in dry masonry walls by this procedure, and may not be attempted.

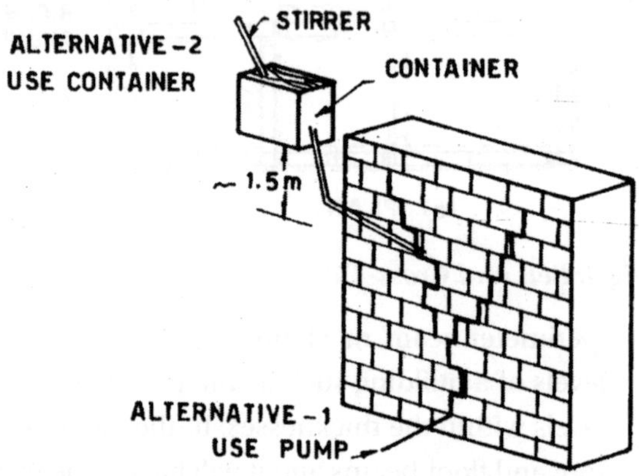

Fig.5:Grouting of cracks

(b) *Stitching of stone wythes:* For new constructions, the 'through' stones or 'bonding' elements are recommended to be provided at the rate of one per m2 area of the wall. For restoration work, however, it is considered adequate that stitching elements may be provided at one level in each storey at about mid-height between the floor and the reinforced concrete band, that is, about 1.2 m above the floor and a horizontal distance of not more than about 1.5 m. For installing the bonding element, a hole will be made through the wall by removing one external stone and one opposite internal stone. After cleaning the hole, it is to be filled with concrete while embedding a 8 mm dia bar hooked at both ends in the transverse to the wall length. The length of this bar will be kept 5 cm less than the wall thickness (see Fig. 6). These may be omitted where cross walls exist at right angles to the external walls.

(c) *Treating bulged walls:* The grouting of the hollow will be done after installation of bonding elements whose number could be increased locally if the bulge area is large. The grouting mortar could in this case be cement-fine sand in the ratio of 1:3. Fallen masonry should be reconstructed using 1:6 cement and mortar and installing 'through' stones or bonding element, as the reconstruction proceeds. See Fig. 3.

(d) *Installing reinforced cement belts:* This will involve the following steps (see Figs. 4 and 7).

(i) Making through hole in the walls, one stone in size, just below the roof or floor, one each at the ends of spans at a spacing not more than 5 m, through which the cross-ties may be installed. Additional holes will be made at the centre of these spans for providing additional shear-connection between the outer and inner bands.

(ii) Installation of cross ties through the holes and filling all holes in internal walls with concrete 1:2:4. The cross ties are placed at two points in each direction dividing the length or width of the building in three parts, each less that about 5 m in length. The ties consist of two high-strength-deformed bars 10 mm dia.

(iii) Installation of belt steel on the outside and inside faces of the external walls. This will consist of two bars 10 mm dia on each face which can be placed in position by driving concrete nails in the stones at about 0.9 m apart at the correct level, supporting the bars on the nails and tying with them by wire. It may be stated that the quantity of band steel required here was calculated for the span between the ties using a seismic coefficient of 0.2.

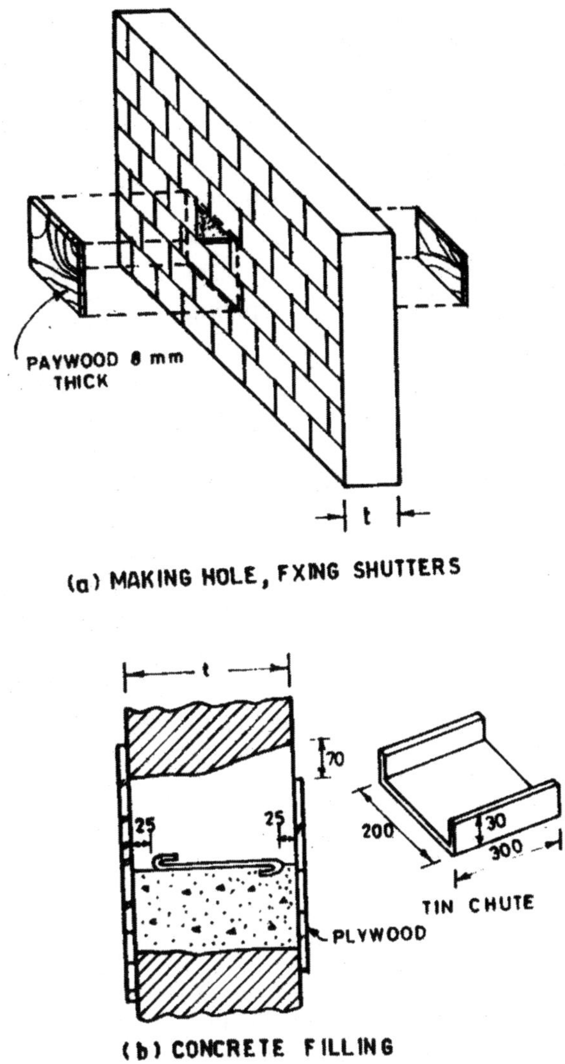

Fig.6: Providing Bonding Elements

(iv) Inserting U-shaped stirrups, 8 mm dia, two from outside and two from inside in each hole so as to go round the 10 mm dia longitudinal bars and be tied to them.

(v) Tying of galvanised wire mesh (wire 1 mm dia, mesh 25 mm x 25 mm) to the main bars in between the holes.

(vi) Nailing of wooden strips 4 cm thick and 7.5 cm wide into the joints of the stone wall to form the edges of band-concrete so as to achieve a 30 cm wide band.

(vii) Filling of all holes in external walls with 1:2:4 concrete, and applying cement concrete 1:2:4, using aggregate passing 10 mm size, by trowels like plaster to form the concrete plate. The mesh will hold the concrete and prevent its shrinkage cracking.

The longitudinal bars will have an overlap of 60 cm wherever necessary and bend around external corners by at least 30 cm each to give a total overlap of 60 cm.

(e) *Splicing existing wooden band beam:* This may be done by nailing iron flats.

(f) *Safeguarding parapets:* Reduce the height of parapets to about 45 to 60 cm and cast on top 40 mm thick concrete reinforced with welded wire mesh throughout.

The total cost of repair and strengthening was estimated and came to less than 20% of the reconstruction cost of these stone buildings. The procedure was

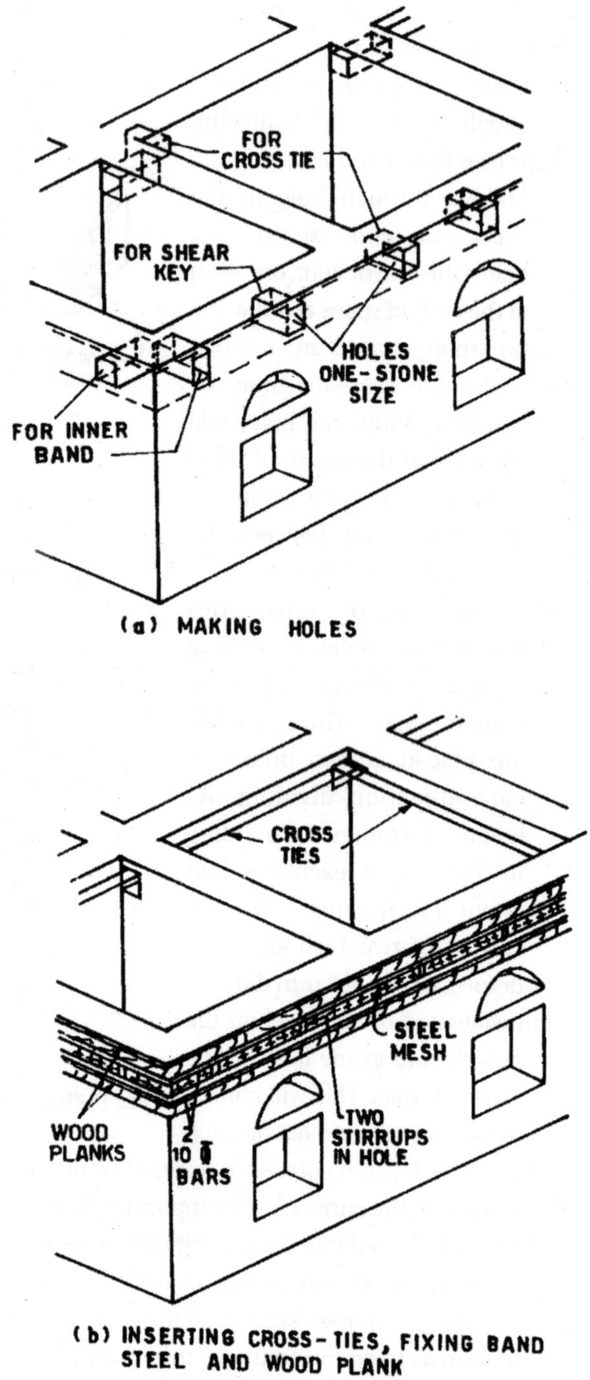

Fig.7: Installing belts and cross-ties

demonstrated in the field by carrying out the seismic strengthening of the post office building at Dhamar.

24.6 Conclusion

The scheme of repair and seismic strengthening presented above would achieve stitching of the inner and outer stone wythes in the walls to avoid their delamination during an earthquake, and will more than restore the lost strength by grouting the cracks and rebuilding fallen wythes. Provision of reinforced concrete belts in each storey just below the roof and floor and installation of steel cross-ties at the same level will effectively integrate all walls like a multicellular boxed unit with great resistance against horizontal inertia forces. Also walls will get horizontal bending strength to resist the horizontal inertia of roof and floor. Thus no-collapse seismic strengthening to about MSK VIII will be achieved. ■

24.7 References

1. Arya A.S., L.S. Srivastava and S.P. Gupta, "Damages During Dhamar Earthquake of Dec. 13, 1982 in Yemen Arab Republic", Bull ISET, Vol. 21, No.4, Dec. 1984, pp.115-162.
2. Basic Concepts of Seismic Codes, Vol. 1, Part II, Non-Engineering Construction, IAEE Committee (ii), IAEE, Tokyo, 1980 pp.33-76.

Index